PHYSICAL-CHEMICAL MECHANICS OF DISPERSE SYSTEMS AND MATERIALS

PROGRESS IN
COLLOID AND INTERFACE SCIENCE

Series Editors
Reinhard Miller and Libero Liggieri

PHYSICAL-CHEMICAL MECHANICS OF DISPERSE SYSTEMS AND MATERIALS

Eugene D. Shchukin
Andrei S. Zelenev

CRC Press
Taylor & Francis Group
Boca Raton   London   New York

CRC Press is an imprint of the
Taylor & Francis Group, an **informa** business

CRC Press
Taylor & Francis Group
6000 Broken Sound Parkway NW, Suite 300
Boca Raton, FL 33487-2742

First issued in paperback 2021

ISBN 13: 978-1-03-223940-8 (pbk)
ISBN 13: 978-1-4665-6709-2 (hbk)

Library of Congress Cataloging-in-Publication Data

Shchukin, E. D. (Evgenii Dmitrievich)
  Physical-chemical mechanics of disperse systems and materials / Eugene D. Shchukin and Andrei S. Zelenev.
      pages cm. -- (Progress in colloid and interface science)
  Includes bibliographical references and index.
  ISBN 978-1-4665-6709-2 (alk. paper)
  1. Colloids. 2. Surface chemistry. 3. Fluid mechanics. 4. Chemical engineering. 5. Chemistry, Physical and theoretical--Mathematics. I. Zelenev, Andrei S. II. Title.

QD549.S495 2015
541'.345--dc23
                                                                    2015021254

Visit the Taylor & Francis Web site at
http://www.taylorandfrancis.com

and the CRC Press Web site at
http://www.crcpress.com

Contents

SECTION II   Surface Phenomena in the Structures with Phase Contacts and in Continuous Solid Bodies

Foreword

The authors of this book, Eugene D. Shchukin and Andrei S. Zelenev, represent the well-known Russian school of colloid and interface science, although both have been residing and working in the United States since the early 1990s. Dr. Shchukin has been a professor at Johns Hopkins University in Baltimore, Maryland, while continuing to teach and supervise research at the Colloid Chemistry Division at the Chemistry Department of Moscow State University, Moscow, Russia and at the Institute of Physical Chemistry of the Russian Academy of Science, Moscow, Russia. Dr. Zelenev has been pursuing a career as an industrial scientist holding a number of research and managerial positions with international chemical companies. This book is the authors' second joint project aimed at introducing the results and accomplishments of the Soviet and Russian school of colloid and interface science to the English-speaking audience.

Physical-chemical mechanics is a fundamental interdisciplinary area of science addressing the reciprocal effects of physical-chemical surface phenomena in disperse systems and materials and their mechanical behavior. Although it is based on colloid chemistry and material science, it is principally different from both of these disciplines. Colloid chemistry addresses the properties of matter in the dispersed state, and material science deals with properties of bulk materials. Conversely, physical-chemical mechanics is focused on the processes of transformation of disperse systems with a broad range of particle size from nanometers to millimeters into a material, as well as on the inverse processes of the transformation of materials into individual particles. The former takes place in various processes involving particle bridging, pressing, sintering, hydration hardening and sol-gel transitions, while the latter are encountered in material dispersion and failure by wear, friction, milling, various mechanical treatment, etc. The description of processes of material degradation and disintegration utilizes the principles and methods of material science as they apply to the description of mechanical strength. Although these two groups of processes are seemingly opposite to each other, they are united by the universal role of surface phenomena, such as adsorption of surface-active substances and media-induced lowering of interfacial free energy. These phenomena constitute a physical-chemical basis for the formation and healing of microscopic surfaces in the course of origination, development, and destruction of contacts between the particles. Contact interactions are the cornerstone subject of physical-chemical mechanics. Controlling contact interactions provides one with the means to influence numerous industrially important processes, such as those of increasing the strength of catalysts, drilling for oil and gas, papermaking, and many others.

Theoretical and experimental investigation of contact interactions is the leading theme of this book, while the chapters addressing the subjects of colloid and surface science and of material science are written with a level of detail that is necessary for understanding contact interactions and the transformation of individual particles into a material, and contrarily the conversion of a material into individual particles. Physical-chemical mechanics covers the broad spectrum of objects and systems, from milli- to nano-sized, and establishes the fundamental basis for controlling and tuning the properties of such systems and the processes taking place in them.

Physical-Chemical Mechanics of Disperse Systems and Materials contains seven chapters. Section I, with four chapters, presents the basics, starting from surface forces and the contact of particles with liquids. Chapter 2 is dedicated to adsorption phenomena, accumulation of surface-active molecules at various interfaces, and the importance of surfactant's adsorption on the contact between particles. The bulk properties of particle dispersions in liquids are discussed in Chapter 3 in terms of coagulation processes and the rheological behavior. Chapter 4 describes in a comprehensive way the stability of disperse systems and emphasizes the Rehbinder effect as an important mechanism in stable colloidal systems. Section II consists of three chapters. Chapter 5 provides an introduction to the methodology of mechanical testing; Chapter 6 describes in detail the structures with phase contacts, emphasizing numerous industrially important applications, such as hydration hardening of mineral binders; and Chapter 7 focuses on the essence and role of interfacial phenomena in the deformation and fracture of solids with different types of chemical bonding, with special attention to the role of active media, that is, on the Rehbinder effect.

We are very happy that this book, unique in its topic and way of dealing with disperse systems and materials, is published as Volume 6 in the series Progress in Colloid Interface Science.

Reinhard Miller
Potsdam, Germany

Libero Liggieri
Genoa, Italy

Preface

This book provides a comprehensive description of physical-chemical mechanics, which is a discipline that uniquely bridges solid-state physics and materials science and colloid and surface chemistry. While colloid and surface science deals mainly with the *properties of a substance in a dispersed state*, and solid-state physics/material science are concerned mainly with the *properties of a substance making up a solid phase*, the scope of physical-chemical mechanics is different from the traditional studies of the properties of substances in that it consists of the investigation of complex *physical-chemical interfacial phenomena* taking place (1) in the course of the transition of a dispersed system into a material and/or, conversely, (2) in the course of the dispersion of a bulk material. This book is based on a critical assessment of results published in the literature, mostly in Russian, over the past four decades, and on the materials used in the lecture course on physical-chemical mechanics developed and taught by Professor Eugene D. Shchukin over many years at the Chemistry Department of Moscow State University in the USSR, and then in Russia, and at Johns Hopkins University in Baltimore, Maryland. The author has also taught the course on physical-chemical mechanics at the University of Sofia, Bulgaria, at Havana University in Cuba, at the Institute of Surface Chemistry (YKI) in Stockholm, Sweden, and at the Helsinki University of Technology in Finland. Many illustrations encountered throughout the book are based on the slides used in those lecture classes.

Professor Eugene D. Shchukin was a long-time collaborator and close friend of Professor Peter Alexandrovich Rehbinder (1898–1972), the founder of physical-chemical mechanics. Over many years, he worked closely with Rehbinder at the Laboratory of Physical-Chemical Mechanics at the Frumkin Institute of Physical Chemistry in Moscow and at the Chemistry Department of Lomonosov Moscow State University and continued to pursue active research in the area following Rehbinder's death. In the course of his work, Professor Shchukin has served as an advisor, co-advisor, and direct participant on numerous PhD and doctor of science thesis research projects. Some of the results presented in the book were generated as part of such thesis research. Those studies are not readily accessible to the English-speaking scientific community but are critical for understanding the underlying principles behind the formation of contacts and structures and common processes leading to wear, fracture, and failure, as well as the ambient media effects in these processes.

Since 1991, Professor Shchukin has resided in Baltimore, Maryland, where he has continued with his work in physical-chemical mechanics. The work conducted in Baltimore included the studies on contact interactions between cellulosic fibers in the presence of polyelectrolytes and surfactants, also included in this book. This was a joint project between Professor Eugene Shchukin, Dr. Igor Vidensky, and Dr. Andrei S. Zelenev, who is the second author of this book. Dr. Zelenev is a graduate of the Chemistry Department of Moscow State University and of Clarkson University (Potsdam, New York), where he earned his PhD in physical chemistry under the supervision of Professor Egon Matijević. Following graduation, Dr. Zelenev pursued a career in industrial research focusing on developing specialty chemicals for pulp and paper and later for oil and gas production. While belonging to the younger generation of scientists, Dr. Zelenev has a long history of collaboration with Professor Shchukin. The authors previously jointly coauthored the textbook *Colloid and Surface Chemistry* (Elsevier, 2001), which laid the path for preparing this book.

The present book is composed of the seven chapters that cover the material used as the basis for the lecture course on physical-chemical mechanics and basic material essential for understanding the content from the areas of colloid and surface chemistry, strength of materials, rheology, and tensors. Such coverage makes the book suitable for readers who do not have extensive knowledge and expertise in any of these areas. The book also devotes a lot of attention to the experimental methods used in physical-chemical mechanics and the relevant instruments, many of which were built and developed over the years by collaborators of Professor Shchukin. Where appropriate and relevant, the authors have included a somewhat detailed description of specific significant studies. These studies are included in the corresponding chapters as essays. Some examples include studies on fluorinated systems, the damageability of glass, the strength of catalysts, etc. Dr. Andrei Zelenev has brought together and organized all of this very different material into a continuously flowing text. In order to maintain proper emphasis and the integrity of the presentation, the authors have occasionally used the same illustrations in different chapters.

The authors acknowledge the contribution from several critical participants of this project. They are indebted to Dr. Aksana M. Parfenova, who has prepared many others original illustrations used by Professor Shchukin in lecture classes. The authors also thank Harald Hille, a retired United Nations translator and terminologist, who read through the manuscript to check for infelicities in the English text, a language of which none of the authors is a native speaker, although they have worked as scientists in the United States for many years. Finally, the authors are indebted to Dr. Kristina D. Kitiachvili, who has provided invaluable help with the editing and prepublication organization of the manuscript.

Authors

Eugene D. Shchukin graduated in physics from Moscow State University (MSU, 1950) and earned a PhD (1958) and a doctor of science (1962) in physics and mathematics. The main positions he has held include the following: research professor at Johns Hopkins University (1994–2006) and professor emeritus (since 2006); professor (1967–2002), chairman (1973–1994), distinguished professor (since 1995) at the Colloid Chemistry Department at MSU, Institute of Physical Chemistry of the USSR Academy of Sciences, Head of Laboratory of Physical Chemistry Mechanics (1967–2003), and advisor (since 2003).

Dr. Shchukin's principal research interests are physical-chemical mechanics of disperse systems and materials, colloid and interface science, surfactant effects at various interfaces, stability of disperse systems, particles interactions and structure formation, materials science and engineering, physical chemistry of solid state, stability and damageability of solids and their surfaces in active media, control and applications in technology and environmental engineering.

He is a member of various Russian, U.S., and other international scientific committees and advisory and editorial boards. He was vice president of the International Association of Colloid and Interface Scientists during 1986–1988.

Dr. Shchukin was elected to the Russian Academy of Pedagogical Science (1965; recently Academy of Education), to the U.S. National (1984), Royal Swedish (1988), and USSR (1990, Russian) Engineering Academy; Founding Member of the Russian Academy of Natural Sciences (1990).

Dr. Shchukin is a recipient of the following prizes and awards: Lomonosov Prize from the MSU (1972), Lenin Prize from the Soviet Union (1988) (Equivalent to National Medal of Science in the United States), Rehbinder's, Russian Academy of Sciences (1998), and a Gold Medal from the Russian Academy of Education (2008).

Andrei S. Zelenev graduated from the Department of Chemistry at Moscow State University, Moscow, Russia, (MSc, 1993), and from Clarkson University in Potsdam, New York (PhD, 1997). Since 1998, he has been working as an industrial scientist and researcher at Nalco Company (staff scientist) and CESI Chemical, a Flotek Company (research scientist and research manager), and currently with Fritz Industries (R&D manager, Stimulation).

Dr. Zelenev's professional interests include industrial applications of colloid and surface science, pulp and paper, oil and gas production, coagulation and flocculation, lyophobic and lyophilic colloidal systems, surfactant phase behavior, interaction of surfactants with surfaces, microencapsulation, particle deposition and aggregation, particle and surfactant transport in porous media, wetting and spreading, development of novel experimental methods for studying colloidal systems, and physical-chemical mechanics. Dr. Zelenev is an inventor on four issued U.S. patents and five pending patent applications, coauthor of 22 scientific publications, and coauthor of the textbook *Colloid and Surface Chemistry* (Elsevier, 2001).

Introduction

Looking at the title of this book, *Physical-Chemical Mechanics of Disperse Systems and Materials*, the first and very important question one might ask is what does it mean. Each of the six terms has a distinct and well-defined meaning, and together they define the scientific discipline originally introduced by Peter A. Rehbinder. The essence of physical-chemical mechanics is schematically illustrated by the chart shown in Figure I.1. Over many years, this chart was used by Professor Shchukin in lecture classes on the subject.

Let us try to understand which principal factors are the same for both disperse systems and the resulting materials.

First, there is the universal importance of mechanical properties defined in a broad sense as the ability of a material to resist mechanical action and the interest of both science and industry to control those properties. Depending on the need, one may seek to increase the resistance to mechanical action, thus ensuring strength and durability, or one may want to decrease that resistance in order to improve one's ability to form and machine a material.

Second, we deal with the real physical structure of a material, with the numerous defects associated with it, that is, micro- and ultra-micro-heterogeneity, defined by the connectivity between its many particles. Those connected particles form the basis for the structure. That implies that there is a universal nature of the disperse state with its highly developed internal and external surfaces.

Third, we consider the critical role played by various chemical and physical-chemical phenomena taking place at these surfaces and specifically the role of surface-active components present in the ambient medium.

Fourth, all of the *mechanical* processes of the deformation and fracture of a solid are of interest, and those involving the rupture and reformation of contacts between individual particles, as they constitute the processes of the rupture and restructuring of chemical bonds; that is, these are the *physical-chemical processes.*

Fifth, both disperse systems and the materials associated with them are united in the same cycle of transformations taking place in nature and technology. The universal nature of these mutual transitions is illustrated on the left-hand and right-hand squares shown in the diagram in Figure I.1 and reflect Rehbinder's principle of "achieving strength via damage."

The discussion of the transformation, rupture, and reformation of the bonds between the particles, and of the influence of the medium on these processes, constitutes the main subject of the first part of the book, which addresses structures with "weak" coagulation contacts between the particles, and the second part, which is devoted to the description of structures with "strong" phase contacts and of continuous materials. It would be useful to define these two broad classes of systems here.

Free-disperse systems are typically represented by dilute sols having low concentrations of particles, while the connected-disperse systems are those in which one typically encounters high particle concentrations, such as concentrated sediments or pastes (Figure I.2).

When the contacts between the particles in a free-disperse system are established, the transition of the system into a connected-disperse state takes place. This transition is associated with the development of a spatial network of particles in which the cohesive forces between the particles forming a network are sufficiently strong to resist thermal motion and the action of external forces. As a result of the transition, the system acquires a set of new structural-mechanical (rheological) properties that characterize the ability of the system to resist deformation and separation into individual parts. That is, the system acquires *mechanical strength*, which is the principal and universal characteristic of all solid and solid-like materials. For many materials, their mechanical strength defines the conditions of their use.

Although it may sound intuitively obvious that one would want to get durable and long-lasting materials, strength is not always the feature that is sought. Some examples where that is the case

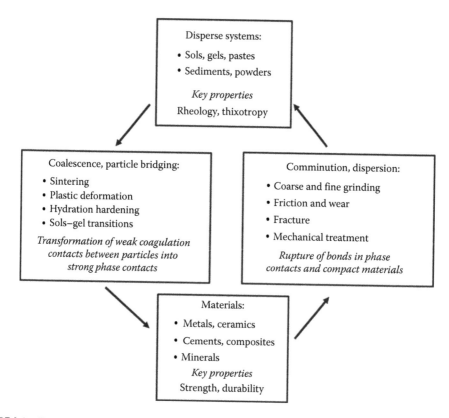

FIGURE I.1 The main objects of physical-chemical mechanics and the relationships between them.

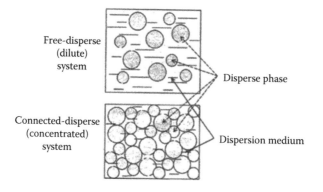

FIGURE I.2 Free-disperse systems and connected disperse systems.

include various processes involving the formation of disperse systems, the packing of powders, the densification of suspensions or in the treatment of metals in the course of grinding, cutting, or pressing. In all of these applications, the main goal is to decrease mechanical resistance to treatment. At the same time, a decrease in strength may result from specific use conditions, and one needs to pay special attention to such factors in the numerous and dangerous processes of wear and failure in construction materials and the parts of various machines and mechanisms, etc. In order to seek ways to prevent and minimize potential damage, a good understanding of the underlying mechanisms is needed.

The understanding of structure formation in disperse systems, including the investigation of the mechanical properties of various structured systems, the optimal conditions for their formation, treatment, and fracturing (with a special emphasis on the role of active media), is one of the central subjects in the study of the physical-chemical mechanics of disperse systems and materials.

Physical-chemical mechanics originated in the 1930s and 1940s and in the 1950s became a separate and unique discipline bridging colloid and interface science with material science. Critical contributions to this area were made by various academic and industrial research groups working in the former USSR, specifically by scientists belonging to the scientific school established by Peter Alexandrovich Rehbinder (1898–1972), the discoverer of the Rehbinder effect. Numerous early and highly significant studies in the area conducted over the span of several decades are reviewed and referenced throughout the book.

Physical-chemical mechanics addresses a great variety of different objects and applications. These include various disperse systems, such as pastes, powders, and suspensions, encountered in various areas and applications; various materials used in modern technology; and such natural objects as minerals, soils, and bone tissue. This great variety emphasizes the universal nature of the disperse state of matter and the universal importance of mechanical properties.

Among the variety of disperse systems, fine disperse systems with particles in the range of 1–100 nm, commonly referred to as *nanosystems*, play an important role. Achieving a fine degree of dispersion and the highest possible uniformity is the prerequisite for obtaining strong and durable materials. At the same time, the high surface area of such materials provides for the highest effectiveness of various heterogeneous chemical processes involving mass and heat transfer.

The highest possible degree of dispersion cannot be achieved by simple comminution and grinding, that is, by dispersion processes. It can only be achieved by condensation methods. The latter involve the nucleation of a new phase in the initial homogeneous system at high levels of supersaturation.

Due to the large number of contacts between the particles, one encounters difficulties in stabilizing and molding fine disperse colloidal systems. The cohesive forces existing between the particles in such systems constitute the manifestation of physical-chemical interactions. By investigating the relationship that exists between the mechanical properties of disperse systems and materials, their structure, and interfacial phenomena, one seeks to find new ways to tune and control the properties of materials. This is achieved by using a controlled combination of mechanical action (including vibration) and physical-chemical means of controlling the interfacial phenomena—for example, by using surfactants.

Within the scope of physical-chemical mechanics, various approaches are used to describe the mechanical properties of various liquid-like and solid-like bodies and materials. These include the methods of macro- and microrheology, and molecular dynamic experiments, allowing one to approach the problem at the molecular dimension. The combination of these approaches provides one with the means to analyze the properties of real disperse systems and with methods for controlling them. Special attention is devoted to the Rehbinder effect, that is, to the adsorption-related influence of the dispersion medium on the mechanical properties of solids.

The structure formation that takes place in disperse systems is the result of spontaneous, thermodynamically favorable processes of particle cohesion, leading to a decrease in the free energy of the system, such as particle coagulation or substance condensation at the points of particle contact. The development of spatial networks of various types is the foundation for the ability of a disperse system to transform into a material. As a result of such a transformation, the system acquires new characteristics and properties that are completely different from those in the original state.

Material strength, P_c (N/m), is the principal characteristic of a given material, determining the ability of a material to resist fracture under the action of external stresses. In disperse systems of *globular type* (Figure I.3), the strength P_c is determined by the combination of cohesive forces acting in the particle–particle contacts and the number of such contacts per unit area, that is,

$$P_c \approx \chi p_1$$

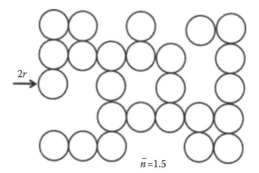

$\bar{n} = 1.5$

FIGURE I.3 The model of a globular disperse structure.

Both p_1 and χ can be obtained experimentally and estimated theoretically, as will be thoroughly discussed in the book.

The value of χ is determined by system geometry, primarily by the particle size (radius r, if the particles are spherical) and the particle packing density. In the case of moderately dense structures, such as those with a primitive cubic lattice with the coordination number of 6, one can approximately write that

$$\chi \approx \frac{1}{(2r)^2},$$

which provides one with the means to establish the spectrum of possible χ values in real systems. In particular, for $2r \approx 100$ μm, $\chi \sim 10^3$–10^4 contacts per cm^2; for $2r \approx 1$ μm, $\chi \sim 10^7$–10^8 contacts per cm^2; and for $2r \approx 10$ nm, $\chi \sim 10^{11}$–10^{12} contacts per cm^2.

This description holds true for disperse systems of the globular type, in which a continuous backbone is formed due to the cohesion of the individual particles in the course of transformation of a free disperse system into a connected disperse system. In such systems, the backbone formed is the main carrier of the strength. At the same time, there are also other types of systems, for example, those with a cellular structure (solidified foams or emulsions). Such structures are typical in polymeric systems and may form in the course of new phase formations by condensation in mixtures of polymers. An individual approach also needs to be employed in the description of the mechanical properties of structures with anisometric particles, due to the specifics of the cohesive forces in such systems. In addition to porous structures, we also consider various compact microheterogeneous structures, such as mineral rocks, modern composite materials, and natural materials such as bone and wood.

Depending on the nature of the forces involved in the particle cohesion, the contacts can be subdivided into *coagulation contacts* and *phase contacts*, as schematically shown in Figure I.4. Coagulation contacts are discussed in detail in the first section of the book. In these contacts, particle–particle interactions are limited to direct "touching," either directly, or via the remaining gaps filled with the dispersion medium (Figure I.4). The strength of the coagulation contacts is determined

(a) (b) (c)

FIGURE I.4 Contacts between particles: (a and b) coagulation and (c) phase.

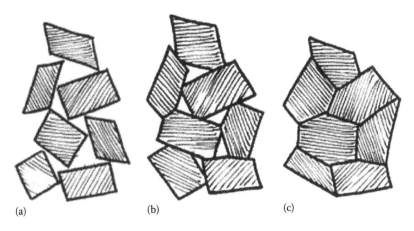

FIGURE I.5   Structures with (a) coagulation contacts, (b) phase contacts, and (c) a continuous body.

by the surface forces of the intermolecular interactions and for spherical particles with radii of 10 nm, 1 μm, and 1 mm, which is on the order of 10^{-9}–10^{-10}, 10^{-7}–10^{-8}, and 10^{-4}–10^{-5} N, respectively.

Structures based on coagulation contacts (Figures I.4 and I.5) are typically characterized by low strength and mechanical reversibility manifested in their ability to spontaneously restore their previous structure after mechanical degradation (*thixotropy*, see Chapter 3).

The presence of an essentially similar liquid medium and of surfactant adsorption influences the magnitude and nature of the surface forces and may result in weakened cohesion in the contacts by two to three orders of magnitude. In a *lyophilized*, highly concentrated system in which the particles are brought into mechanical contact, this is revealed through lower resistance to deformation, τ^*, and results in a plasticizing of the system (see Chapters 2 and 3). When the disperse phase concentration is low, lyophilization leads to the preservation of the colloidal stability of a free-disperse system, that is, the resistance of the system to coagulation (see Chapter 4).

The second section of the book principally deals with different types of contacts—the so-called *phase contacts*. These contacts are substantially stronger than coagulation contacts and are mechanically irreversible. The transition from coagulation contacts to phase contacts corresponds to a transition from a simple "touching contact" (due to weak van der Waals forces) to the "accretion" of particles over a significant area (Figures I.4 and I.5). The accretion process is associated with the formation of a bridge of the new phase. This takes place in various processes involving the mutual plastic deformation of particles subjected to compressive forces, in sintering, the hydration hardening of mineral binders, and in sol-gel transitions.

In phase contacts, the cohesion between the particles is caused by short-range cohesion forces, which propagate over the area, significantly exceeding the dimensions of an elementary cell. There are about 10^2–10^3 interatomic bonds involved in the cohesion, and the contact surface may be compared to the region of a grain boundary in polycrystalline material. Once phase contact has been established, the transition from one bridged particle to the other can be accomplished within the same phase (Figure I.4), which is the reasoning behind the terminology.

For a large number of materials, the minimal value of the strength of such contacts is in the range between 10^{-8} and 10^{-6} N. As the area increases, the strength of the phase contacts also increases, reaching ~ 10^{-4}–10^{-3}. The strength of structures with phase contacts covers a range from 10^4 to 10^8 N/m² or even broader. In contrast to coagulation contacts, phase contacts are fractured irreversibly. Structures with phase contacts and the corresponding materials are discussed in detail in Chapter 6.

Consequently, we can see that physical-chemical mechanics is the interdisciplinary area of science that employs the principles and concepts of colloid and surface chemistry on the one hand and of physics and material science on the other.

At the same time, physical-chemical mechanics should not be viewed as a simple combination of particular fields of these areas of science. The main subject of colloid and surface chemistry concerns the properties of a substance in the disperse state, which is a microheterogeneous state characterized by highly developed interfaces. Material science and physics mainly address the properties of a substance in the state of a solid phase. This state is characterized by the absence of highly developed interfaces. Physical-chemical mechanics connects these two principally different areas by addressing the physical-chemical interfacial phenomena involved in the mutual transformation of a disperse system into a material, or in the conjugate processes of the dispersion of solids, rather than by focusing on the substance properties. Within the scope of physical-chemical mechanics, the general thermodynamic, kinetic, and structural regularities describing the formation of cohesive contacts or the onset of damage as principally surface phenomena are identified. Physical-chemical mechanics especially focuses on the determining role of initial surface defects in the origination and further development of damage in solids. The presence of the active components of the medium, of adsorption, or of contact with the related liquid phases may significantly influences the processes of particle cohesion or the formation of cracks. Depending on the conditions, the latter can be both facilitated and retarded. Among such physical-chemical processes studied by physical-chemical mechanics, special attention is devoted to the media-induced lowering of the strength of solids due to the lowering of surface energy, known as the *Rehbinder effect*, which is discussed in detail in Chapter 7.

Section I

Coagulation Contacts and Structures

1 Surface Forces and Contact Interactions

The properties of a contact between the particles of a solid phase depend on a combination of the properties of that solid phase, the media in the gap between the particles, and the interactions between the solid phase and the medium. To a great extent, these interactions are due to intermolecular forces acting at the interface between phases. Since *surface forces* are one of the main subjects discussed in modern books on colloid and surface science, we will restrict ourselves only to a brief review of the subject relevant to physical-chemical mechanics.

1.1 PHYSICAL-CHEMICAL PHENOMENA AT INTERFACES

Before turning to a discussion on particular interactions, it is worth reviewing the basic principles of the thermodynamics of surface forces and the concept of surface (or interfacial) free energy and of surface tension in particular.

On an intuitive level, the concept of surface tension becomes apparent in the description of Dupré's original experiment, in which one observes the stretching of a soap film, either in a bubble or on a frame (Figure 1.1). Equilibrium is established when a force, $F = 2d\sigma$, is applied to the moving boundary, where d is the frame width, σ is the *surface tension*, and the numerical coefficient 2 implies that the film has two sides.

Displacing the boundary by Δl increases the film area by $2d\Delta l$ and requires that work equal to $F \Delta l$ is expended (Figure 1.1). Per unit of the newly formed surface, this work is $F \Delta l / 2d\Delta l = F/2d = \sigma$. Consequently, the meaning of σ now is *specific surface free energy* (units of work per unit surface area). In a thermodynamic sense, the term "free" means that this (equilibrium) process takes place at constant temperature and volume. The identity of those two notions, that is, of the surface tension and the surface free energy, is true only for "common" (unassociated) liquids. For solids, the situation is more complex [1–3].

A rigorous definition of the specific *surface* free energy of a body as the excess of free energy at an interface between the condensed phase and air (or vacuum), as well as the definition of specific *interfacial* free energy at the interface between two condensed phases, can be given in terms of Gibbs's thermodynamics of surface phenomena [2,4]. Often the terms "surface free energy" and "interfacial free energy" are used interchangeably, but, as the discussion that follows will show, it is best to separate them.

As shown in Figure 1.2, let us choose a prism with a unit cross section of $S = 1$, along a direction 1 perpendicular to the interface between phase 1 and phase 2, for example, perpendicular to the interface between a liquid (1) and its vapor (2). In a region adjacent to the geometrical interface between those two phases (the so-called dividing surface), we can select a transition layer of finite thickness (*physical discontinuity surface*), with properties differing from those of both phases 1 and 2. The boundaries of this layer are positioned at distances δ' and δ'' from the dividing surface, and consequently, the effective thickness of the discontinuity surface is $\delta = \delta' + \delta''$. Let the density of the surface free energy f (units of work per unit volume) outside the discontinuity surface be equal to f' within phase (1) and f'' within phase (2). The distribution of the surface free energy density along the z-axis is illustrated by a plot of $f(z)$ in Figure 1.2 within the physical discontinuity surface $f(z)$ and is different from both f' and f''. In principle, the transition from f' to f'' may be smooth, but it is more

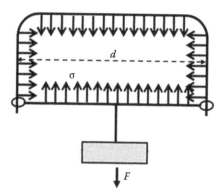

FIGURE 1.1   Schematic illustration of Dupré's experiment, which helps to understand the concept of surface tension.

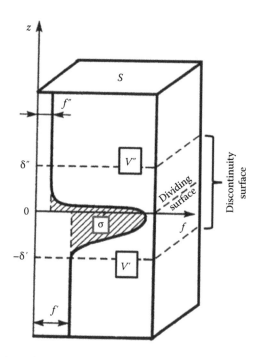

FIGURE 1.2   The thermodynamic definition of the surface free energy according to Gibbs.

commonly "tonguelike," as seen in Figure 1.2. To characterize this "tongue" quantitatively as the free energy density excess over f' and f'', one needs to extrapolate f' and f'' to the dividing surface. According to Gibbs, this excess defines the value of the specific (per unit area) interfacial free energy, or surface free energy, in the case of the interface between a vapor and a vacuum:

$$\sigma = \int_{-\delta'}^{0} \left[f(z) - f' \right] dz + \int_{0}^{\delta''} \left[f(z) - f'' \right] dz \tag{1.1}$$

Typically, when far away from the critical conditions, δ is on the order of interatomic (intermolecular) distances, b, that is, on the order of angstroms (1 Å = 10^{-10} m). As the system nears its critical point corresponding to the complete miscibility or complete mutual solubility of the constituent phases, the value of δ may reach hundreds or thousands of angstroms.

Gibbs's approach to the determination of σ is valid for any interface: liquid/gas (L/G), liquid/liquid (L1/L2), solid/gas (S/G), solid/liquid (S/L), or solid/solid (S1/S2). The methods are used to either measure or indirectly estimate the interfacial free energy, which may vary significantly depending on the type of interface [3,5].

Typical measured values for the surface and interfacial tension, σ, for liquid phases (in mN/m = dyn/cm = mJ/m^2) are as follows [6–11]:

Liquefied Noble Gases	Single Digits	
Hydrocarbons	20–25	(300 K)
Alcohols	30–35	(300 K)
Water	72	(300 K)
Water/vapor interface near critical point	Approaches 0	
Water/hydrocarbon interface	50	(300 K)
Water/hydrocarbon interface with t-butanol	50–0	(300 K)
Molten salts	Hundreds	
Liquid metals	Hundreds or thousands	

It would be worthwhile to immediately turn one's attention to a discussion of the molecular origin of the surface free energy and the surface tension, and the principal difference of the latter from the state of a rigid solid body deformed by stretching.

To do this, let us follow Laplace and consider a spherical gas holder shell (i.e., a balloon filled with gas) with a radius R and an excessive pressure, P, inside (Figure 1.3). Let us position a cone of small volume with a small angle at the apex, 2ϕ, in such a way that the cone's apex matches the center of the sphere and the base matches the surface of the shell. The length of the base is $a = R\phi$, and the length of the base perimeter is $2\pi a = 2\pi r\phi$, while the area is $\pi a^2 = \pi R^2 \phi^2$. The pressure force exerted on this surface is $F = PR^2\phi^2$. If the stretching elastic stresses acting in the shell are equal to σ^{elast} [N/m], then the force F is balanced by the component of a stretching stress acting in the direction of the radius R (axis of the cone), namely, $\sigma^{elast}\phi$, acting along the perimeter $2\pi r\phi$. This force is given by $F = 2\pi R\phi\, \sigma^{elast}\phi$. The force balance relationship, $F = P\pi R^2\phi^2 = 2\pi R\phi\sigma^{elast}\phi$, yields the Laplace equation for a sphere

$$P = \frac{2\sigma^{elast}}{R} \tag{1.2}$$

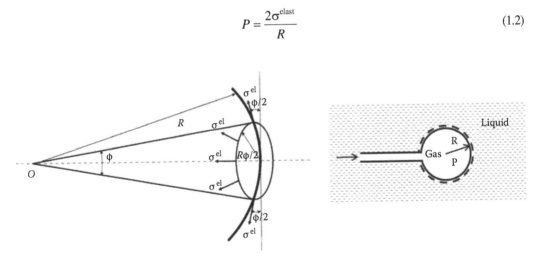

FIGURE 1.3 Derivation of the Laplace law for a rigid shell and for a bubble.

The aforementioned equation constitutes Laplace's law for a sphere. For a cylinder, the expression is similar, except that it does not have a numerical coefficient 2.

Instead of considering a balance of forces, one may also derive Laplace's law by using the *variation of work* approach. Let Δp be the excessive pressure as compared to the surrounding atmospheric pressure inside a bubble with radius r and a surface tension of σ. The change in the bubble radius in the vicinity of the equilibrium by δr is associated with the work of expansion in the system

$$W_{\text{exp}} = \Delta p \delta V = \Delta p \delta \left(\frac{4}{3} \pi r^3 \right) = 4 \pi r^2 \Delta p \delta r \tag{1.3}$$

As a result, the free energy of the system is lowered by the amount of this work. At the same time, an increase in bubble size results in an increase in the surface area associated with an increase in the surface free energy, $\sigma \delta(4 \pi r^2) = 8 \pi r \sigma r$. The overall free energy variation around the equilibrium is zero

$$\delta W = -4 \pi r^2 \Delta p \delta r + 8 \pi r \sigma \delta r = 0$$

which readily yields the Laplace equation

$$\Delta p = p_\sigma = \frac{2\sigma}{r} \tag{1.4}$$

This "continuum" approach does not clearly reveal a molecular-level difference between an *elastic* deformation of a solid shell and the *nonelastic* behavior of a liquid surface. The essence of this difference is as follows: Stretching a solid body (within the limits of ideal elasticity) causes the stretching of all interatomic (intermolecular) bonds within the volume of the body in the direction of the applied stretching force. This process is reversible—both thermodynamically and mechanically. As a result of such stretching, an increase in dimensions (length, volume, surface area) takes place. This increase is not large, as determined by a high value of Young's modulus, and the increase in surface area is stipulated strictly by the change in the distance between the surface atoms with their relative position remaining unchanged. This scenario is completely different from what happens in the course of the deformation of a liquid at moderate pressures, such as that caused by viscous flow. In that case, a change in the surface area neither results in a change in the intermolecular distances nor does it cause any change in volume. Instead, it is caused by molecules entering the surface from the bulk of the liquid phase. This process is associated with the restructuring of the intermolecular bonds. These two scenarios for the increase in surface area taking place in a solid body and a liquid are illustrated in Figure 1.4.

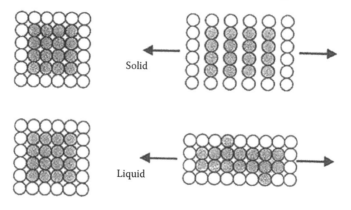

FIGURE 1.4   Elastic deformation of a solid body and viscous flow in liquids.

The following simple example illustrates quantitative relationships within a dense pack of spherical particles, for example, atoms. In a metal, an atom in the bulk is surrounded by 12 neighbors, while an atom within the surface layer is surrounded by only 9 neighbors. This deficiency of three interatomic bonds results in ~25% lower cohesive energy at the surface as compared to that in the bulk. The energy of cohesion in the bulk of a body is *negative,* because it is counted from the zero reference point when mutually attracting particles cluster together from infinity to form a solid body. The lack of any bulk energy of cohesion per unit area is exactly the *positive* excess associated with the surface layer, referred to as the specific surface free energy, σ.

This scheme corresponds to the well-known Stefan's approach to the estimation of the surface energy of solid phases. The density (energy per volume) of the energy of cohesion, G_c, in the bulk of a condensed phase may be approximately estimated as the heat of sublimation per molar volume, H_{subl}/V_m, or as the Young's modulus, E, or as the so-called molecular pressure, K (see [12–14] for details).

Within the volume of a discontinuity surface of thickness δ, which is on the order of the interatomic distance b, the lack of cohesive energy is qbG_c, where q is the ratio of the so-called boundary coordination number, z_b, to the bulk coordination number, z, within the bulk of a phase. The boundary coordination number z_b is characterized as the number of "missing bonds" (or missing neighbors) between surface atoms. For instance, in the case of face-centered close packing, $z = 12$, and the value of z_b for the atoms in the (111) plane is 3, with $q = 1/4$ (as above). In this way, we have introduced a possibility for getting an estimate of the free surface energy of the condensed phases, σ, as qbG_c. Such an approximate estimate allows a comparison with other approaches and clearly illustrates the general conclusion, namely, that for solids having high strength and high melting points, such as transition metals, oxides, carbides, and nitrides, the values of σ are the highest, ($\sim 2–3 \times 10^3$ mJ/m²), while organic liquids have low σ (tens of mJ/m²), and the values of σ for liquefied gases are even lower.

When one compares the thermodynamic description of a surface with that of a rigid shell, the following principal difference between these two cases needs to be revealed. In the case of a rigid body, the entire potential energy of elastic deformation is determined by the work performed on the body. In the case of a liquid surface, stretching is also associated with the transfer of heat. The total surface energy ("internal surface energy"), ε, at a given temperature T includes, in addition to the mechanical work, σ, also hidden heat, $\varepsilon = \sigma + \eta T$, where η is specific (per unit area) surface entropy, which can be established from the temperature dependence of the surface tension, $\eta = - d\sigma/dT$ (Figure 1.5). For a free surface, the surface entropy is positive and, for most liquid phases, is around 0.1 mJ/m² K.

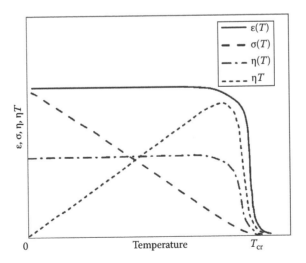

FIGURE 1.5 Temperature dependence of free energy (σ), internal energy (ε), entropy (η), and latent heat (ηT). (Redrawn from Shchukin, E.D. et al., *Colloid and Surface Chemistry*, Elsevier, Amsterdam, the Netherlands, 2001.)

1.1.1 THREE-PHASE CONTACT LINE: WETTING

Before we start a discussion on the thermodynamics of a contact between particles, it is worthwhile to briefly address the phenomena taking place at the three-phase contact line, and in particular, wetting and capillary forces acting within a liquid meniscus. We will also briefly summarize the principal methods of surface tension measurement.

Following Young, let us examine the contact between a solid phase and a liquid phase in ambient air, that is, a drop of liquid sitting on a solid substrate, as shown in a 2D representation in Figure 1.6. The three interfaces share a common perimeter, which is a linear boundary perpendicular to the plane of the drawing. The drop is at equilibrium when a balance is achieved between the three surface tension forces acting along the perimeter, as depicted by the vectors in Figure 1.6. The equilibrium between those three forces requires that the algebraic sum of the projections of the three vectors is equal to 0, which readily yields Young's equation

$$\sigma_{SL} + \sigma_{LG}\cos\theta = \sigma_{SG}$$

where the contact angle, θ, is measured inside the liquid. Depending on whether $\theta < 90°$ or $\theta > 90°$, one talks about wetting or nonwetting, respectively (Figure 1.6).

Young's equation can also be obtained by a variation approach, that is, by considering variations in the free energy of the system: the algebraic sum of the small variations in the three components of the interfacial energy should equal to zero when we encounter small deviations from the equilibrium. This approach is schematically summarized in Figure 1.7. By shifting the three-phase contact point A by a small distance AA′, we introduce small variations in the three components of the free interfacial energy, the sum of which is zero

$$\delta F_s = \delta S_{SL}\sigma_{SL} + \delta S_{LG}\sigma_{LG} + \delta S_{SG}\sigma_{SG} = AA'\sigma_{SL} + AA'\sigma_{LG}\cos\theta - AA'\sigma_{SG} = 0$$

which readily yields the Laplace law.

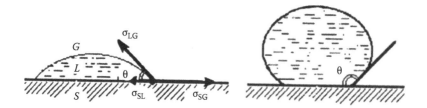

FIGURE 1.6 Young–Laplace representation of a three-phase contact line between solid, liquid, and gas phases. Cases of wetting and nonwetting liquids.

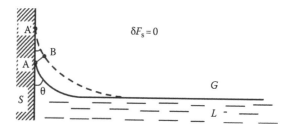

FIGURE 1.7 The schematic illustration of the derivation of Young's equation by the variation method: $\delta F_s = 0$.

We have addressed the situation of a rigid solid surface when a single contact angle is formed. In a contact involving two liquid phases and a gas phase (e.g., a drop of fat on water surface), two contact angles are formed. For each of these angles, the equations describing the equilibrium must take into account the individual horizontal and vertical components. The same argument can also be applied to the solid–liquid–gas interface between a gas, a liquid, and a yielding material, such as a polymer with a low Young's modulus.

It is also worth emphasizing that the interatomic bonds are not fully compensated at the three-phase contact line. This results in a free energy excess and a *linear tension*, æ, acting along the perimeter of a three-phase boundary. This linear tension can be either positive or negative and does not exceed 10^{-4} dyn/cm. While the linear tension can in most cases be neglected, it plays an essential role in the case of very small droplets, particularly in nucleation.

When a three-phase contact line is formed by a solid phase and two liquid phases, a *selective wetting* of the solid phase by one of the liquids takes place. Usually, there is competition between the polar phase (e.g., water) and the nonpolar phase (e.g., hydrocarbon or "oil") in the wetting of the polar and nonpolar solid surfaces. By convention, in selective wetting, the contact angle, θ, is measured into the more polar phase. The solid surface is referred to as *hydrophilic* ("oleophobic") when it is predominantly wet by water ($θ < 90°$), and *hydrophobic* ("oleophilic") when it is predominantly wet by a nonpolar liquid ($θ > 90°$), as illustrated in Figure 1.8.

The concept of oleophobicity is applicable only to the case of selective wetting by *two* liquids: in air, hydrocarbons exhibit good wetting toward nearly all solid materials. The terms "surface hydrophilicity" and "surface oleophilicity" were introduced by Freundlich to indicate the physical-chemical similarity between the solid and liquid phases or more precisely the affinity of a solid phase toward a liquid, whether oil or water. Both of these terms go back to a common notion of "lyophilicity," which in literal translation from the Greek means "liking to dissolve." In the opposite case, one would use the term "lyophobicity" to stress dissimilarity between the phases (the word itself literally means "fearing to dissolve").

Selective wetting (or nonwetting) is governed by the difference in the values of $σ_{SL1}$ and $σ_{SL2}$. Here, it would be worthwhile addressing the general concept of interfacial free energy, $σ_{12}$, associated with the interface between any two condensed phases, including the contact between two solids.

Following Dupré's approach, one can consider a column having a unit cross-sectional area consisting of two phases in contact with each other, phases 1 and 2 (Figure 1.9).

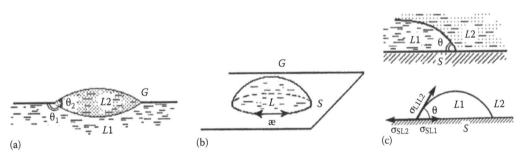

(a)　　　　　　　　(b)　　　　　　　　(c)

FIGURE 1.8 Interface between two liquids and a gas (a); linear tension, æ, (b); selective wetting: $θ > 90°$, hydrophilic (oleophobic) surface; $θ > 90°$, hydrophobic (oleophilic) surface (c).

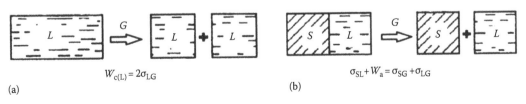

$$W_{c(L)} = 2σ_{LG}$$

$$σ_{SL} + W_a = σ_{SG} + σ_{LG}$$

(a)　　　　　　　　　　　　　　　　(b)

FIGURE 1.9 The definition of work of adhesion (b) and work of cohesion (a).

Separating the phases requires that the adhesion at the interface is overcome, that is, that the work of adhesion, W_a, is consumed. As a result of the separation of phases 1 and 2, the interface with energy σ_{12} vanishes, and two new surfaces with surface free energies σ_1 and σ_2, respectively, are formed. The energy balance of this process is given by Dupré's equation and can be written as follows:

$$\sigma_{12} + W_a = \sigma_1 + \sigma_2 \tag{1.5}$$

This rule for the case of contact between a solid and a liquid phase can be written as

$$\sigma_{SL} + W_a = \sigma_{SG} + \sigma_{LG} \tag{1.6}$$

By combining Young's equation with Dupré's equation, one gets

$$W_a - \sigma_{SG}\cos\theta = \sigma_{LG}$$

or

$$W_a = \sigma_{SG}\left(1 + \cos\theta\right) = \sigma_{LG} + W_w \tag{1.7}$$

where the *work of wetting*, W_w, is given by

$$W_w = \sigma_{SG}\cos\theta = \sigma_{SG} - \sigma_{SL} \tag{1.8}$$

Dupré's scheme may be complicated by the possibility of the adsorption of liquid phase molecules on a bare solid substrate. Such adsorption can be neglected for hydrophobic solid surfaces.

Young's and Dupré's equations enable one to compare the values of σ_{SL}, σ_{LG}, σ_{SG}, W_a, and W_w and thus to assess the degree of physical-chemical similarity ("philicity") of the phases in contact. The larger the similarity between the phases (e.g., between the surface of the ionic crystal and water), the higher the value of W_a and the lower the value of $\sigma_{12} = \sigma_{SL}$. In the limiting case of completely identical phases (corresponding to continuous substance, either solid or liquid), $\sigma_{12} = 0$, and $W_a = 2\sigma_1 = 2\sigma_2 = W_{c1} = W_{c2}$, where W_c is the *work of cohesion* inside a given condensed phase. For a liquid, $W_{c(L)} = 2\sigma_{LG}$. The work of cohesion corresponds to the work necessary for the formation of two identical surfaces having a unit area in a thermodynamic equilibrium process. In the opposite case of two "very dissimilar" phases in contact, such as a drop of mercury on paraffin, where the contact angle reaches 150°, the interfacial energy, σ_{12}, is high, and the work of adhesion is low.

The ratio $(\sigma_{SG} - \sigma_{SL})/\sigma_{LG}$ may serve as a dimensionless characteristic of cohesion between phases. When the value of this ratio ranges between −1 and +1, $(\sigma_{SG} - \sigma_{SL})/\sigma_{LG} = \cos\theta$, that is, it corresponds to the measured contact angle. This ratio may not assume values of less than −1, while $(\sigma_{SG} - \sigma_{SL})/\sigma_{LG} > 1$ yields $\sigma_{SG} - \sigma_{SL} > \sigma_{LG}$, which corresponds to a complete spreading of a liquid phase over the solid substrate, that is, an equilibrium contact angle is not established under these conditions. The work of spreading, W_{sp}, can be written as $W_{spr} = \sigma_{SG} - \sigma_{SL} - \sigma_{LG}$. For the work of spreading, one can also write that $W_{spr} = W_{a(SL)} - W_{c(L)}$. This expression is also true for the case of the spreading of one liquid over the surface of another liquid. The condition when $W_{a(SL)} = \frac{1}{2} W_{c(L)}$ corresponds to the situation when $\theta = 90°$. We will use these equations further during the discussion of the molecular dynamics of wetting.

Of special interest is the case of contact between solid and liquid metals of similar chemical nature, such as zinc and mercury and aluminum and gallium. In these cases, the work of adhesion between the solid and liquid phases remains high, but the interfacial tension, σ_{SL}, may be very low, reaching about 10% of σ_{SG}. Such lowering in the interfacial tension is the reason for liquid-metal embrittlement, which is discussed in detail in Chapter 7.

The wetting of solid surfaces by liquids is greatly influenced by the state of the solid surface and particularly by its microgeometry, that is, surface roughness. The topography of a surface can be approximated with a series of microgrooves of depth H and width d; $H = (d/2) \tan \chi$, where χ is the angle between the idealized flat surface and a side wall of the groove (Figure 1.10).

If the surface is rough, the real surface area, S_{real}, is greater than that of an idealized surface, S_{ideal}. The ratio of the real surface area to the area of its projection onto an idealized flat surface is referred to as the *coefficient of roughness*, k_r:

$$k_r = \frac{S_{real}}{S_{ideal}} = \frac{d/\cos\chi}{d} = \frac{1}{\cos\chi}$$

Roughness results in an increase in the true surface area of the solid, which in turn increases the input from the solid–liquid and solid–gas interfaces into the energy of wetting. According to Derjaguin, the expression for the work of adhesion in the case of contact between a liquid and a real solid surface should be written as

$$W_a = k_r(\sigma_{SG} - \sigma_{SL}) + \sigma_{LG}$$

and the averaged (the "effective") value of the cosine of the contact angle is

$$\cos\theta_{ef} = \frac{k_r(\sigma_{SG} - \sigma_{SL})}{\sigma_{LG}} = \frac{\sigma_{SG} - \sigma_{SL}}{\sigma_{LG}\cos\chi} = \frac{\cos\theta}{\cos\chi}$$

The aforementioned equation shows that surface roughness improves the wetting of a solid surface by a liquid (the value of θ_{ef} decreases) but makes nonwetting worse (the value of θ_{ef} increases). The condition where $\chi = \theta$ is sufficient for wetting to turn into spreading. This effect is used in such processes as soldering and gluing: prior to applying glue or solder, the solid surfaces are treated with sand paper, which in addition to removing the impurities makes the surfaces rough.

A very important application of nonwetting is the attachment of rock particles to air bubbles in flotation. This is schematically shown in Figure 1.11 and will be discussed in more detail in Chapter 2.

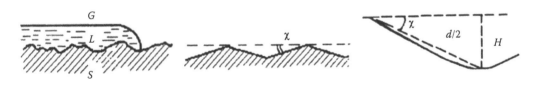

FIGURE 1.10 The effect of surface roughness on wetting.

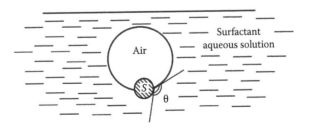

FIGURE 1.11 Nonwetting during the elementary act of flotation.

Finally, let us briefly address some of the principal methods that are used to measure surface tension (free surface energy) in liquids. Nearly all of these methods rely on the Laplace equation:

$$P = p_\sigma = \frac{2\sigma}{r}$$

(1.9)

which is the main equation for capillarity, describing the *capillary pressure* over curved surfaces [3,5,7].

1.1.2 CAPILLARY RISE METHOD

In this method, the rise of a liquid in a capillary of radius r is examined (Figure 1.12). The capillary rise force, $F\sigma = \pi r^2 p\sigma = \pi r^2 \, 2\sigma/r = 2\pi r\sigma$, acting from the side of a concave hemispherical wetting ($\theta = 0$) meniscus with a radius equal to the radius of the capillary, is compared to the weight of a column of liquid with height H, $F_g = \pi r^2 H\rho g$, where g is the acceleration of gravity and ρ is the density of the liquid (corrected for the density of the ambient air). Fluid rises in the capillary until equilibrium is established, so that $F\sigma = F_g$, which readily yields σ

$$\sigma = \frac{1}{2} r\rho g H$$

(1.10)

In the case of incomplete wetting, characterized by a nonzero value of the contact angle θ, the earlier expression is written as

$$\sigma = \frac{r\rho g H}{2\cos\theta}$$

(1.11)

1.1.3 SESSILE DROP METHOD

This precise and universal method utilizes the deformation of a relatively large drop of a liquid resting on a solid substrate under gravity, as shown in Figure 1.13. As a result of deformation,

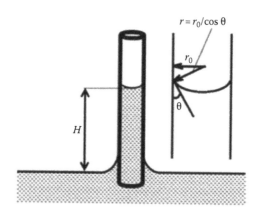

FIGURE 1.12 Schematic illustration of capillary rise method.

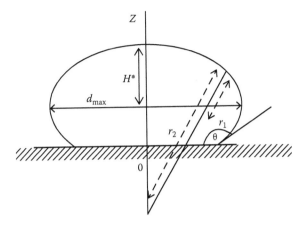

FIGURE 1.13 Schematic illustration of sessile drop method.

the droplet's surface has a nonspherical shape, and the Laplace pressure at any point on the surface is determined by main radii of curvature, r_1 and r_2

$$p_\sigma = \sigma \left(\frac{1}{r_1} + \frac{1}{r_2} \right) \tag{1.12}$$

The numerical integration of the Laplace equation enables one to estimate the value of σ at a given liquid density as a function of the droplet's maximum diameter, d_{max}, and the distance from the plane of maximum diameter to the top of the droplet, H^* (Figure 1.13), both of which can be directly measured. The results of a numerical integration of the Laplace equation are utilized in the software supplied with modern contact angle instrumentation and can also be found in published tables [15]. The sessile drop method can be used to determine the surface tension of melts of metals with high melting points: the shape of a molten metal droplet placed into an isolated high-temperature heater can be obtained from the x-ray image.

1.1.4 Spinning Drop Method

This method is used to determine the interfacial tension between two liquids, $\sigma_{L1/L2}$, and enables one to measure interfacial tensions as low as 10^{-3} mN/m. In the spinning drop method, a drop of a less dense fluid is placed into a tube containing a more dense fluid. Then, the tube is made to spin rapidly, and the centrifugal forces elongate the droplet of the less dense fluid into a cylinder along the axis of rotation (Figure 1.14). The interfacial tension can then be determined from the geometry of the elongated drop, and in the case when the shape of the drop can be approximated by a cylinder, Vonnegut's equation can be used

$$\sigma = \frac{\omega^2 \Delta \rho r^3}{4} \tag{1.13}$$

where
  ω is the angular rotation speed
  $\Delta \rho$ is the density difference between the liquids
  r is the diameter of the elongated droplet

The spinning drop method is commonly used in petroleum-related applications for determining the interfacial tension at the crude oil/aqueous solution interface in the presence of different surfactants.

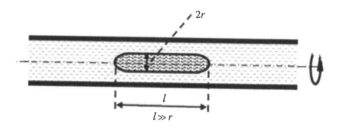

FIGURE 1.14 Schematic illustration of spinning drop method: the equilibrium shape of the rotating drop.

All of the methods described earlier can be classified as *static*, because all of the measurements are conducted under conditions of stable equilibrium. This condition corresponds to the *minimum* in potential energy, which in turn corresponds to the balance between the total surface energy and the energy in a field of gravitational or centrifugal force. There are also methods in which the measurements of surface tension are conducted under conditions of metastable equilibrium, corresponding to the *maximum* of the free energy of the system. Such methods can be classified as *semistatic*.

An example would be the classic method of *maximum bubble pressure*, in which a bubble of air is squeezed through a capillary immersed into the liquid (Figure 1.15). As the pressure of the air flowing through the capillary, P, is increased, the bubble radius, r, decreases, and the capillary pressure is balanced by the applied pressure, P, up to the moment when the bubble assumes a hemispherical shape with radius r equal to the radius of the capillary, r_0. The capillary pressure then decreases, and the bubble rapidly detaches from the capillary. The condition of a metastable equilibrium corresponds to the *maximum* bubble pressure,

$$P_{max} = p_\sigma = \frac{2\sigma}{r} \qquad (1.14)$$

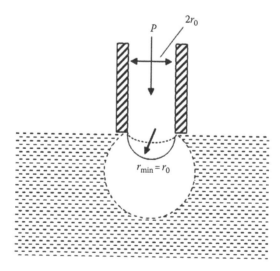

FIGURE 1.15 Schematic illustration of maximum bubble pressure method: change in the curvature radius of the growing bubble.

In the aforementioned equation, the pressure P is corrected for the hydrostatic pressure ρgH, where H is the depth of immersion of the capillary. This method can also be utilized to determine the interfacial tension, $\sigma_{L1/L2}$.

Another quite popular and rather simple method is the *Du Noüy ring detachment* method (Figure 1.16) in which the force needed to pull the ring through the interface is measured. That force is given, within the first approximation, by $F = 2 \times 2\pi r_r \sigma$, where r_r is the ring radius. If more precise measurements of the surface tension are needed, then proper corrections need to be introduced. This method can also be used to measure the interfacial tension between two liquids and is especially useful for interfacial tensions on the order of several mN/m.

One of the simplest techniques that can be used to measure surface tension is the so-called drop weight method (Figure 1.17). In this method, the weight of a droplet separating from a pipette tip with a radius r_0 is determined by direct weighing of 50–100 droplets, and the surface tension is estimated from the relationship $mg = 2\pi r_0 \sigma$. This method provides only a crude estimate of the

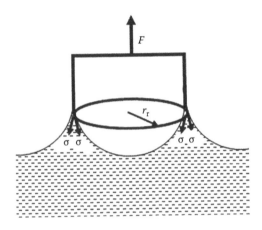

FIGURE 1.16 Schematic illustration of Du Noüy ring detachment method.

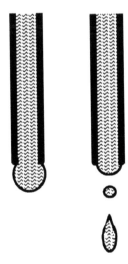

FIGURE 1.17 Schematic illustration of drop weight method.

surface tension, because a drop separating from a pipette tip has a rather complex geometry and the separating droplet is often accompanied by several smaller ones, the so-called satellite droplets.

A third group of methods is the so-called dynamic methods, such as the oscillating jet, capillary wave, or oscillating drop/bubble methods. These methods are typically based on the evaluation of periodically applied interfacial stresses followed by interface relaxation and provide the means for determining the dilational rheology of the liquid–gas and liquid–liquid interfaces.

A stand-alone *static* method is the popular *Wilhelmy plate* method (Figure 1.18). In this method, a completely wetted platinum plate is brought into contact with a liquid surface, and a pull force is applied to the plate. Equilibrium is achieved when that force, corrected by the buoyancy force acting on the immersed part of the plate, is balanced by the surface tension, that is, $F + dbH\rho g = 2(d + b)\sigma$. The force F is measured with a sensitive dynamometer, which typically forms the core of modern surface tension meters.

It is much more difficult to carry out experimental studies of the surface free energy of solids. Two *direct* measurement methods are the *cleavage method* and the *zero-creep method*. The cleavage method developed by Obreimow can be used on brittle solid materials, especially those with a clearly defined layer structure, such as mica. In this method, the crystal is split along the cleavage plane (Figure 1.19), and a measurement of the cleavage force, F, enables one to estimate the work of cohesion, $W_c = 2\sigma$. In the opposite case of ductile solids, such as metals, the Tamman-Odin *zero-creep method* can be employed to measure the surface tension at temperatures close to the melting point. In this method, the material of interest (e.g., aluminum foil) is cut into strips of width d, onto which weights of different magnitudes are mounted (Figure 1.20). At loads $F > d\sigma$ sample, elongation takes place, while at loads $F < d\sigma$ sample, contraction is observed. The surface free energy can be estimated from the load F_0 resulting in zero creep, that is, $\sigma = F_0/d$. Note that the expression for the surface free energy calculation does not contain a numerical coefficient 2. The reasons for that can be more easily visualized if instead of foil a thin cylindrical wire is used in the zero-creep method. The Laplace law for nonspherical surfaces yields

$$p_\sigma = \sigma\left(\frac{1}{R_1} + \frac{1}{R_2}\right) \tag{1.15}$$

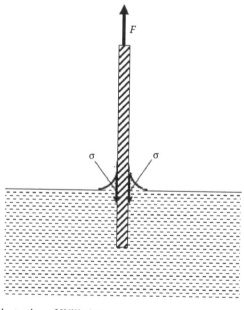

FIGURE 1.18 Schematic illustration of Wilhelmy method: the force balance equilibrium.

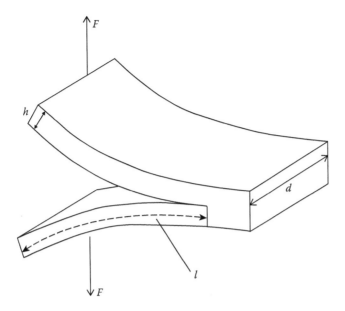

FIGURE 1.19 Schematic illustration of Obreimow's single-crystal cleavage method.

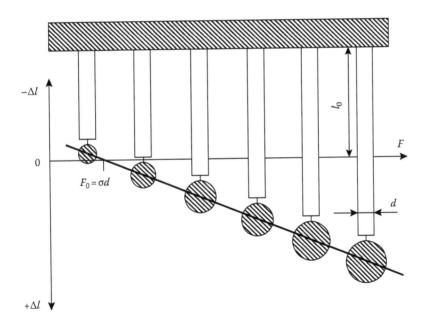

FIGURE 1.20 Schematic representation of the zero-creep method.

In the case of a cylindrical surface, $R_2 = \infty$, which immediately yields $p_\sigma = \sigma/R_1$. The equilibrium state of "zero creep" is established at the balance between the applied stretching force, F, a counteracting contracting force acting along the contour, $2\pi R\sigma$, and a force resulting from a positive Laplace pressure, $\pi R^2\sigma/R$, namely, $2\pi R\sigma - \pi R\sigma = \pi R\sigma$. Similarly, in the case of an aluminum strip, only "half of the perimeter" is involved in the equation.

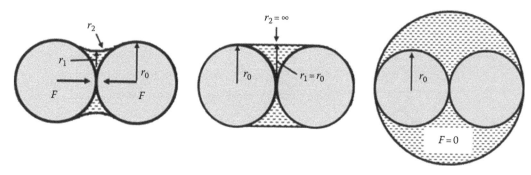

FIGURE 1.21 Estimation of the capillary attractive force between two spherical particles in the presence of a liquid meniscus. (Redrawn from Shchukin, E.D. et al., *Colloid and Surface Chemistry*, Elsevier, Amsterdam, the Netherlands, 2001.)

The surface curvature, capillary pressure, and wetting reveal themselves in the properties of the meniscus between solid surfaces. In the case when the meniscus is formed between spherical surfaces of radius r_0 under complete wetting conditions ($\theta = 0$), a capillary attractive force (or meniscus contraction force) appears between the solid surfaces. This force has two components: the contracting force acting along the perimeter, which is always attractive, $F_1 = -2\pi r_1 \sigma$ (the minus sign reflects the attraction), and the capillary force $F_2 = \pi r_1^2 (1/r_1 + 1/r_2) \sigma$, resulting from the action of the capillary pressure, $p_\sigma = (1/r_1 + 1/r_2) \sigma$, on the area πr_1^2 ($r_1 > 0$). The capillary force F_2 can be either repulsive or attractive, depending on the radii, and hence on the amount of liquid in the meniscus between the solid surfaces (Figure 1.21). Indeed, F_2 is repulsive when $p_\sigma > 0$, which occurs when $|r_2| > r_1$, and attractive when $p_\sigma < 0$, which is the case when $r_2 < 0$, and $|r_2| < r_1$. The net force F acting between the two surfaces is then

$$F = F_1 - F_2 = \pi r_1 \sigma \left(1 - \frac{r_1}{r_2} \right) \tag{1.16}$$

For a cylindrical meniscus, $r_1 = r_0$, $r_2 = \infty$, and the contractive force $F = \pi r_0 \sigma$ (and not $2\pi r_0 \sigma$). When the particles are completely immersed in the liquid, $r_1 = r_2 = r_0$, and $F = 0$. The opposite limiting case of a nearly dry meniscus, that is, when $r_1 \to 0$ and $r_2 \to 0$ ($r_1 > 0$ and $r_2 < 0$), yields $r_1^2 = 2r_0 |r_2|$ (from the Pythagorean theorem) and $F \to 2 \pi r_0 \sigma$. Under incomplete wetting conditions, $F = \pi r_1 \sigma \cos \theta (1 - r_1/r_2)$. These types of *meniscus* attractive forces between particles are frequently encountered in nature and in many practical applications. These forces play a critical role in the structural and rheological properties of soils and the ground.

Of particular interest is the case of a "nonwetting" meniscus, that is, the formation of a bubble (cavity), when $\cos \theta < 0$, such as between hydrophobic (methylated) surfaces immersed in water [16–18]. The description of this case is analogous to the one mentioned earlier, if the contact angle θ (in this case, measured outside the meniscus) is replaced by an ($\pi - \theta$) angle inside the cavity, but with one principal difference: The formation of a wetting meniscus implies a particular volume of a liquid, while in the case of a cavity, the liquid volume depends on the external pressure. The analysis of meniscus formation upon the contact of particles in a nonwetting liquid has been studied and has been discussed in detail in a study by Shchukin et al. [18]. Here, we will review the major findings of this study and use them as an example of the quantitative description of phenomena that are essential in physical-chemical mechanics.

We will focus first on theoretical considerations. Let us consider the case of two solid particles immersed in a liquid that forms a contact angle $\theta > 90°$ with their surface. When these particles are brought into contact, the liquid will spontaneously flow out from a narrow gap separating the particle surfaces. As a result, a cavity filled with vapor from the escaping liquid is formed.

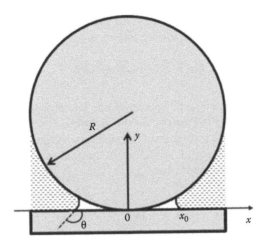

FIGURE 1.22 The cavity formed between particles placed into contact in a nonwetting liquid. (From Yushchenko, V.S. et al., *J. Colloid Interface Sci.*, 96, 307, 1983.)

Around this cavity, a meniscus with an equilibrium radius equal to x_0 is formed. This situation is schematically illustrated for the case of contact between a spherical particle and a flat plate in Figure 1.22.

In the absence of the diffusion of dissolved gases, the pressure P inside the cavity is equal to the pressure of the saturated vapor of a liquid. Because of the difference between this pressure, P, and the atmospheric pressure, P_0, the sphere will be pressed against the flat planar surface with a force p^s equal to the product of the pressure gradient $\Delta P = P - P_0$ acting over the area $S = \pi x_0^2$ of the plane bounded by the wetting perimeter, that is,

$$p^s = \pi x_0^2 \Delta P \tag{1.17}$$

The other component of the capillary pressing force, p^l, originates from the action of the surface tension of the liquid, σ, exerted on the perimeter of the wetting, $l = 2\pi x_0$, at an angle of θ:

$$p^l = 2\pi x_0 \sigma_L \sin\theta \tag{1.18}$$

The net adhesive force between the particles in a contact formed in a nonwetting liquid, p_{SLS}, is given by the sum of the molecular adhesive force between the particles, p_{SS}, acting in the gas phase and the capillary attractive force, $\Delta p = p^s + p^l$:

$$p_{SLS} = p_{SS} + \Delta p \tag{1.19}$$

Strictly speaking, the force p_{SS} is the molecular adhesive force in an atmosphere of saturated vapor. However, because the adsorption of vapors of a nonwetting liquid on solid surfaces is insignificant and does not influence adhesion between particles, one can nearly always assume that the value of the force p_{SS} is the same as that in air.

The force acting on the wetting perimeters, l, on each of the contacting surfaces in the direction of their displacement is the reason behind the recess of the fluid from the gap. This force F equals $-2l\Delta\sigma_S$, where $-\Delta\sigma_S$ is the force per unit length of a perimeter. That force results from the lowering of the surface free energy owing to a replacement of the solid–liquid interface having the surface energy of σ_{SL} with the solid–gas interface having the surface free energy of σ_{SG}. Young's equation yields

$$\Delta\sigma_S = \sigma_S - \sigma_{SL} = \sigma_L\cos\theta \tag{1.20}$$

where σ_L is the surface tension at the liquid–gas interface. The ratio of the force $-2l\Delta\sigma_S$ acting on the perimeters of the area of a "ring" formed by these perimeters, $S' = 2\pi xz$, is equal to the capillary pressure, P_c. The "ring" is, in fact, a projection of the meniscus on the direction of its displacement. If we approximate the spherical surface in the vicinity of the contact with a paraboloid of revolution that has a generatrix $z = x^2/2R$, and remembering that $-2l\Delta\sigma_S = -4\pi x\Delta\sigma_S$, and $S' = \pi x^3/R$, we can write that

$$P_c = \frac{4R\Delta\sigma_S}{x^2} \tag{1.21}$$

The liquid stops receding when the capillary pressure (which decays rapidly with the increase in x) is balanced by the counteracting pressure at the surface of the growing meniscus. The equilibrium meniscus radius $x = x_0$ is then obtained from the condition $P_c = \Delta P$:

$$x_0 = 2\left(\frac{R\Delta\sigma_S}{\Delta P}\right)^{1/2} \tag{1.22}$$

The substitution of the aforementioned expression for x_0 into Equation 1.17 allows one to exclude x_0 and ΔP and to express ΔN as a function of $\Delta\sigma_S$ and R:

$$\Delta p = -4\pi R\Delta\sigma_S \tag{1.23}$$

Equation 1.23 is a well-known limiting relationship for the capillary attractive force, which for the general case of convex particles of arbitrary shape can be written as [18]

$$\Delta p = -2\pi R_g\Delta\sigma_S \tag{1.24}$$

where R_g is the function of the curvature and mutual orientation of the contacting particles. In the case of a contact between a sphere and a plate, $R_g = 2R$, and $R_g = R$ for two equivalent spheres of radius R.

Let us now discuss the experimental part and address the formation of a cavity upon the contact of surfaces in a nonwetting liquid, which can be observed in the following simple experiment in which it is sufficient to immerse an empty round-bottom glass test tube into a jar filled with mercury. When the test tube contacts the bottom of a jar, a clear round window appears around the point of contact and becomes very visible, if one looks through the top of the empty tube. Mercury does not wet the glass, so it spontaneously recedes away from the gap between the two glass surfaces, forming an *empty* meniscus containing a mercury vapor. A variation of this experiment has been used for simultaneous quantitative measurements of *contact forces* as well as of the diameter of a cavity. In a modified experiment, a meniscus is formed between a fused glass sphere and the flat wall of a glass cuvette filled with mercury (Figure 1.22). Mercury has been used as a nonwetting fluid for the following four reasons. First, the high surface tension of mercury, in combination with a large obtuse contact angle, ensures a high negative value of the surface energy of wetting, due to which the cavity has a rather large diameter that can easily be measured. Second, the force Δp has high values significantly exceeding p_{SS}, so that the variation in p_{SS} does not contribute to an uncertainty in the value of Δp. Third, mercury does not dissolve gases, which enables one to assume that ΔP is approximately equal to the atmospheric pressure. This approximation is justified because the pressure of the saturated vapor of mercury in the cavity is negligibly small and the correction for the hydrostatic pressure of mercury above the cavity is also small (it has been demonstrated that the height of the layer of mercury was about 1 cm, and the correction for hydrostatic pressure was around 1% and could be neglected). Fourth, one can make an effective observation of the cavity because mercury is not transparent.

Before performing measurements, the glass cuvette was cleaned with hot chromic acid, rinsed thoroughly with water, and dried in the oven at 110°C. Distilled mercury with a surface tension around 450 mJ/m^2 was used in all experiments. The contact angle formed by a drop of mercury on the wall of the glass cuvette was 115°, as determined with a goniometer. The cavity diameter was measured with a micrometer mounted on the eyepiece of a microscope. The contact forces were measured as described earlier [16,19]. The methodology of such measurements will be discussed in detail in Chapter 2.

The determined values of the adhesive forces between a sphere and a plane, like the forces between two spheres determined earlier, did not depend on the application of a preliminary compression force, which was varied in a broad range between 10^{-1} and 10^3 μN, but in contrast to the forces between two spheres, they showed a significantly higher variability. The adhesive forces in the experiments carried out with a sphere having a diameter $2R = 1.68$ mm were predominantly around 500 μN, but in some instances, values on the order of tens of micrometers, or even no adhesion at all, were observed. This spread of values is likely related to the surface roughness of the glass cuvette walls, in contrast with the ideally smooth surface of fused spheres. The dependence of the adhesive forces on the surface geometry is given by the Derjaguin equation [20,21]

$$p_{SS} = \pi R_g F_{SS} \tag{1.25}$$

In Equation 1.25, the geometric parameter R_g is expressed in terms of the curvature radii of the contacting surfaces in the same way as in Equation 1.24, and F_{SS} is the specific (per cm^2) surface free energy of adhesion between two flat and parallel molecularly smooth surfaces,

$$F_{SS} = 2\sigma_S - \gamma_{SS} \tag{1.26}$$

where
  σ_S is the specific surface free energy ("surface tension") of a solid particle at the interface with air,
  γ_{SS} is the specific free energy associated with the interface between contacting solid surfaces

The latter variable is analogous to a grain boundary in a polycrystalline material. In the case when the contacting particles have a molecularly smooth surface, the value of R_g can be estimated using macroscopic curvature radii. In the case of a contact between rough surfaces, the values of R_g should be estimated on the basis of the effective local curvature radii of the nonuniformities, which will result in smaller value of p_{SS} than expected for smooth surfaces.

The value of F_{SS} obtained from the measurements of the adhesive forces between two fused spheres was around 100 mJ/m^2. The results showed good reproducibility with a standard deviation of 10%. This value can be used in Equation 1.25 to estimate the value of p_{SS} for the case of a meniscus formed between a plate and a sphere. Those values are summarized in Table 1.1, which shows that the estimated values are close to the highest experimentally observed values of p_{SS} in the experiment with a cuvette wall and a sphere. Therefore, the contact between a sphere and a smooth planar surface produces large adhesive forces.

As follows from the Equation 1.25, the contact between the rough areas of the contacting surfaces yields forces with values lower than the calculated ones. For contacting surfaces that have nonuniformities

TABLE 1.1

Adhesive Forces (µN) and Cavity Diameters (mm) in Contacts between a Sphere and a Plane Immersed in Mercury

Surface	Unmodified Glass		Methylated Glass
	$(\theta \sim 115°, F_{SS} \sim 100 \text{ mJ/m}^2)$		$(\theta \sim 115°, F_{SS} \sim 100 \text{ mJ/m}^2)$
$2R$	1.68	0.935	2.42
p_{SLS}, experiment	1900–2450	1000–1250	4000–4800
p_{SS}, experiment	30–550	0–250	30–300
$p_{SS} = 2\pi R F_{SS}$	530	290	300
$\Delta p \approx p_{SLS} - p_{SS}$	1900	1000	4500
	(1400–2400)	(750–1250)	(3700–4800)
$2x_0$, experiment	0.151 ± 0.005	0.116 ± 0.005	0.24 ± 0.010
$p^s = \pi x_0^2 \, \Delta P$	1700 ± 120	1060 ± 90	4520 ± 370
$p^l = 2\pi x_0 \, \sigma_L \sin \theta$	190 ± 20	150 ± 20	220 ± 30
$\Delta p = p^s - p^l$	1980 ± 140	1210 ± 140	4740± 400
$2 x_0 = \theta (R\sigma_S \cos\theta / \Delta P)^{1/2}$	0.160 ± 0.020	0.117 ± 0.015	0.258 ± 0.010

on the order of a micron or submicron scale, the adhesive forces can be lower by several orders of magnitude as compared to the case of smooth surfaces.

Measurements carried out in mercury indicated that the appearance of a cavity takes place immediately upon the approach of the sphere to the wall under the action of very small forces, not exceeding several µN, which is in line with the earlier conclusion regarding an activation-free formation of the cavity, that is, under these conditions, there is no action of positive linear tension. The perimeter of a contact is easily visible through the eyepiece of a microscope and has the shape of a perfect circle. The values of the measured adhesion forces p_{SLS} and of the meniscus diameters $2x_0$ are reported in the two left-hand columns of Table 1.1. While confidence intervals for the $2x_0$ values indicate primarily the accuracy of measurements, the confidence intervals for the p_{SLS} are characteristic of the true spread of the observed force values. For example, for a sphere with $2R = 1.68$ mm, the measured force values were typically on the order of ~2400 µN, but values as low as 1900 µN have also been observed.

In agreement with Equation 1.19, the adhesive force between the solid surfaces in mercury should be equal to the sum of the adhesive force in air, p_{SS}, and the capillary contraction force, Δp. For this reason, the spread in the values of p_{SS} should be included in the spread of the p_{SLS} values. Surface roughness may have some impact on the value of Δp due to hysteresis phenomena taking place on the wetting perimeter. However, in contrast to the values of p_{SS}, which can be distributed over a broad range covering one or two orders of magnitude, the influence of the surface roughness on the value of Δp may only reveal itself through small variations in the measured values of the perimeter l and the area s and can't be large. This is indeed confirmed by the good reproducibility of the x_0 values. It can be seen in Table 1.1 that the confidence intervals in p_{SLS} are indeed similar to those in p_{SS}.

The maximum values of p_{SLS} correspond to instances when the contact involves smooth surfaces (maximum p_{SS}). The values of Δp, calculated from the maximum values of p_{SLS} and p_{SS}, are also summarized in Table 1.1. The values of Δp cannot fall outside of the range given in the parentheses because, at the top, they are limited by the maximum values of p_{SLS} and at the bottom—by the difference between the minimum value of p_{SLS} and the maximum value of p_{SS}.

The values of p^s and p^l, estimated from Equations 1.17 and 1.18, using the experimentally determined cavity diameter (Table 1.1), indicate that in these systems, the pressure difference contributes approximately 90% to the total value of the capillary attractive force, Δp. The remaining 10%

comes from the action of the surface tension acting along the perimeter. The capillary attractive force, obtained by the summation of p^s and p^l, is in good agreement with the measured one.

Table 1.1 indicates that the measured cavity diameters, $2x_0$, are in good agreement with those calculated from Equation 1.22. Those diameters are on the order of 10% of the particle diameters. The values of $2x_0$ and Δp, calculated using the limiting relationships (Equations 1.22 and 1.23), deviate from those calculated rigorously [18] by no more than a few percent, and the error associated with using Equations 1.22 and 1.23 does not exceed the experimental errors.

Similar experiments were also conducted with the methylated surfaces obtained by the exposure of glass to dimethylsilane vapor [16,20]. The 140° contact angle formed by a drop of mercury on methylated glass was much higher than the corresponding angle on the original glass. As shown in Table 1.1, in the case of methylated surfaces, there was good agreement between the experimentally determined and the calculated values.

The Derjaguin equation for the force of the molecular adhesion between solid particles in a liquid medium can be written as

$$p_{SLS} = \pi R_g F_{SLS} \tag{1.27}$$

where

$$F_{SLS} = 2\sigma_L - \gamma_{SLS} \tag{1.28}$$

A direct liquid-free contact between the particles immersed in a liquid is possible if the difference $\Delta\sigma_s = \sigma_S - \sigma_{SL}$ between the interfacial tensions at particle–air and particle–liquid interfaces is smaller than the specific free energy of adhesion between the particles in air, F_{SS}. In this case, the difference between the values of γ_{SLS} and γ_{SS} is negligible, and then in agreement with Equations 1.26 and 1.28, $F_{SLS} = F_{SS} - 2\Delta\sigma_s$, which upon substitution in Equation 1.27 yields

$$p_{SLS} = \pi R_g (F_{SS} - 2\Delta\sigma_S) \tag{1.29}$$

or, if one takes into account Equation 1.25,

$$p_{SLS} = p_{SS} - 2\pi R_g \Delta\sigma_S \tag{1.30}$$

So, if the liquid wets the solid surface ($\theta < 90°$), the cavity does not form, and Equations 1.29 and 1.30 are valid. According to Equation 1.30, in the case of a wetting liquid, the forces of molecular adhesion in a liquid phase, p_{SLS}, should be lower than the forces of molecular adhesion in air. For nonwetting liquids ($\theta > 90°$), one should expect the formation of a cavity, an increase in the adhesive force in a liquid, as compared to that in air, by the value of the capillary contraction force, Δp, which is predicted by Equation 1.19. However, as one can see, Equation 1.19 becomes identical to Equation 1.30 upon the substitution of the expression for Δp given by Equation 1.24.

It is thus evident that, within a first approximation, Equation 1.30 is valid for both wetting and nonwetting liquids, with a meniscus either present or absent. It is, however, worth pointing out that for macroscopic particles, this is valid only in the case of molecularly smooth surfaces. In this case, the equations for the molecular adhesive forces and for the capillary contraction force both contain the same macroscopic value of R_g. The situation is different for rough surfaces. Namely, the value of R_g in the expression for the molecular forces may be determined by the radii of microheterogeneities between which the contact is formed, while the value of R_g in the expression for the capillary adhesion force may be determined by the macroscopic radii of the particles. Consequently, particles with a microscopically rough surface may not reveal noticeable adhesive forces, neither in air nor upon immersion in wetting liquids, but may adhere to each other in nonwetting liquids. Indeed, in many of the described experiments, the adhesion in air was nearly nonexistent, while in mercury, a strong adhesion was observed.

It has been shown previously [5,21–27], and will be further shown in Chapter 2, that Equations 1.29 and 1.30 hold very well for molecularly smooth methylated spheres in various wetting and nonwetting liquids (water, organic solvents, mixtures of the two, and in surfactant solutions). The experimentally established increase in adhesive force in nonwetting liquids in comparison with that in air by the value $-\pi R_g \Delta \sigma_S$ cannot by itself serve as a proof of the formation of a cavity. Indeed, Equations 1.29 and 1.30 predict the same increase in adhesive force, even in the case when a cavity in nonwetting liquids is not formed for some reason.

As will be shown in Chapter 2, measurements of the adhesive forces, p_{SS}, between two spheres yield values of the specific energy of adhesion, $F_{SS} \sim 100$ mJ/m^2 for ordinary glass and ~ 40 mJ/m^2 for methylated glass. The same results for F_{SS} can be obtained from the calculations using the maximum numbers for the sphere-plate system reported in Table 1.1. In the hydrophobic methylated glass–water system, the value of $2\Delta\sigma_S$ is approximately -40 mJ/m^2, and since F_{SS} is ~ 40 mJ/m^2, the adhesive force between two methylated spheres in water is approximately twice as high as in air, which has been demonstrated experimentally [18].

The analysis of interactions between particles in a nonwetting liquid has resulted in the introduction of the term "hydrophobic (lyophobic) interaction" into the literature on colloid science [24]. This term initially was used to describe the interactions between nonpolar segments of individual molecules in water. Hydrophobic interaction is the reason for the observed strong aggregation taking place in aqueous dispersions of nonpolar particles. The interaction between particles in the case of nonwetting by a dispersion medium can be encountered in numerous practical applications, such as flotation or the behavior of slag particles in metal melts with high surface tension. The results discussed earlier clearly indicate that the formation of a cavity between particles in nonwetting liquids may be viewed as an example of a hydrophobic interaction. While in the described case the particles were in direct contact, a more versatile and more complex case is the analysis of the attractive force (and the work of separation) across the gap as a function of interparticle distance, in particular in the presence of a nonwetting meniscus (see Chapter 4).

1.1.5 MOLECULAR DYNAMICS OF WETTING

The use of molecular dynamic modeling of wetting was first described in a crucial 1976 work by Yushenko and Shchukin [25,26], also presented at the plenary lecture of the "Seventh International Symposium on Surfactants" (Moscow, USSR, 1976), where the presentation was accompanied by a short film illustrating the findings.

While a thermodynamic description of interfacial phenomena provides means for their macroscopic description, the use of molecular dynamic simulation enables one to study these phenomena on a molecular level. One of the first and most interesting developments in this direction was the analysis of the behavior of a drop of liquid on wetting and nonwetting solid surfaces. As a result, it was possible to reveal molecular insights into the mechanisms of wetting and nonwetting on a qualitative level and to relate these findings to the thermodynamic characteristics of these systems.

In molecular dynamic simulation, a 2D system giving a pictorial illustration of the arrangement of molecules at various time intervals was produced by computer. A solid substrate was modeled by a box, the bottom of which contained a layer of molecules fixed on a substrate. The attraction of the mobile molecules to the fixed ones corresponded to the interaction of a substrate with a semi-infinite 2D crystal built out of molecules belonging to the main component of a fluid. The other walls of the box were the reflecting walls. The liquid was represented by 19 mobile molecules, which was sufficient for observing the studied phenomena. The interaction between the molecules was described by the Lennard-Jones "6–12" potential. The parameters of the interaction potential of the molecules belonging to the main component of a fluid (ε_{LL}, r_0) and their mass (m) were assigned a value of 1. Mean kinetic energy (i.e., kT) was ~40% of a unit energy per molecule (ε_{LL}), and the duration of each experiment was 120 units of time, (r_0 [m/ε_{LL}]$^{1/2}$). For argon, this corresponded to 50 K (2D argon is liquid at this temperature) and 3×10^{-10} s. As a metric characterizing wetting and spreading, one used the number of intermolecular bonds established between molecules in a liquid (N_{LL}) and between molecules in a liquid and those in a solid phase (N_{SL}), that is, the number of pairs of molecules located close to each other. The value of $r = 1.5$ was used as a boundary of the first coordination sphere.

The results of the molecular dynamic simulation are shown in Figure 1.23. On a lyophilic surface, spreading of a microdroplet took place when the parameters $\varepsilon_{SL} = 1$ and $r_0 = 1$, that is, when the work of adhesion was equal to the work of cohesion. Under these conditions, the adsorption layer constitutes a stable state. The real structure of the substrate significantly influenced the mechanism and kinetics of the propagation of the molecules of a liquid as compared to the case of a structureless barrier. In the system under consideration, the molecules moving along the surface overcame barriers that were on the order of ~60% of a unit of energy (i.e., 1.5kT). Isolated molecules were primarily involved in the oscillatory motion, and only a few "hot" ones were capable of translational motion. The need to spend energy in order to separate a molecule from the rest of the mobile molecules made surface diffusion more difficult. In addition to that, in the present model, the substrate molecules were fixed and were thus unable to provide any additional momentum to the diffusing molecules, which resulted in a slower migration. The propagation of the liquid in the system took place mainly via an "overflow," that is, by a corporative move of many molecules. Spreading resulted in a decrease in the number of

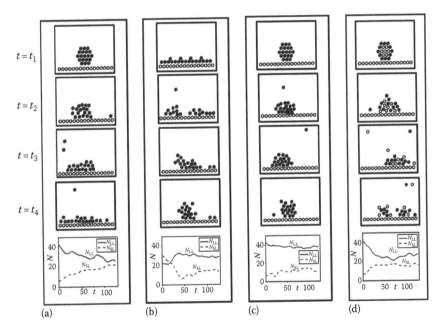

FIGURE 1.23 Spreading of a microdroplet on a lyophilic substrate (a); receding of a liquid on a hydrophobic substrate (b); oscillations of a microdroplet on a lyophobic substrate (c); wetting influenced by a surfactant (d). Diagrams are arranged vertically according to the time elapsed since initial time, $t = t_1$, $t_1 < t_2 < t_3 < t_4$. The bottom diagrams illustrate the change in N_{LL} and N_{SL} for each case as a function of time (the curves consist of merged data points averaged over a time period of 1.5 units. (Redrawn from Shchukin, E.D. and Yushchenko, V.S., *Kolloidnyi Zh.*, 39, 331, 1977.)

bonds between the liquid molecules from 42 to ~25 and in an increase in N_{SL} from 6 to ~22, as shown in the bottom chart in Figure 1.23a.

On a lyophobized surface (Figure 1.23b and c), the microdroplet was stable. This condition corresponded to $\varepsilon_{SL} = 0.5$. The microdroplet was formed when the liquid both advanced and receded. If the microdroplet was artificially transformed into the adsorption layer (Figure 1.23b, $t = t_1$), the molecules spontaneously gathered to reform the microdroplet (Figure 1.23b, $t = t_4$). Similar to the spreading case, the main mechanism here was a cooperative movement of molecules, but a weaker interaction between the mobile molecules and a substrate led to faster surface diffusion, which played a larger role in the case of a lyophobic substrate, as compared to the case of a lyophilic one. It is worth pointing out that even in such a small system, the instability of the adsorption layer may cause its separation into parts (Figure 1.23b, $t = t_2$) that further reconnected (Figure 1.23b, $t = t_3$). In the case when an already formed droplet was placed on a lyophobic surface (Figure 1.23c, $t = t_1$), the process of a limiting spreading quickly came to an end (Figure 1.23c, $t = t_2$ and $t = t_3$). The main time of the experiment corresponded to the establishment of a macroscopic equilibrium. Macroscopically, the drop formed a contact angle close to 90°, which was expected since the energy of adhesion was half of the work of cohesion of a liquid. On a microscopic scale, as visualized by the molecular dynamic experiment, there was no "established equilibrium." The methods enabled one to observe a live picture of the motion of molecules and a continuous change in the droplet shape. One could directly observe surface fluctuations, such as Mandelshtam waves, evaporation and condensation, and the exchange of matter between a droplet and the unsaturated adsorption layer. The thermodynamic analysis of the latter was done by Frumkin [28]. In contrast to a microdroplet consisting of 37 particles, no spontaneous dispersion was observed in the present case. This may be explained by a steep rise in the surface tension as the droplets decrease in size.

Figure 1.23d illustrates a series of experiments in which the role of surfactants on wetting was investigated. In these experiments, molecules of a second component, a surface active additive, were placed into the middle of a droplet, as shown by the open symbols in Figure 1.23d. This second component was a reasonably weak wetter that displayed surface activity only at the gas–liquid interface. The surfactant was modeled using the following parameters of the potential: $\varepsilon_{AA} = 0.5$, $\varepsilon_{AL} = 0.75$, $\varepsilon_{AS} = 0.5$, where "A" stands for "additive" (i.e., surfactant molecules). The molecular dynamic experiment revealed the exit of the surfactant molecules from the bulk of the droplet to the surface, which is a realistic picture of adsorption dynamics. The lowering of the surface tension of the liquid resulted in an improvement in wetting, that is, of the lyophilization of the system. Simultaneously, the splitting of a droplet became easier, and one observed the establishment of a dynamic equilibrium between the spontaneous dispersion and coalescence of microdroplets, as shown in Figure 1.23d, $t = t_3$, $t = t_4$.

The results presented earlier have clearly indicated that a molecular dynamic simulation enables one to observe all of the basic physical-chemical phenomena in the processes of wetting and spreading and enables one to conduct a rigorous analysis of the molecular mechanisms of various interactions taking place at the interfaces. The method of molecular dynamics offers a unique opportunity to study the mechanisms of the interfacial phenomena at a microscopic level. The equilibrium state of interfacial phenomena can be investigated using statistical methods, such as a Monte Carlo simulation [29].

1.2　THERMODYNAMIC CHARACTERISTICS OF A CONTACT BETWEEN PARTICLES

Now, we can address the subject of the thermodynamics of a contact between solid particles. We will start by examining a *thin film* in a gap formed between solid surfaces; such a film serves as a carrier of the properties of the contact [19,21,23,30–36]. First, we will examine the properties of a

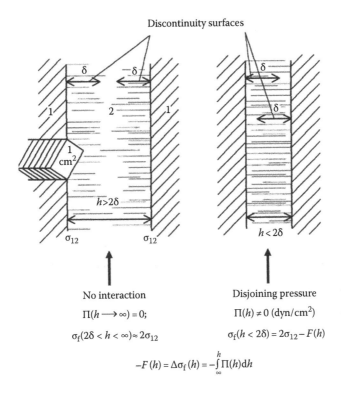

FIGURE 1.24 Disjoining pressure $\Pi(h)$ and the free energy of interaction $\Delta\sigma_f(h)$ in the film.

film between two parallel planar surfaces. In this way, we will be able to determine the essential parameters that are invariant with respect to particle size and shape. Let us select a column with a unit cross section and generatrix perpendicular to the gap (Figure 1.24). The interlayer *film* can be liquid, gaseous, or even solid.

Because of the difference between the properties of the film and the properties of the interfaces and adjacent phases, the body occupying the volume of the column has an excess of free energy, referred to as the *free energy of the film*, σ_f (mJ/m² = mN/m). There is no interaction between the surfaces if the gap h between them is large, (e.g., $h > 2\delta$), where δ is the effective thickness of the Gibbs discontinuity surface. In this case, the surfaces do not feel each other's presence. In this case of a film between two identical bodies, $\sigma_f = 2\sigma$, where $\sigma = \sigma_{12}$ is the interfacial free energy at each of the boundaries. When the surfaces are brought closer together at a distance $h < 2\delta$, the discontinuity surfaces overlap, and the surfaces start to interact with each other. According to Derjaguin, the force of this interaction per unit area can be viewed as a *disjoining pressure*, $\Pi(h)$, which is regarded as positive in the case of repulsion and negative in the case of attraction. When the surfaces are allowed to approach each other from infinity to a separation distance h, work equal to $-\int_{\infty}^{h}\Pi(h)\,dh$ is performed. The minus sign indicates that this work has the same sign as the disjoining pressure. When the surfaces approach each other in an equilibrium process at $T = $ const, this work is performed by external forces and corresponds to an increase in the energy of the film

$$\Delta\sigma_f(h) = \int_{-\infty}^{h}\Pi(h)\,dh \qquad (1.31)$$

which is viewed as a specific (per unit surface area) *free energy of interaction*. Based on how they are defined, the free energy of film, σ_f, the interfacial energy σ, and the free energy of interaction, $\Delta\sigma$, are related to each other via the following linear equation:

$$\sigma_f(h) = 2\sigma + \Delta\sigma_f(h) \tag{1.32}$$

Since σ = const, we can write that

$$\Pi(h) = -d\Delta\sigma_f/dh = -d\sigma_f/dh \tag{1.33}$$

In the literature, one can encounter different types of designations. For instance, γ is frequently used in place of σ, and $F(h)$ is frequently used to designate the free energy of interaction. We have used these notations earlier in this chapter. We will use the symbol $\Delta\sigma_f(h)$ for the free energy, so that we can emphasize the dependence on the distance. At the same time, for the case of attraction at the immediate point of contact at $h = h_0$, where $\Delta\sigma_f(h) < 0$, we will use the absolute value and a designation that $|\Delta\sigma_f(h_0)| = |F(h_0)| = F > 0$. In some instances, we will also utilize another expression that has been used in the literature, $|\Delta\sigma_f(h_0)| = |\Delta\sigma_f| = \Delta\sigma_f > 0$. In the text (not in the equations), we may also omit the minus sign in front of $\Delta\sigma_f$ in cases when this does not cause confusion.

As will become evident from the discussion that follows, essentially only the value of $\Delta\sigma_f(h)$ in Equation 1.32 is of interest to us. For this reason, $\Delta\sigma_f(h)$ can be viewed as a primary characteristic describing film properties and the interactions between surfaces. Unfortunately, for solid planar parallel surfaces, such measurements are nearly impossible: the experimental integration of $\Pi(h)$ between macroscopic surfaces requires that the surfaces remain *flat and parallel* to the precision of fractions of an angstrom in the course of measuring very small forces. While this is impossible for solid surfaces, such measurements are quite possible and indeed broadly utilized in the investigation of *liquid films:* emulsion films, foam films, and wetting films. In all of these cases, a *flat and parallel* state can be maintained because of the high mobility of the interfaces.

The following approach, originally described by Derjaguin et al. [21], lays out the path to the experimental determination of the free energy of interaction between solid surfaces. A comparison of the values of the free energy of interaction per unit area, $\Delta\sigma_f(h)$, between flat–parallel surfaces and the force $p(h)$ between two spherical particles of radius R of the same solid phase in the same medium separated by the same distance h, indicates that the two values are proportional to each other, namely,

$$p(h) = \pi R\Delta\sigma_f(h) \tag{1.34}$$

Generally speaking, the linear dependence shown by Equation 1.34 is evident *a priori*: the coefficient relating $p(h)$ (mN = mJ/m) and $\Delta\sigma_f(h)$ (mJ/m²) must have the dimension of length, and the only such parameter is the particle radius. At the same time, to identify the numerical coefficient, that is, π, one needs a rigorous approach. To do that, we utilize the following scheme for calculating the force $p(h)$ from the known $\Pi(h)$ dependence. The approach illustrated here is a simpler one than that utilized by Derjaguin.

Let us consider rings of radius ρ and thickness dz located at a distance $h + z$ from each other, as shown in Figure 1.25. The force of interaction between the parts of the surface of the particles belonging to these rings, $dp(z + h)$, can be written as

$$dp = 2\pi\rho d\rho\Pi(h + z) \tag{1.35}$$

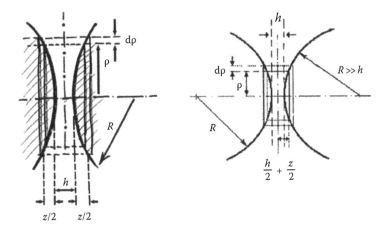

FIGURE 1.25 The derivation of Derjaguin's theorem.

Using the Pythagoras' theorem, one obtains that $\rho^2 = 1/2\ z\ (2R - 1/2z)Rz$, and $2\rho d\rho = Rdz$, which in turn yields

$$p(h) = \int dp(h)dh = \int_{-\infty}^{0} \pi R\Pi(h + z)dz = -\pi R\int_{-\infty}^{h} \Pi(z)dz = \pi R\Delta\sigma_f(h) \tag{1.36}$$

$$\Delta\sigma_f(h) = \frac{p(h)}{\pi R} \tag{1.37}$$

It is assumed here that the particle surface is molecularly smooth and that the gap $h \ll R$; "infinity" as the integration limit has the meaning of distances comparable to h. The particles may not necessarily be spherical: in the general case, instead of π, one can introduce a dimensionless coefficient dependent on the main curvature radii of both particles, $k = k(R_1', R_2', R_1'', R_2'')$. At the same time, there is a limitation associated with the use of the Pythagoras' theorem: both surfaces must conform to a second-order equation, that is, must be quadratic.

Overall, the Derjaguin theorem is a universal one. It can be applied to the interaction forces of any physical-chemical nature (including nonmonotonous ones) acting between particles of different chemical nature in a variety of media (including a vacuum) (Figure 1.26).

For hydrophobic particles in air,

$$p_1^d = \pi R F^d = \frac{AR}{12h_0^2}$$

$$\sigma_{12} = \sigma^d = \frac{A}{24\pi h_0^2} \sim \frac{5 \times 10^{-13}}{75 \times 3 \times 10^{-16}} \sim 20 - 25 \text{ erg/cm}^2$$

(A is the Hamaker constant)

The expression for the work of adhesion is also valid for the case of a direct contact (i.e., for the case of attraction) between particles, that is, when $h = h_0$. In this case, $\Delta\sigma_f(h) < 0$ and $p(h_0) < 0$, but since we are using absolute values, it is possible to write that $|p(h_0)| = p_1 > 0$ and $F = |\Delta\sigma_f(h_0)| = -\Delta\sigma_f(h_0)|F(h_0)| = \Delta\sigma_f = p_1/\pi R > 0$. One must make clear how the distance h_0 is defined within a direct contact. For example, it is possible to take h_0 as the distance between the centers of atoms

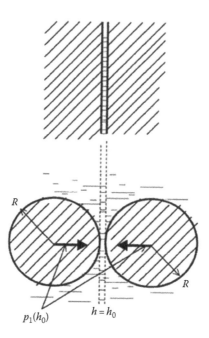

FIGURE 1.26 The application of Derjaguin's theorem to the determination of the free energy of interaction in the direct equilibrium contact, that is, work of adhesion (or cohesion), $|\Delta\sigma_f(h_0)| = |F(h_0)| = F$.

(molecules) in the surface layers of two particles, but in the end, that value is close to interatomic (intermolecular) distances on the order of several angstroms. Such a clarification is essential when these expressions are used to calculate interaction potentials.

Before we start addressing the experimental estimation of the interaction free energy in the contact, $-\Delta\sigma_f(h_0) = F$, it is worthwhile to compare the role that forces of various nature play in the contact interactions.

Depending on the nature of the matter, the cohesive forces can be ionic, covalent, metallic, or molecular. Molecular forces can in turn be subdivided into orientation forces (between molecules containing rigid dipoles), induction forces (between a dipole and a polarizing molecule), and dispersion forces (between molecules that do not have permanent dipole moments, but that can polarize each other). Only dispersion forces are long range, remain largely unscreened by the substance, and therefore play a critical role in the attraction between particles. Those features form the basis for the special recognition of dispersion forces in colloid science: these forces are often opposed to all other (nondispersion) forces. Let us now briefly address the physics of dispersion interactions.

A typical way to characterize interactions between two molecules each with the polarizability α, separated by the distance r, is to use the same Lennard-Jones (or "6–12") potential as was used in the section on molecular dynamic simulation:

$$\varphi(r) = -\frac{\alpha_L}{r^6} + \frac{b}{r^{12}} \tag{1.38}$$

The second term in Equation 1.38 describes the so-called Born repulsion observed at the shortest separation distances and caused by direct contact between the electron shells of atoms. The first term describes the attraction at distances that are rather large compared to molecular dimensions (Figure 1.27). The parameter that characterizes dispersion interactions is the *London constant*, α_L,

$$\alpha_L = \tfrac{3}{4} h\nu_0\alpha^2 \tag{1.39}$$

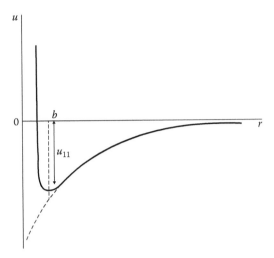

FIGURE 1.27 On the evaluation of dispersion interactions.

where $h\nu_0$ is the quantum of the minimum mutual influence of molecules, which can be obtained from molecular spectroscopy. Mutual perturbation in the state of the molecules corresponds to the appearance of absorbance bands—the so-called anomalous dispersion. In the near-UV region, the value of $h\nu_0$ can be on the order of several units, for example, 5×10^{-19} J. Polarizability has the dimensions of a volume, and for a single molecule, it is a fraction of a molecular volume, for example, for a molecule with a volume $v_m \sim 10^{-28}$ m^3, the value of $\alpha \sim 10^{-29}$ m^3. This rather crude estimate yields $\alpha_L \sim 4 \times 10^{-77}$ J m^6. Dispersion forces are additive.

Let us now estimate the dispersion (molecular) component of the free energy of interaction per unit area in a plane–parallel gap between two identical semi-infinite bodies in a vacuum, $\Delta\sigma_f^d(h)$ (Figure 1.27). To do this, one needs to sum (i.e., integrate over the volume) the potential $\frac{3}{4} h\nu_0 \alpha^2 / r^6$ in the column with a volume V_1 with a unit cross section on one side of the gap and the entire volume V_2 on the other side of the gap. If these unit volumes contain n molecules with polarizability α, and the forces are assumed to be additive, one gets

$$\Delta\sigma_f^d(h) = \int dV_1 \iiint dV_2 \frac{\alpha_L n^2}{r^6} \tag{1.40}$$

For the estimation of $\Delta\sigma_f^d(h)$, let us use a simple approach: quadruple integration with respect to r increases the power of r by four orders of magnitude. Using the integration limits, one then gets 0 and $-1/h^2$. Then, one can write that

$$\Delta\sigma_f^d(h) \approx -\frac{\alpha_L n^2}{h^2} \tag{1.41}$$

A rigorous solution produces only a dimensionless coefficient of $\pi/12$. Multiplying both numerator and denominator by π, one obtains

$$\Delta\sigma_f^d(h) = -\frac{\pi^2 \alpha_L n^2}{12\pi h^2} = -\frac{A}{12\pi h^2} \tag{1.42}$$

where $A = \pi^2 \alpha_L n^2$ is the *Hamaker constant*. The minus sign reflects the universal nature of the dispersion forces, as forces of attraction. With $h\nu_0 \sim 5 \times 10^{-19}$ J, $v_m = 10^{-28}$ m^3, and $n = 10^{28}$ m^{-3}, one has $A \sim 5 \times 10^{-20}$ J. Indeed, the numbers between 10^{-19} and 10^{-20} J are the typical values of the Hamaker constant for dispersion interactions in a vacuum for substances such as water, hydrocarbons (nonpolar molecules of which may be involved only in dispersion interactions), glass, and quartz.

If the gap between the surfaces approaches zero, forming a direct "ideal" contact, that is, there is no film, then $\sigma_f = 0$, and $|\Delta\sigma_f(h_0)| = 2\sigma = W_c$, which yields the work of cohesion in the condensed phase. The value of the parameter $h = h_0$ remains somewhat undefined. To generate a crude estimate for h, it is possible to work backward from the relationship for the Hamaker constant, that is, $A/12\pi h_0^2 = 2\sigma$. For hydrocarbons, typical values of σ are around 20–25 mJ/m², from where we can obtain h_0: $(A/24\pi\sigma) = (1.5-2) \times 10^{-10}$ m, that is, h is on the order of 1.5–2 Å, which corresponds to interatomic distances.

By subdividing all interactions into dispersion and nondispersion interactions, Fowkes [12–14] had assumed that the dispersion and nondispersion components of the surface free energy are additive, and that their summation yields the net value of the surface free energy, that is, $\sigma = \sigma^d + \sigma^n$. The dispersion component of the surface free energy, σ^d, is determined by the substance density and can vary for different substances by an order of magnitude. In particular, for hydrocarbons, $\sigma^d \sim$ 20–25 mJ/m², while for metals, σ^d can be on the order of hundreds of mJ/m². Conversely, the nondispersion component of the surface free energy covers a much broader spectrum: from 0 for nonpolar hydrocarbons to several thousands of mJ/m² for high-melting-point transition metals and covalent compounds. The dispersion components of neighboring phases may compensate each other at the interfaces, in which case, the main portion of the uncompensated interactions and, consequently, the value of σ is in the main determined by the nondispersion interactions.

The next step in our discussion of the thermodynamics of a contact is related to the consideration of a condensed (mostly liquid) film present in the gap between the solid phases. If the solid phase (phase 1) is characterized by the Hamaker constant A_1 and the liquid (phase 2) is characterized by the Hamaker constant A_2, the interaction between phase 1 and 2 is characterized by the Hamaker constant A_{12}. A thermodynamic analysis (the traditional cycle) then tells us that the interaction between the solid surfaces of phase 1 via the film of phase 2 is characterized by the *complex Hamaker constant, A*:

$$A^* = A_1 + A_2 - 2A_{12} \tag{1.43}$$

According to Fowkes, the simplest way one can approximate the value of A_{12} is to use the geometric mean of the values A_1 and A_2: $A_{12} \approx (A_1 A_2)^{1/2}$. Then,

$$A^* \approx (A_1^{1/2} - A_2^{1/2})^2 \tag{1.44}$$

Equation 1.44 is a reasonably "smooth" function, which is an indication that dispersion interactions compensate each other to a significant extent, even when there is a large difference in the nature of the contacting phases. For example, if both A_1 and A_2 are around 10^{-19} J and differ from each other by, for example, 20%, then $A^* \sim 10^{-21}$ J; for a 5% difference, $A^* \sim 2.5 \times 10^{-22}$ J, and for a 2% difference between A_1 and A_2, A^* is lowered to 10^{-23} J. We will use these estimates later in the characterization of the contact interactions in lyophilic and lyophobic systems.

Based on the proportionality between A and σ, Fowkes has offered an estimate of the dispersion component of the free interfacial energy by analogy with A^*, namely, as

$$\sigma_{12}^d \approx \left(\sqrt{\sigma_1^d} - \sqrt{\sigma_2^d}\right)^2 \tag{1.45}$$

and overall

$$\sigma_{12} \approx \left(\sqrt{\sigma_1^d} - \sqrt{\sigma_2^d}\right)^2 + \sigma_{12}^n \tag{1.46}$$

If one of the contacting phases is nonpolar (e.g., solid phase 1), then $\sigma_1^d = \sigma_1 = \sigma_s$, and the interfacial energy is determined only by the nondispersion component of the other phase (e.g., liquid), $\sigma_{12}^n = \sigma_2^n = \sigma_1^n$, and we thus arrive at the following important relationship

$$\sigma_{SL}^d \approx \left(\sqrt{\sigma_S^d} - \sqrt{\sigma_{2L}^d}\right)^2 + \sigma_L^n \tag{1.47}$$

We will come back to a more detailed discussion of these relationships in Section 4.2.

1.3 CONTACT INTERACTIONS BETWEEN SOLID SURFACES OF DIFFERENT NATURE IN VARIOUS LIQUID MEDIA

1.3.1 METHODS USED TO MEASURE CONTACT FORCES

In the discussion that follows, we will present the results of the direct measurements of the contact forces, that is, of the cohesive forces between particles of different nature in liquid media varying in polarity. These results were generated by the original method utilizing a magnetoelectric device as a very sensitive dynamometer. The method and the description of the device are described in detail in [21–23,29,35]. A basic schematic diagram illustrating the principles of measuring small contact forces is shown in Figure 1.28. First, two particles immersed in a liquid medium are brought into direct contact. They can then be further kept in contact with an additional compressive force f applied over a given period of time. The compression is achieved by passing a direct current via the frame of a galvanometer in a particular direction. Upon reversing the current direction, the contact between the particles can be broken, and the force necessary to break the contact is proportional to the magnitude of the applied current. The precision of determining the strength of a contact, p_1, by this technique is 1×10^{-8} N (1×10^{-3} dyn). The main difficulty associated with such measurements in liquid media is the effect of the wetting meniscus, which can be overcome by using L-shaped mounts (Figure 1.28).

The principal advantage of the method utilizing a galvanometer core is the low mechanical compliance of the rigid suspension of the galvanometer frame. This enables one to measure force with very high precision using a reasonably simple calibration. This is what is indeed needed in the direct measurements of *contact* interactions. This scheme is not suitable for conducting studies that involve the investigation of long-range forces. These studies involve the measurement of distances, and a stiff dynamometer with a high compliance is needed. This principle is indeed utilized in the surface force apparatus, which is restricted to the use of curved mica sheets with a molecularly smooth surface. An alternative technique that is not restricted to the use of mica is based on the use of a capacitor as a distance measuring device and is described in Sections 4.1 and 5.2.

Let us now turn our attention to the results of the contact force measurements obtained with the device illustrated in Figure 1.28. In these studies, spherical particles with radii $R = 1$–3 mm were used. These particles had molecularly smooth surfaces of three types as follows:

1. Fused glass or quartz with a hydrophilic surface, contact angle θ, does not exceed a few degrees.
2. Methylated fused glass in which the surface is covered with a dense layer of chemisorbed CH_3 groups. Such glass has a hydrophobic surface producing contact angles on the order of 100°–105°.
3. Fluorinated fused glass that has a surface even more hydrophobic than the methylated one.

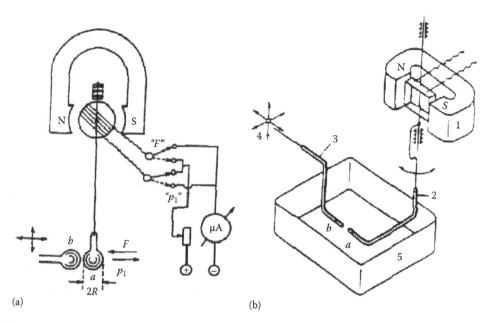

(a)   (b)

FIGURE 1.28 A diagram illustrating the method for measuring the contact forces in a liquid medium between solid particles (a) and a schematic illustration of the corresponding instrument (b); 1, magnetoelectric galvanometer; 2, holder (L-shaped tube, 1 mm in diameter); 3, holder connected to a manipulator; 4-manipulator; 5, immersed in cuvette. (From Shchukin, E.D. et al., *Colloid and Surface Chemistry,* Elsevier, Amsterdam, the Netherlands, 2001.)

The measurements of the force p needed to separate particles were conducted in air, in water, and in hydrocarbon (heptane). The free energy of interaction, $\Delta\sigma_f$, was determined from Equation 1.48:

$$\Delta\sigma_f = \frac{p}{\pi R} \tag{1.48}$$

From now on, we will talk about *contact interactions*—the measurement of the force necessary to rupture an equilibrium unit contact, $p = p(h_0)$, and the determination of the value of $\Delta\sigma_f$ at $h = h_0$. As was mentioned before, both $p(h_0)$ and $\Delta\sigma_f(h_0)$ (negative in a thermodynamic sense) are represented by their absolute values, $|p(h_0)| = p_1$ and $|\Delta\sigma_f(h_0)| = \Delta\sigma_f = F = p_1/\pi R$, respectively. Since the results of the measurements are compared to the values of the free surface and interfacial energies, it is common to also report the values with the coefficient ½: $\frac{1}{2}|\Delta\sigma_f(h_0)| = \frac{1}{2}\Delta\sigma_f = \frac{1}{2}F = p_1/2\pi R$.

The validity of Equation 1.48 can be verified experimentally by establishing the proportionality between p_1 and R for particles of different sizes of a given nature immersed in a given medium. The experimentally determined spread of data did not exceed 10% and, in most cases, was on the order of a few percent. The results of actual measurements are reported in Table 1.2.

Data in the table mentioned before indicate clear and very large differences for the following two types of results:

1. For hydrophilic surfaces in water and for methylated surfaces in hydrocarbon, the cohesive force is very weak. It is on the brink of what can be experimentally determined, and the value of ½F does not exceed several units of 10^{-2} mJ/m^2.
2. In all other cases, the value of ½F is on the order of tens of mJ/m^2, that is, higher by three to four orders of magnitude (the special case of fluorinated surfaces is discussed further down).

TABLE 1.2

Cohesive Force between Particles of Different Natures in Various Media, $\frac{1}{2}F = \frac{1}{2}p_1/\pi R$, (mJ/m²)

Particles \ Medium	Air	Heptane	Water
Glass	~40	25	≤0.01
Methylated glass	22[a]	≤0.01	40
Fluorinated glass	28	5	50

[a] 22 mJ/m² ≈ σ^d ≈ σ_c.

In discussing these results, let us first analyze the case of methylated surfaces in air. This case can be regarded as a "rather simple" one for two reasons. First, there are only dispersion interactions present in this case, as the methyl groups effectively screen the nondispersion interactions of glass. Second, there is no adsorption from the air on a hydrophobic surface. Because of this, in Equation 1.49,

$$\frac{1}{2}\sigma_f = \sigma_{12} + \frac{1}{2}\Delta\sigma_f(h_0) = \sigma_{12} - \frac{1}{2}\Delta F \qquad (1.49)$$

In addition to the value of $\frac{1}{2}F$, one can also obtain an estimate for two other terms. The value of σ_f is small, because in the absence of an adsorption film, the contact can be regarded as direct, almost an *ideal* one. Although σ_f is small, it is still nonzero, as it would have been in the case of a complete cohesion of two surfaces, such as in the coalescence of two liquid droplets. There are two issues that introduce some uncertainty. First, the screening of the glass with methyl groups may be incomplete. Second, the surface of particles (fused glass) may not necessarily be a molecularly smooth spherical surface. For this reason, the value of σ_f assumes finite positive values on the order of tenths and even units on mJ/m². This is further confirmed by the observed spread in the $\frac{1}{2}F$ values for various specimens: from 18 to 22 mJ/m². The sum $\frac{1}{2}\sigma_f - \frac{1}{2}\Delta\sigma_f(h_0) = \frac{1}{2}\Delta\sigma_f + \frac{1}{2}F = \sigma_{12}$ now yields an approximate estimate for the surface free energy, σ_{12}. The reliability of this estimate is biased by the mentioned uncertainty in σ_f and the 10% variation in the experimental values of p_1 and F. For the nonpolar solid phase, we therefore find a quantitative estimate of $\sigma_{12} = \sigma_{SG} \approx 20$ mJ/m² with the precision down to a few mJ/m².

One has to emphasize here the principal importance of this result: we have provided a thermodynamically based estimate of the surface free energy of the solid phase, which for a long time was considered impossible, because of the immobility of the surface. However, it is the *opposite* that is true: it is the nature of the surface (rigid and smooth) that has made these measurements possible. At the same time, these measurements yield a quantitative estimate for the *depth of the primary potential energy minimum* in the distance dependence of the free energy of interaction of these solid surfaces (see Section 4.1).

The value of $\sigma_{SG} \approx 20$ mJ/m² estimated for a solid hydrocarbon is in good agreement with values known for liquid hydrocarbons, in which $\sigma_{SL} = 18$–24 mJ/m².

In the case of methylated surfaces in a nonpolar liquid medium, the interactions are also stipulated only by the dispersion components of the cohesive forces, but the situation in this case is principally different. The cohesive force for macroscopic millimeter-sized beads is on the brink of what is experimentally measurable, and the corresponding value of the free energy of interaction does not exceed 10^{-2} or even 10^{-3} mJ/m². This means that $\frac{1}{2}\sigma_f \approx \sigma_{12} = \sigma_{SL}$, and these measurements by themselves do not provide an estimate for the state of an equilibrium contact, regardless of whether the contact contains a residual adsorption layer or a film. Nevertheless, as in the previous case, it is possible for one to estimate σ_{SL}. The dispersion component, $\sigma_{12} = \sigma_{SG}$, is very small. Using Fowkes' approach, $\sigma_{SL}^d = \left(\sqrt{\sigma_S^d} - \sqrt{\sigma_L^d}\right)^2$, one can see that even a difference of 5 units, assuming $\sigma_S \sim 25$ and $\sigma_L \sim 20$ (for hydrocarbon) mJ/m², we find that $\sigma_{SL} \sim 0.3$ mJ/m². At the same time, the difference between σ_S and σ_L of 2 units results in a drop in σ_{SL} down to 10^{-2} mJ/m². Obviously, the possibility of an incomplete screen out of the nondispersion forces brings a certain level of uncertainty into these estimates. However, it is evident that the interfacial energy $\sigma_{12} = \sigma_{SG}$ is very low, and thus σ_f is low as well.

In contrast to the methylated surface/air case, in this system, the equilibrium state in the contact can't be estimated, along with the possibility of an incomplete screening of the nondispersion forces and a possible surface nonideality. It makes no difference whether the residual film is modeled as a saturated monolayer, a unsaturated monolayer, or a polylayer of a liquid phase. Consequently, the value of σ_f is low, not because the contact is ideal, but because the value of σ_{12} is low.

Here, we have a typical example of a *lyophilic system*, in which the particles (and their surfaces) of the solid dispersion phase and the liquid dispersion medium are quite similar in their physical-chemical properties. The term lyophilic (from Greek, "likes to dissolve") was introduced by Freundlich.

For nonpolar surfaces, the value of the free energy of interaction is proportional to the Hamaker constant A^* (and *not* to σ_{12} in the general case, because $\sigma_{12} \sim \frac{1}{2}F$ only in the case when $\sigma_f \ll \sigma_{12}$). For this reason, a lowering of F in the liquid phase by three to four orders of magnitude in comparison with air corresponds to a comparable decrease in the value of A^* in a similar medium (over air or vacuum). Using these considerations, instead of having $A \sim 10^{-20}$–10^{-19} J, one gets $A^* \sim 10^{-22}$–10^{-23} J. We will further show that the latter value can be viewed as a quantitative estimate of the *lyophilization* of the system (Section 4.3).

The second example of a lyophilic system described in Table 1.2 is the example of glass beads in water. In this case, the cohesive force in a contact, as well as the value of F, is also very small, and $\frac{1}{2}\sigma_f \approx \sigma_{12} = \sigma_{SL}$. However, in this case in addition to the dispersion interactions, the nondispersion interactions are also compensated at the interface. In order to evaluate the extent of this compensation, one needs additional information. The contact angle between the glass and the water may differ significantly from 0 and may reveal a substantial hysteresis: the equilibrium receding angle is usually very low, while the advancing angle may reach 30° more. Water molecules may form an adsorption layer at the hydrophilic surface of the glass, and the 2D pressure in the layer would lower the surface energy of the glass. The nature of the glass surface itself is not very well defined in this case: during ageing of the glass specimen in air, its surface becomes hydroxylated to a significant extent, in contrast to that of the glass specimen heated in a vacuum at high temperatures. In the latter case, the glass surface does not contain a water adsorption layer. The lowering of the surface free energy of the quartz surface due to the presence of a hydration layer has been verified experimentally in a number of studies [37]. For these reasons, appropriate corrections need to be introduced into Young's and Dupré's equations for W_a and W_w. It is important to emphasize here that the values for the work of adhesion and the work of wetting can be estimated from the measured values of σ_{LG} and θ ($W_a = \sigma_{LG}[1 + \cos\theta]$; $W_a = \sigma_{LG}\cos\theta$) only for the case when there is no adsorption at the solid surface, that is, mainly for nonpolar hydrocarbon surfaces.

Similar to the case described earlier, the primary reason for a low value of σ_f may be the low value of σ_{12} rather than the ideality of the contact.

At the same time, the similarity between the solid and liquid phases, as evidenced by good wettability and high values of the work of wetting and the work of adhesion, suggests that the interfacial energy $\sigma_{12} = \sigma_{SL}$ is much lower than the σ_{SG} of the same glass specimen in air. In Chapter 2, we will further address contact interactions in the case of surfactant adsorption.

In Table 1.2, the value of F for the hydrophilic glass/air system is the highest among all of the other values reported in that table. Since σ_f is not negative, the value of $\sigma_{12} = \sigma_{SG}/(\frac{1}{2}F) = 40–60$ mJ/m^2. Because this surface is strongly hydrophilized, the value of its surface free energy may approach the value of the surface tension of water. The difference $2\sigma_{SG} - F$ may be related to a rather high value of σ_f (in this case as well as in the case of glass in water). Indeed, the adsorption of water vapor from the atmosphere results in a deviation of the gap from an ideal, that is, *perfect* one. Consequently, the cohesion is significantly weaker, and the value of σ_f may reach, for example, units or even tens of mJ/m^2. It is of interest to obtain an estimate for the cohesive force in a contact between hydrophilic glass surfaces, under the assumption that this force is the capillary attractive force associated with the presence of a residual meniscus in the contact zone (see Section 1.1). The maximum value of this capillary attraction force is achieved right before meniscus "drying" reaches $2\pi R\sigma_{LG} \sim 2\pi(1 - 1.5) \times 10^{-3}$ m ~ 70 mN/m $\sim (50–70) \times 10^{-2}$ mN. The latter value agrees well with the results of the initial measurements of the cohesive force in a contact.

The interaction between hydrophilic surfaces in a nonpolar liquid (unmodified glass/heptane) can be viewed as a behavior "intermediate" between that of the same surfaces in air ("most nonpolar media of all") and in water. In the expression $\frac{1}{2}\sigma_f = \sigma_{12} - \frac{1}{2}\Delta\sigma_f$, the known last term, $-\frac{1}{2}\Delta\sigma_f \sim 20–30$ mJ/m^2, and the value of the interfacial energy, $\sigma_{12} = \sigma_{SG}$, can be independently estimated. The dispersion components at this interface may be regarded as mutually compensated: the value of $(\sqrt{\sigma_S^d} - \sqrt{\sigma_L^d})^2$ is only on the order of ~ 0.3 when $\sigma_S^d - \sigma_L^d = 5$ and is on the order of ~ 0.01 when $\sigma_S^d - \sigma_L^d = 2$ mJ/m^2. Utilizing the same approach as used in the previous two cases, the nondispersion component of the hydroxylated glass specimen may be estimated as being close to that of water, that is, $\sigma_{H_2O}^d \sim 70 - 20 \sim 50$ mJ/m^2. This corresponds to good wetting: the contact angle of the oil on the glass is low, and the value of the work of wetting, $W_w = \sigma_L \cos\theta$, is approximately 20 mJ/m^2, which is indeed indicative of dispersion components. Consequently, the interfacial energy can be estimated as 50 mJ/m^2. The difference $2\sigma_{SL} - \Delta\sigma_f \sim 2 \times 50 - 2 \times (20–30)$ yields $\sim 40–60$ mJ/m^2, which can be assigned to σ_f. This contact is rather far from being a perfect one—the gap may contain residual film or an adsorption layer of a hydrocarbon. In this case, one needs to also account for the role played by the meniscus in the corresponding surface tension. This does not lead to any contradictions.

It is interesting to note that a good wetting (in this case, of glass by oil in air), that is, the apparent "oleophilicity," does not necessarily mean that the system of the polar particles in the nonpolar media is lyophilic. Cohesion is lower than in air, but it is still rather high, that is, approximately, $\frac{1}{2}\Delta\sigma_f \sim 20–30$ mJ/m^2. This means that the system is lyophobic, not lyophilic! This peculiarity is revealed in the following case.

In the case of methylated particles in water, the cohesion is twice as strong as in air: $\frac{1}{2}\Delta\sigma_f \sim 35–40$ mJ/m^2. One might have expected to see fairly low values of σ_f upon the complete displacement of the water from the zone of contact and the formation of a "nonwetting" meniscus (Section 1.1). Indeed, the contact angle of the water on the methylated glass surface is similar to the contact angle on solid hydrocarbons and reaches $100°–105°$. This yields $\cos\theta \sim -\frac{1}{4}$, and thus the work of wetting is negative: $W_w \sim \sigma_{H_2O}^d \cos\theta \sim -(15 - 20)$ mJ/m^2. Most likely, the residual nondispersion interactions result from an incomplete screening by the nonpolar monolayers. A molecular-level surface nonuniformity plays a role as well.

In any case, the contact should not contain any traces of water, which is expelled due to the dissimilarity of the contacting phases. Here, one can appreciate the meaning of the term "lyophobicity" (meaning "fearing to dissolve" in Greek), also introduced by Freundlich. The term "lyophobicity" is not proper to use in characterizing the cohesion of particles in air, because there is no liquid present. Nevertheless, the term "phobicity" may be used in a very broad sense with respect to any media, including both hydrophobicity and hydrophilicity, as well as oleophobicity, when one refers to fluorinated surfaces.

The next step in a thermodynamic analysis of the interactions between the surfaces in different media is the comparison of the results of free energy measurements corresponding to a gradual

TABLE 1.3

Free Energy of Film for Hydrophobic Particles in Various Media (in Air $\frac{1}{2}F_S = 22$ mJ/m^2)

Medium	$\frac{1}{2}F_{SL}$	$-\frac{1}{2}\Delta F = \frac{1}{2}(F_S - F_{SL})$	$W_w = \sigma_L \cos\theta$	$\sigma_{SL} = 22 - \sigma_L \cos\theta$
Water	40.5	−18.8	~ −17	~39
Ethylene glycol	16.1	5.7	≈8	≈14
Ethanol	1.6	20.2	≥20	≤2
Propanol	0.1	21.7	≥20	≤2
Heptane	≤0.01	21.8	≥20	≤2

transition from complete lyophobicity to complete lyophilicity. Let us here refer to the results of the cohesive force measurements between hydrophobic methylated spherical particles in water, alcohols, and hydrocarbons. In the first and last case, we will be replicating results that have already been presented earlier in this chapter.

The data in Table 1.3 show that, with an increasing degree of lyophilization, the value of $|\Delta\sigma_f| = F$ gradually decreases down to very low values. Gradual lyophilization is achieved with a sequential transition from polar liquids, such as water and ethylene glycol, to weakly polar and nonpolar liquids, such as alcohols and heptane.

We have already provided approximate estimates for the absolute value of the surface and interfacial energy in various systems using the thermodynamic relationship containing three parameters, $\frac{1}{2}\sigma_f = \sigma_{12} - \frac{1}{2}F$, and one directly measurable parameter, $F = \Delta\sigma_f$. Here, we encounter a very typical situation: in systems of thermodynamic equations, the number of unknown parameters is always larger than the number of equations relating these parameters. Consequently, finding a solution (in a general sense, establishing a rigorous pairwise functional relationship between two variables) requires additional information about the relationship between the parameters. This information comes from models or theoretical considerations, rather than from thermodynamics. The most commonly encountered relationships involve the proportionality of the parameters when the values of variables are small or actual experimental data. For this reason, in addition to the values of the free energy of interaction, $\Delta\sigma_f$, and its change upon the transition from water to heptane, Table 1.3 also contains the independent experimentally determined values for the work of wetting, $W_w = \sigma_L \cos\theta$. The latter has been determined from the values of the surface tension of the fluid and the contact angles formed by the same fluid on methylated surfaces. The use of methylated surfaces makes it unnecessary for one to account for the effect of adsorption on the work of wetting. While we do have information on the *additional* parameter, the work of wetting, W_w, we still are not able to determine the absolute values of the parameters rigorously. Nevertheless, we can estimate their *differences*. On the one hand, we do have the change in the free energy of interaction, $\Delta\sigma_{f(SG)} - \Delta\sigma_{f(SL)}$, upon the transition from air to liquid. On the other hand, the values of the work of wetting, $W_w = \sigma_L \cos\theta$, characterize the surface free energy of the solid phase lowering in a given medium relative to air. These two sets of data are in agreement within the 10% variation observed for $\Delta\sigma_f$. By employing another independent consideration, that is, the concept of *perfect* contact between the hydrophobic surfaces in air: $\sigma_{f(SG)} \sim 0$ and $\frac{1}{2}\Delta\sigma_{f(SG)} = \frac{1}{2}F_{SG}$, we can use the difference $\frac{1}{2}\Delta\sigma_{f(SG)} - W_w = \sigma_{SL}$ as an approximation and get a justified estimate for σ_{SL}. It is worth remembering here that for an equilibrium contact, $\Delta\sigma_f(h_0)$ is always < 0, while we use the designation $|\Delta\sigma_f(h_0)| = \Delta\sigma_f = F > 0$.

The next step in the development of these ideas is the transition from a "line spectrum" to a "continuous spectrum," that is, the use of aqueous solutions of polar and of some weakly polar fluids as liquid media. Figure 1.29 shows the values of $\frac{1}{2}|\Delta\sigma_f(\phi)|$ for methylated surfaces immersed in aqueous solutions of alcohols, where ϕ is the volume fraction of the alcohol [21,23]. The values of *half* of $|\Delta\sigma_f(\phi)| = \frac{1}{2}F$ are reported for easier comparison with the values of the surface tension.

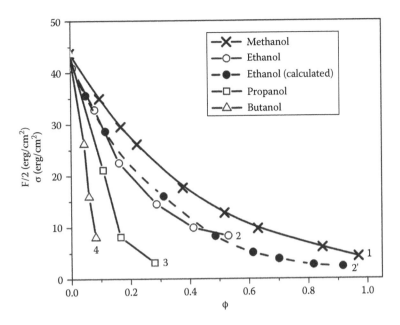

FIGURE 1.29 Free energy isotherms for the interaction of hydrophobic surfaces in aqueous solutions of alcohols (1–4); curve 2′ shows calculated values of σ_{12} for solutions of ethanol. (Redrawn from Shchukin, E.D. and Amelina, E.A., *J. Dispers. Sci. Technol.*, 24, 377, 2003.)

Depending on the polar or nonpolar nature of the additive, that is, on the degree of lyophilization in the system, different slopes of the decaying portions of the curves were observed. The values of $\Delta\sigma_f$ decrease very quickly for butanol (4) and gradually for methanol (1). The data shown in Figure 1.29 illustrate a general rule applicable to the entire homologous series of alcohols: the concentration of alcohol needed to achieve the equivalent lowering of $\Delta\sigma_f$ decreases by a factor of 3–3.5 upon each subsequent increase in the hydrocarbon chain by a CH_2 group (i.e., change from methanol to ethanol and further to propanol and butanol). In the case of butanol, the curve in Figure 1.29 is limited by the solubility of that alcohol in water. These empirical general trends are indeed well known in colloid science as the *Duclaux–Traube's rule*. The latter can be explained rigorously in the thermodynamic analysis of the surface activity and the contribution of methyl groups to the driving force of adsorption (see Section 2.1). Indeed, $\Delta\sigma_f$ curves shown in Figure 1.29 are very similar to the surface tension isotherms of the same solutions.

At the same time, the same figure shows independent experimental results for the work of wetting of methylated surfaces, $W_w = \sigma_L \cos\theta$, by aqueous solutions of ethanol. These data are presented in the form of the difference $\frac{1}{2}\Delta\sigma_{f(SG)} - W_w(\phi)$. The fact that these two independent data sets coincide confirms the conclusion that polar fluids are expelled from the gap between hydrophobic surfaces and hence confirms that it is valid for one to regard this difference as the interfacial free energy, $\sigma_{12} = \sigma_{SL}$.

As the lyophilicity of the system increases upon the transition to a high concentration of nonpolar alcohols, the conclusion regarding low values of σ_{SL} (and $\sigma_{f(SL)}$) holds true but does not allow one to draw any certain conclusions with regard to the nature of a residual fluid potentially remaining in the gap.

Substantial progress in the understanding of contact interactions on the molecular level can be achieved with the use of molecular dynamic simulation. It would be worthwhile to emphasize here again some specifics of this method.

1. The method of molecular dynamics (numerical modeling of the dynamic type) combines both theoretical and experimental approaches. The method specifies the interaction potentials of molecules (atoms), the temperature as a chaotic distribution of velocities, and the laws of motion (in our case, the integration of Newton's laws for each molecule). At the same time, molecular dynamic simulation is *an experiment*, which enables one to observe the behavior of molecules (atoms) with a resolution of up to 10^{-14} s in the time domain and up to 10^{-12} m in the space domain, which can't be realized in any other experiment.

2. This method enables one to seek an answer to a fundamental question: Down to what size levels (of particles, gaps between surfaces, etc.) can one extrapolate macroscopic thermodynamic concepts? This can be established by comparing the thermodynamic parameters, such as surface free energy, obtained from macroscopic experiments, to those estimated from molecular dynamic simulations.

3. This method gives one unique means for comparing the nature and probability of different mechanisms of the processes taking place on the molecular level.

Because of the high computing power of modern computers, a molecular dynamic experiment can involve the simultaneous investigation of the motion and interaction of millions of molecules (atoms). In Section 4.2, we will learn in greater detail one of the studies in which the breaking of the contact between two small particles was observed under conditions of complete lyophilicity, allowing for a spontaneous penetration of liquid into the contact zone between the particle and the surface of the same nature.

REFERENCES

1. Adamson, A. 1979. *Physical Chemistry of Surfaces.* Moscow, Russia: Mir.
2. Rusanov, A. I. and F. C. Goodrich (Ed.). 1980. *The Modern Theory of Capillarity.* Leningrad, Russia: Himiya.
3. Rusanov, A. I. and V. A. Prohorov. 1994. *Interphase Tensiometry.* Saint-Petersburg, Russia: Himiya.
4. Gibbs, J. W. 1982. *Thermodynamics: Statistical Mechanics.* Moscow, Russia: Nauka.
5. Shchukin, E. D., Pertsov, A. V., Amelina, E. A., and A. S. Zelenev. 2001. *Colloid and Surface Chemistry.* Amsterdam, the Netherlands: Elsevier.
6. Holmberg, K. (Ed.). 2002. *Handbook of Applied Surface and Colloid Chemistry*, Vol. 2. New York: Wiley.
7. Hartland, S. (Ed.). 2004. *Surface and Interfacial Tension: Measurements, Theory, and Applications.* CRC Press/Taylor & Francis Group.
8. Cosgrove, T. (Ed.). 2005. *Colloid Science: Principles, Methods and Applications.* Oxford, U.K.: Blackwell Publishing.
9. Somasundaran, P. (Ed.). 2006. *Encyclopedia of Surface and Colloid Science.* CRC Press/Taylor & Francis Group.
10. Birdi, K. S. (Ed.). 2009. *Handbook of Surface and Colloid Chemistry*, 3rd edn. CRC Press.
11. Riviere, J. C. and S. Myhra (Eds.). 2009. *Handbook of Surface and Interface Analysis: Methods for Problem-Solving,* 2nd edn. CRC Press, New York.
12. Lyklema, J. 1991–2000. *Fundamentals of Interface and Colloid Science*, Vols. 1–4. Orlando, FL: Academic Press.
13. Hunter, R. J. 1991. *Foundations of Colloid Science*, Vols. 1–2. Oxford, U.K.: Clarendon Press.
14. Adamson, A. W. and A. P. Gast. 1997. *Physical Chemistry of Surfaces*, 6th edn. New York: Wiley.
15. Matijević, E. (Ed.). 1969. *Surface and Colloid Science,* Vol. 1. New York: Wiley.
16. Amelina, E. A., Yaminskiy, V. V., Syunyaeva, R. Z., and E. D. Shchukin. 1982. Cohesion of heterogeneous particles in air and liquid. *Kolloidnyi Zh.* 44: 640–644.
17. Yaminskiy, V. V., Yushchenko, V. S., Amelina, E. A., and E. D. Shchukin. 1982. Meniscus formation during particles contact in a non-wetting liquid. *Kolloidnyi Zh.* 44: 956–963.
18. Yushchenko, V. S., Yaminskiy, V. V., and E. D. Shchukin. 1983. Interaction between particles in a non-wetting liquid. *J. Colloid Interface Sci.* 96: 307–314.

19. Yaminskiy, V. V., Yusupov, R. K., Amelina, E. A., Pchelin, V. A., and E. D. Shchukin. 1975. Surface tension at the solid-liquid boundaries. *Kolloidnyi Zh.* 37: 918–925.
20. Yaminskiy, V. V., Pchelin, V. A., Amelina, E. A., and E. D. Shchukin. 1982. *Coagulation Contacts in Disperse Systems*. Moscow, Russia: Himiya.
21. Churaev, N. V., Derjaguin, B. V., and V. M. Muller. 1987. *Surface Forces*. New York: Plenum Publishing Corporation.
22. Shchukin, E. D. and E. A. Amelina. 2003. Surface modification and contact interaction of particles. *J. Dispers. Sci. Technol.* 24: 377–395.
23. Shchukin, E. D. 1996. Some colloid-chemical aspects of the small particles contact interactions. In *Fine Particles Science and Technology*, E. Pelizzetti (Ed.), pp. 239–253. Amsterdam, the Netherlands: Kluwer Academic Publishers.
24. Shchukin, E. D. 2002. Surfactant effects on the cohesive strength of particle contacts: Measurements by the cohesive force apparatus. *J. Colloid Interface Sci.* 256: 159–167.
25. Shchukin, E. D. 1976. Mechanisms of action of surfactants on various interphase boundaries. *Proceedings of VII International Congress on Sufractants*. Part 1:15–53. Moscow, Russia.
26. Shchukin, E. D. and V. S. Yushchenko. 1977. Molecular dynamics of wetting. *Kolloidnyi Zh.* 39: 331–334.
27. Yushchenko, V. S., Grivtsov, A. G., and E. D. Shchukin. 1977. Stability and dynamics of droplets on a solid surface. *Kolloidnyi Zh.* 39: 335–338.
28. Frumkin, A. N. 1938. About the phenomena of wetting and adhesion of bubbles. *Zh. Fiz. Khim.* 12: 337.
29. Ferrenberg, A. M. and R. H. Swendsen. 1989. Optimized Monte Carlo data analysis. *Phys. Rev. Lett.* 63: 1195–1198.
30. Somasundaran, P., Lee, H. K., Shchukin, E. D., and J. Wang. 2005. Cohesive force apparatus for interactions between particles in surfactant and polymer solutions. *Colloids Surf.* 266: 32–37.
31. Israelashvili, J. N. 1985. *Intermolecular and Surface Forces*. Orlando, FL: Academic Press.
32. Shchukin, E. D. and V. V. Yaminskiy. 1985. Thermodynamic factors in the sol-gel transition. *Colloids Surf.* 32: 19–55.
33. Yushchenko, V. S., Edholm, O., and E. D. Shchukin. 1996. Molecular dynamics of the particle/substrate contact rupture in a liquid medium. *Colloids Surf.* 110: 63–73.
34. Claesson, P. M., Ederth, T., Bergeron, V., and M. W. Ruthland. 1996. Techniques for measuring surface forces. *Adv. Colloid Interface Sci.* 67: 119–183.
35. Christensen, H. K. and P. M. Claesson. 2001. Direct measurements of the force between hydrophobic surfaces in water. *Adv. Colloid Interface Sci.* 91: 391–436.
36. Shchukin, E. D. 1997. Development of P. A. Rehbinders' studies on factors of strong stabilization of disperse systems. *Kolloidnyi Zh.* 59: 270–284.
37. Zdziennicka, A., Szymczyk, K., and B. Jańczuk. 2009. Correlation between surface free energy of quartz and its wettability by aqueous solutions of nonionic, anionic and cationic surfactants. *J. Colloid Interface Sci.* 340: 243–248.

2 Adsorption of Surfactants and Contact Interactions

The data presented in the previous chapter clearly indicate the universal importance of investigating particle cohesion under various conditions for establishing the scientific basis for explaining (and controlling) the mechanical properties of disperse systems in various natural and industrial processes. An important aspect of such investigations is the use of surface-active substances (surfactants), which at a low bulk concentration accumulate at the interfaces and radically change their properties. Before addressing specific results pertinent to the studies of contacts between particles of various natures in various surfactant solutions, let us briefly summarize the concepts of the adsorption of surfactants, primarily of the thermodynamics of adsorption.

2.1 ADSORPTION PHENOMENA AT INTERFACES

At constant temperature, T, and volume, V, the condition of stable equilibrium corresponds to the minimum of free energy. Due to the uncompensated interactions, there is an excess of free energy associated with the surface (or interface). In the absence of external forces, such as gravity, the minimum of free energy at constant T and V is reached at the lowest value of the surface free energy, σS, where S is the surface area of a body. At a constant value of the specific surface energy, σ, the minimum in energy for liquid surfaces is reached at the minimum surface area, which explains the spherical shape of small droplets, bubbles, or the meniscus in a capillary. For crystalline solids, this corresponds to the minimum of the sum of the product of the surface area of faces and their specific surface free energy (Curie–Wulff theorem).

In the case when one or both of the neighboring phases contain a component capable of lowering the interfacial energy σ, the minimum in the surface energy is reached when this component is concentrated within the interfacial layer, that is, when *adsorption* takes place. It must be emphasized here that the concentration of surface-active component at the interface and the lowering of the free surface energy are mutually related: the spontaneous concentration of a surfactant takes place because it can cause a lowering of free energy, and conversely, the surfactant lowers the free energy because it is concentrating at the interface.

Adsorption as a physical-chemical parameter can be defined in a number of ways. We will focus here on the same principle as we used in defining the surface energy, that is, the principle of surface excess, originally introduced by Gibbs [1–4].

Figure 2.1 shows a column that consists of two neighboring volumes of two-component phases, for example, an aqueous solution of *n*-hexanol, $C_6H_{13}OH$ (phase δ) in equilibrium with its own vapor phase (phase δ'). The square cross section represents the unit of surface area. The horizontal axis shows the concentrations in these two phases: of water and water vapor, $c_1(z)$, and of the additive—*n*-hexanol and its vapors, $c_2(z)$.

Outside of a particular layer of a finite thickness (*the physical discontinuity surface*), the concentrations of water and hexanol in the corresponding phases are constant. These concentrations are, respectively, c_1' and c_2' in an aqueous solution (liquid) and c_1'' and c_2'' in vapor. Within the discontinuity surface, there is a transition between these concentrations: either a smooth one or a positive "tongue-like" one, as shown in Figure 2.1b. In order to determine the excess (or the "difference in the composition" to be more precise) of a given component in a surface layer, we need to use Gibbs's method and extrapolate these four constant values to the dividing surface (the geometric discontinuity surface located within the physical discontinuity surface). The excess (adsorption) of

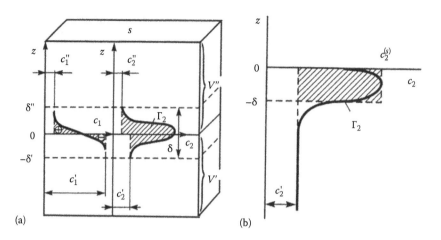

FIGURE 2.1   (a) To the definition of adsorption according to Gibbs. The figure is schematic, not drawn to scale. (b) The "transition tongue" approximation.

component I (1—water, 2—hexane) in moles per unit surface area within the discontinuity surface is given by the following integral:

$$\Gamma_i = \int_{-\infty}^{0}(c_i - c_i')dz + \int_{0}^{+\infty}(c_i - c_i'')dz \approx \int_{\delta'}^{0}(c_i - c_i')dz + \int_{0}^{\delta''}(c_i - c_i'')dz$$

As in the case of the definition of σ, the values of these integrals may depend on the choice of dividing surface position. Without getting into details (see [1,4,5]), we will just emphasize that in the case of a large area of the "tongue" on the graph of $c_2(z)$, there is no need to focus on the actual position of the dividing surface, while in the case of a smooth transition in the plot of $c_1(z)$, the value of integral Γ_1 may be both positive and negative (Figure 2.1). For this reason, it is convenient for one to choose a position of dividing surface that yields $\Gamma_1 = 0$.

At constant temperature and volume, the system is characterized by a single parameter, which in our case is the concentration of additive, (of n-hexanol in a liquid phase), $c_2' = c$. The adsorption of this additive, $\Gamma_1 = \Gamma$, is a function of the concentration, c: $\Gamma = \Gamma(c)$. Establishing this dependence for different interfaces is in fact the main subject of the theoretical physical chemistry of surface phenomena. Both rigorous solutions, originating in the fundamental Gibbs equation, as well as the simpler approximations, leading to the thermodynamic dependence $\Gamma = \Gamma(c)$, have been reported in the literature [1–8]. These approaches are based on the fact that the chemical potentials of each component in the phases coexisting in equilibrium are equal. The latter can be illustrated as follows. In the case of an equilibrium distribution of an additive between phases, the chemical potential of that additive, μ, should be the same everywhere: in a liquid phase (solution with the concentration c [moles per volume]), in a rarified gaseous phase, and in the transition layer represented by the physical discontinuity surface with the surface excess Γ (moles per surface area). The value of σ (J/m²) is the result of *mechanical* work spent on the creation of a unit surface area, while the $\Gamma\mu$ product (mol/m² × J/mol = J/m²) is the *chemical* component of that work. At equilibrium, the variation of the sum $\sigma + \Gamma\mu$ should be equal to zero, that is, $\delta\sigma + \Gamma\delta\mu + \mu\delta\Gamma = 0$. By restricting ourselves to the dependence between the concentration of the additive c and the surface tension at the equilibrium value of Γ (i.e., $\delta\Gamma = 0$), we establish that $\delta\sigma + \Gamma\delta\mu = 0$. Since there is only one independent variable in the system, it is possible to use the full derivative, that is,

$$\Gamma = \frac{-d\sigma}{d\mu} \tag{2.1}$$

Equation 2.1 is the well-known Gibbs adsorption equation describing the adsorption of a single component. The corresponding thermodynamic consideration produces a general equation for the case of multicomponent systems [1,3,4]. Within the limits of the low concentration of the additive in the bulk liquid phase, $c_2' = c$, the chemical potential of the additive is given by $\mu(c) = RT \ln c +$ const, and $d\mu = RT\, dc/c$. These relationships together with Equation 2.1 readily yield Gibbs law in its most common form

$$\Gamma = -\left(\frac{c}{RT}\right)\frac{d\sigma}{dc} \tag{2.2}$$

The meaning of Gibbs law (Equation 2.2) can be formally stated as follows: the surface tension of a liquid phase (or in the general case the interfacial free energy), $\sigma(c)$, decreases with the increase in surfactant concentration c of the adsorbing component, and the rate of this (relative) decrease at constant temperature is described by the value of adsorption, Γ.

Gibbs equation contains three parameters. Being a typical thermodynamic equation, it does not produce solutions for the *surface tension isotherm*, $\sigma = \sigma(c)$, and for the *adsorption isotherm*, $\Gamma = \Gamma(c)$. Here, the term "isotherm" emphasizes constant temperature. Consequently, in order to integrate Equation 2.2, one needs an additional *independent* equation relating the same parameters. This equation can be both theoretical (utilizing some kind of a model, most commonly a molecular one) or experimental. For a liquid interface, the experimental surface tension isotherm, $\sigma = \sigma(c)$, can serve as the second relationship and can allow one to obtain the adsorption isotherm $\Gamma(c)$ from the Gibbs equation. Conversely, for the solid–gas and solid–liquid (S/L) interfaces, experimental measurements yield the adsorption isotherm, $\Gamma(c)$, which together with the Gibbs equation produces the surface free energy isotherm, $\sigma(c)$.

The common universal approximation frequently utilized in thermodynamics is that of ideal solutions, that is, a restriction to the limit of very low concentrations. When the parameters of linear thermodynamic equations have very small values, they are usually proportional to each other. This proportionality provides the necessary additional universal relationship.

Let us demonstrate the application of this approach to the analysis of adsorption from an aqueous solution at the air–liquid interface. The experimentally determined change in the surface tension of solutions, σ, as a function of the additive concentration, c, typically shows linear dependence at low values of c (we will further present an explanation of what specifically can be used as a criterion to judge whether c is low enough):

$$\lim_{c \to 0}\left(-\frac{d\sigma}{dc}\right) = \text{const} = G\,[(\text{mJ/m}^2)/(\text{mol/dm}^3)] \tag{2.3}$$

The constant G characterizes the *surface activity* of the additive. Consequently, at low values of c, the Gibbs equation has the form $-d\sigma = (RT\Gamma/c)dc$, where $RT\Gamma/c = G = \text{const}$ (and $\Gamma/c = \text{const}$ as well). Integration with the initial condition of $\sigma = \sigma_0$ at $c = 0$ yields $\sigma - \sigma_0 = \Delta\sigma = -Gc$, or

$$-\Delta\sigma(c) = \sigma_0 - \sigma = RT\Gamma \tag{2.4}$$

Equation 2.4 can be utilized to present the concept of *surface activity* or *inactivity* (Figure 2.2).

The adsorption Γ, viewed as the excess of a component in the interfacial layer and given by the "tongue" area (see Figure 2.1b), may be presented as the averaged difference between volume-based concentrations of the additive in the surface layer, c^s, and the concentration in the bulk, c, multiplied by the volume (thickness) of the interfacial layer, namely,

$$\Gamma = (c^s - c)\delta \tag{2.5}$$

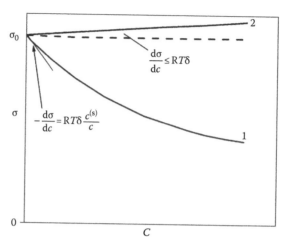

FIGURE 2.2   Surface activity (1) and inactivity (2) in the isotherms of $\sigma(c)$.

which yields

$$-\frac{d\sigma}{dc} = \frac{RTT}{c} = \frac{RT\delta(c^s - c)}{c} \tag{2.6}$$

When the adsorption is positive, that is, when there is a real excess of a component in the surface layer, $c^s > c$, or $c^s \gg c$, and Equation 2.6 becomes

$$-\frac{d\sigma}{dc} = \frac{RTc^s}{c} \tag{2.7}$$

In the aforementioned expression, the ratio c^s/c can be very large, up to several thousands. This is due to the *surface activity*, which is characterized by the decaying linear portion of the isotherm 1 in Figure 2.2. This type of strong surface activity is the general feature of all *surface-active substances* to which we will devote a lot of attention throughout this book. Conversely, under the conditions when the adsorption layer is *deficient* in a given component, the absolute value of c^s-c cannot exceed the value of c, and hence the value of $(c-c^s)/c$ does not exceed 1. In Figure 2.2, this situation is reflected in the small positive slope of isotherm 2, which corresponds to a *surface-inactive* substance. One example of such a system is a NaCl solution: 100% *negative* adsorption simply corresponds to the absence of salt in the surface layer.

Equation 2.4 was experimentally verified by Irving Langmuir and eventually allowed him to propose an explanation of the adsorption layer structure. To further discuss adsorption, we will summarize the essence of Langmuir's method and introduce the concept of *two-dimensional* (or *surface*) *pressure*, π_s (Figure 2.3) [5].

Langmuir and Pockels [5–8] have offered an experimental method for studying surfactant adsorption layers at the air/water interface. Their technique, focused on the use of insoluble or sparingly soluble surfactants, has had a significant impact on the development of colloid science. A shallow rectangular cuvette with hydrophobic walls (paraffin-coated glass or Teflon) is filled with water, which carries an adsorption layer of *an insoluble* surfactant on its surface. Although in these systems the concentration c_2' is very low, it does not prevent the formation of an equilibrium adsorption layer, corresponding to a very high value of the surface activity, G. A moving barrier separates the surface of liquid in the cuvette into two portions, with surfactant being present only on one side of the barrier. Since the surfactant is insoluble, there is no diffusion to the other side of the barrier via the liquid phase.

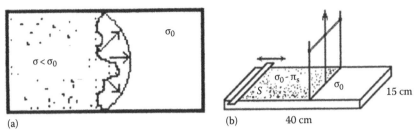

FIGURE 2.3 (a) The concept of two-dimensional pressure, $\sigma_0 - \sigma = \pi_s$ and (b) the Langmuir balance for measuring two-dimensional pressure of water-insoluble surfactants.

The experiment is conducted by first placing on the one side of the barrier a drop of very dilute surfactant solution in a volatile organic solvent (for instance, a drop of 10 mol/dm³ solution of cetyl alcohol in cryoscopically pure benzene) and allowing the solvent to evaporate. Since the surfactant solution concentration in benzene is known, the mass of the surfactant in the adsorption layer is also known—it is typically of the order of fractions of a milligram. When the molecular weight of a surfactant is known, one can estimate the average area occupied by a single molecule, s_m. Repositioning the moving boundary allows one to change the area occupied by the adsorption layer and hence the value of s_m. A sensitive dynamometer connected to the barrier allows one to measure the surface tension gradient existing across the barrier. It is obvious and can indeed be confirmed experimentally that the barrier should slide in the direction of a surfactant-free surface. That is, the barrier slides toward the side containing pure solvent with the surface tension of σ_0 (for a very pure double-distilled water, $\sigma_0 = 72.8$ mJ/m²), which is greater than the surface tension on the other side of the barrier. The difference in the surface tension can be written as

$$\Delta\sigma = \sigma_0 - \sigma = R T \Gamma$$

From this description, it is clear that the adsorption layer exerts "pressure" on the barrier, referred to as the *two-dimensional* (or *surface*) pressure per unit length of the barrier:

$$\pi_s = R T \Gamma = \sigma_0 - \sigma \tag{2.8}$$

Dividing both sides of Equation 2.8 by the value of Γ and by Avogadro's number, $N_A = 6.02 \times 10^{23}$ mol⁻¹, yields

$$\frac{\pi_s}{N_A \Gamma} = kT \tag{2.9}$$

where the value $1/\Gamma = S_m$ is the area occupied in the adsorption layer by a mole of surfactant, while $1/ N_A \Gamma$ is the corresponding area per molecule. Consequently, one can write that

$$\pi_s S_m = RT, \quad \text{and} \quad \pi_s s_m = kT \tag{2.10}$$

This remarkable Langmuir's relationship is, in fact, the equation of the state of a rarified (ideal) two-dimensional "gas" consisting of adsorbed molecules. This equation is analogous to a well-known gas law describing a "conventional" three-dimensional ideal gas.

This theoretically (thermodynamically) derived equation has been verified experimentally for the dilute adsorption layers (where the concentrations are low, not only in the bulk phase but also in the

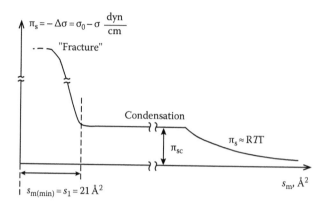

FIGURE 2.4   Characteristic regions typically identified in the $\pi_s(s_m)$ isotherm of insoluble surfactants.

adsorption layer) formed by numerous common organic surfactants, such as fatty acids $C_nH_{2n+1}OH$, fatty amines $C_nH_{2n+1}NH_2$, and many other water-insoluble higher members of homologous series. These particular experiments further extended to include the case of dense adsorption layers (high c^s and Γ values), allowed Langmuir to achieve an understanding of the structure of the adsorption layers at the molecular level, and hence explain the nature of the surface activity.

A typical two-dimensional isotherm of a (surface) pressure, $\pi_s = \pi_s(s_m)$, as a function of area per molecule is shown in Figure 2.4. The right-hand side of the isotherm expression describes rarified adsorption layers in which $s_m \sim 10^3$ Å^2. This portion of the isotherm is similar to the isotherm of a three-dimensional ideal gas. When s_m is lowered, the adsorption layer becomes denser, and the value of π_s increases. When a certain value of $\pi_s = \pi_c$ is reached, the so-called *two-dimensional condensation* takes place. The latter corresponds to the formation of liquid- or solid-like isles in equilibrium with the two-dimensional gaseous phase. The portion of the isotherm associated with the two-dimensional condensation is associated with a constant value of $\pi_s = \pi_c$. When a certain small area per molecule, $s_{m(min)} = s_1$, has been reached, a sharp rise in π_s is observed. For long-chain fatty acids, the value of s_1 is around 21 Å^2. The onset of this rise corresponds to the critical concentration, c^s and adsorption Γ. This region of the isotherm reflects a compression of the two-dimensional condensed phase down to the point at which it loses elastic stability and develops local fractures ("hummocks") at $\pi_s = \pi_{fr}$, which is on the order of several tens of dyn/cm. Langmuir has shown that the critical value of area per molecule, s_1, corresponds to the cross-sectional area of a surfactant molecule in a dense adsorption layer. In such an arrangement, hydrophilic polar groups of surfactant molecules stay immersed in the polar liquid, while hydrophobic hydrocarbon chains are expelled from the aqueous phase into the "most nonpolar medium of all," that is, into the air. This situation is schematically illustrated in Figure 2.5. Similar values of $s_1 = 20$–21 Å^2 obtained for many members of homologous series of fatty acids, fatty amines, and fatty alcohols indicate that this characteristic value corresponds to the cross-sectional area of a hydrocarbon chain and not the polar group. It has also been demonstrated that a similar arrangement takes place at the interface between aqueous and hydrocarbon phases (i.e., at the oil/water interface): polar groups are contained in water, and hydrocarbon chains stay in the nonpolar phase.

The value of $s_{m(min)} = s_1$ numerically yields the value of the limiting adsorption, Γ_{max}, which further yields the thickness of the adsorption layer, δ, that is, the length of the hydrocarbon chain, $\delta = l$.

The estimation of the distance between neighboring –CH$_2$ groups as projected on the axis of a molecule allows one to compare the data obtained from adsorption measurements to the same data independently obtained from the x-ray diffraction analysis [9]. Both methods yield similar results: 1.3 Å in the former case and 1.2 Å in the latter. The value of the valence angle CH$_2$–CH$_2$–CH$_2$ is in good agreement with the known value of 109°. Table 2.1 contains the corresponding $s_{m(min)}$ and l data for stearic and oleic acid, as well as for their triglycerides. These data clearly indicate that the unsaturated oleic acid molecule is "twice-folded" as compared to the stearic acid molecule.

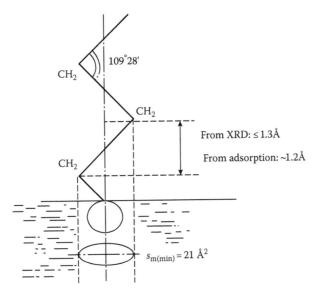

From XRD: ≤ 1.3Å

From adsorption: ~1.2Å

$s_{m(min)} = 21 \text{ Å}^2$

FIGURE 2.5 Langmuir's discovery: the structure of a surfactant adsorption layer.

TABLE 2.1
Values of $s_{m(min)}$ and l for Stearic and Oleic Acid, and of Their Triglycerides

	$s_{m(min)}$, Å^2	l, Å
St–H: $C_{17}H_{35}COOH$	21–22	≤25
Ol–H: $C_{17}H_{33}COOH$	44	11–12
$(C_{17}H_{35}COO)_3C_3H_5$	63	25
$(C_{17}H_{33}COO)_3C_3H_5$	126	13

This specific structure of adsorption layers allows one to explain the phenomenon of surface activity. The molecules of organic surfactants containing both polar and nonpolar features within the same molecule form a bridge between neighboring polar and nonpolar phases. This bridge evens out the transition between two antagonistic phases by compensating otherwise uncompensated interactions across the interface. This leads to a lowering of the free energy excess at the interface, σ, which manifests itself as the observed lowering of surface or interfacial tension. Rehbinder has shown [9,10] that one can formulate the *polarity equalization rule* by stating that the adsorption layer equalizes the polarities of the neighboring phases, decreases the extent of their antagonism toward each other, and as a result lyophilizes the system.

The same considerations regarding the structure of the adsorption layer also hold true for S/L interfaces. At the interface between a polar solid phase (such as mineral salts, oxides, hydroxides, and glasses) and a nonpolar organic phase (e.g., lubricating oils, organic monomers, and oligomers), the polar groups face the surface of a solid phase, while hydrophobic hydrocarbon chains float freely in nonpolar liquids. Conversely, in the case of a hydrophobic nonpolar solid surface (e.g., crude oil paraffins, solidified fats and waxes, plant leaves, animal fur, bird feathers, human skin, and soot particles) and a polar liquid medium (e.g., water), the hydrophobic chains of surfactant molecules face the nonpolar surface, while the hydrophilic groups are immersed into an aqueous phase.

The aforementioned is true for both water-soluble and water-insoluble surfactants. In the first case, one measures the surface tension of the corresponding aqueous solutions, while in the second case,

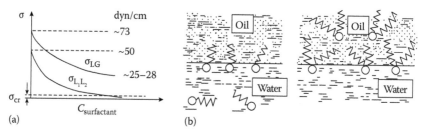

(a)   (b)

FIGURE 2.6 (a) A comparison of the surface tension isotherms of isopropyl alcohol at the aqueous solution–air interface, σ_{LG}, and liquid–liquid interface between aqueous solution and the hydrocarbon phase, σ_{LL}. (b) Isopropanol is a "weakly surface-active" substance soluble in both phases; complete mixing is observed once a certain critical concentration corresponding to σ_{cr} is reached, when complete miscibility of two phases is observed; the "asymmetric nature" of the driving force for the adsorption of a surfactant at the interface from the aqueous and hydrocarbon phases.

the primary experimental method consists of measuring the two-dimensional pressure, $\pi_s(s_m)$, with a Langmuir balance. Some common examples of water-soluble surfactants include soaps formed with monovalent cations, Na^+, K^+, NH_4^+, while water-insoluble surfactants are soaps formed with polyvalent cations, Ca^{2+}, Al^{3+}, etc., and containing a well-developed hydrophobic portion. The water-insoluble surfactants form a class of *oil-soluble additives*, which adsorb on surfaces from a nonpolar liquid phase. In all cases, including those in which adsorption from both polar and nonpolar liquid phases takes place (as, e.g., in emulsions), the polarity equalization rule determines the adsorption layer structure.

Our understanding of a tendency to minimize surface free energy as the general cause of adsorption can be enhanced by emphasizing the role of the chemical potential in the transition of surfactant molecules to the interface. It is the gradient of chemical potential that always determines the direction of mass transfer resulting in the equalization of the chemical potential in all of the phases in contact.

It is worth pointing out that despite the similar arrangement of the hydrophilic and hydrophobic segments of surfactant molecules at the water/hydrocarbon interface, there is a principal difference in the driving force of the adsorption from the aqueous phase and from the liquid hydrocarbon phase (Figure 2.6).

The obvious reason for an "immersion" of the polar surfactant group into the aqueous phase is the work of hydration, that is, a strong nondispersion interaction between the polar head and the molecules of water. The driving force for a transition of hydrocarbon chain C_nH_{2n+1} into a nonpolar medium is different. While a detailed discussion of this rather complex issue is beyond the scope of this book, we will provide the following qualitative explanation. The introduction of a hydrocarbon chain into an aqueous phase disrupts the icelike structure of the water in which dipoles are present in a partially ordered arrangement with a tetrahedral coordination. The reorganization of the water molecules around the hydrocarbon chain in an attempt to restore tetrahedral coordination (i.e., to restore *ordering*) is related to a decrease in entropy. An increase in entropy upon the "expulsion" of the hydrocarbon chain from an aqueous phase into a nonpolar phase is the driving force for adsorption in this case. This applies to the adsorption at the aqueous solution–air interface, aqueous solution–liquid hydrocarbon interface, and to some extent to the adsorption at the interface between an aqueous solution and a solid hydrophobic surface. Similar considerations are also applicable to the formation of surfactant *micelles* in aqueous media. Micelles are spherical nanoparticles formed by the association of the surfactant molecules. In these nanostructures, surfactant hydrocarbon chains are associated together into a core, while polar groups form an interface with the aqueous solution. Micellar cores are capable of dissolving (*solubilizing*) nonpolar phases. At high surfactant concentrations, spherical micelles undergo transformation into cylindrical micelles, form liquid crystalline phases and sponge phases. It is also worth emphasizing here the exceptional role of bilayers formed in aqueous media by phospholipids. In these bilayers, the hydrocarbon tails of phospholipid molecules are associated with each other, while the polar heads are oriented toward the aqueous solution. These bilayer *membranes* are the main building blocks of all living organisms. The specifics of the water structure and the related entropic nature of adsorption constitute one of the main factors of life.

Direct methods of measuring adsorption are essentially restricted to the measurements of adsorption at the solid surface from a gas phase. At sufficiently high specific surface area, S_1 (m²/g), the adsorption can be determined directly as the increase in mass. High-surface-area materials include highly disperse adsorbents with fine pores, such as activated charcoal, zeolites, and various catalysts for which the surface area is on the order of tens and hundreds of square meters per gram. The adsorption on such surfaces from a gas phase can also be determined by measuring a decrease in the gas (vapor) pressure, p, in a closed vessel. The multilayer adsorption of noncorrosive gasses is a commonly used method to determine the surface area of adsorbents on the basis of the Brunauer–Emmett–Teller (BET) theory included in all classic texts on physical chemistry.

The adsorption at the S/L interface between finely dispersed powders and aqueous solutions is measured by the same principle as described earlier, that is, by a decrease in the concentration c in solution, which can be most simply measured by the increase in the surface tension at the air–liquid interface. However, this method lacks accuracy, once the surfactant concentration exceeds the critical micellization concentration. For more accurate determination of surfactant concentrations, spectroscopic or chromatographic–mass spectrometric methods can be used. A review of the methods used to determine the concentration of different surfactants can be found in [5].

It is important to recognize that the measured change in the surfactant concentration due to the adsorption from solutions at the solid surfaces yields the *adsorbed mass*, Γ^*. In order to determine a *specific adsorption* (per unit area), Γ, for a given fine disperse adsorbent, the surface area, s, needs to be determined from the independent measurements from the gas phase (e.g., by the BET method), that is, $\Gamma = \Gamma^*/s$.

For both solid–gas and S/L interfaces, the isotherm of interfacial free energy can be determined from the combination of Gibbs equation and adsorption isotherm.

At the same time, the change in the σ_{SL} value during wetting can be estimated from the work of wetting, $W_w = \sigma_L \cos\theta$, if the adsorption on the hydrophobic surface is negligible, which together with Gibbs equation allows one to estimate the adsorption.

The adsorption at the S/L interface is different from the adsorption at the solid–gas interface in numerous aspects. Yet, the thermodynamics used to describe adsorption at the solid–gas interface are applicable to the S/L interface, and the polarity equalization rule remains valid.

The specifics of adsorption at S/L interfaces are utilized in many different applications, including the removal of toxic substances from wastewaters; liquid chromatography; using the surfactants to direct, alter, and control wetting, which plays a critical role in crude oil recovery; and using polar adsorbents, such and clays and zeolites, to purify a nonpolar medium from oil-soluble surfactants. The mosaic structure of the solid surfaces, and specifically the role of nonuniformities, plays a significant role in adsorption and chemisorption phenomena. There are pronounced differences between these two phenomena. As will be shown later in this book, chemisorption plays a significant role in modifying the strength of contacts between the particles and in influencing the mechanical properties of various materials. Surfactant chemisorption plays a critical role in the hydrophobization of solid surfaces in the aqueous medium, which is employed in mineral flotation.

A peculiar adsorption behavior is observed in the case of adsorption at the solid surface from a mixture of two mutually miscible liquids, A and B (see Figure 2.7). The isotherms may have significantly different shapes depending on the extent of the surface activity (e.g., of liquid B). Case I, represented by a diagonal, corresponds to zero surface activity, which is the case when two liquids have a similar nature: the increase in the surface concentration is the same as the increase in the bulk concentration, which corresponds to $\Gamma = 0$. Case II represents the strong surface activity of liquid B: surface concentration of B and surface excess initially grow rapidly as the mole fraction of liquid B, x_B, is initially increased until a maximum in adsorption is reached. A further increase in x_B results in a lesser accumulation of liquid B in the surface layer; the concentration of B in the bulk increases to a larger extent

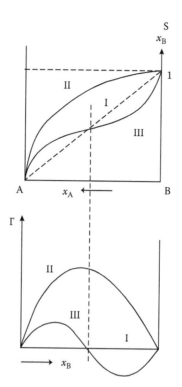

FIGURE 2.7   Adsorption from the mixture of two miscible liquids, A and B, at the solid surface: I, no surface activity of B; II, strong surface activity of B; III, weak surface activity of B.

than the surface concentration and a decrease in adsorption is observed. Case III corresponds to a weak surface activity of liquid B: an initial increase in x_B results in an increase in the surface excess of B, but to a much smaller extent than in case II: similar behavior is now observed at much lower values of x_B. At some point, the surface concentration of B becomes equal to the bulk concentration, and hence $\Gamma = 0$. A further increase in x_B results in a higher bulk concentration than the surface concentration and thus yields $\Gamma < 0$. The understanding of this behavior plays an essential role in the analysis of the effectiveness of the decontamination of liquids by adsorption, for example, by charcoal adsorbents.

Let us provide an example of how one can estimate the change of the surface free energy from the $\sigma(p_v)$ isotherm. This method will be utilized again further in the book. Preweighed samples of microporous magnesium oxide with high surface area are kept in the atmosphere of vapors of the adsorbing component (water vapor at different pressures, p_v) until the equilibrium has been reached. From the gain in weight of the specimen, one can then determine the adsorption $\Gamma(p_v)$ (mol/m²). It is assumed here that the surface area of the specimen is known, for example, from low-temperature nitrogen adsorption (BET) measurements. The parameter p_v plays here the same role as the concentration, c, in Equation 2.2. According to the Gibbs equation, we can write that

$$\Gamma(p_v) = -\left(\frac{p}{RT}\right)\frac{d\sigma}{dp_v}, \quad \text{or} \quad d\sigma = -\Gamma(p_v)\left(\frac{RT}{p_v}\right)dp_v \qquad (2.11)$$

By integrating without the low p_v limitation, one gets

$$-\Delta\sigma(p_v) = \sigma_0 - \sigma(p_v) = RT\int_0^p \Gamma(p_v)\frac{dp_v}{p_v} \qquad (2.12)$$

Further, with some additional data, one can also estimate the value of σ_0.

Depending on the physical-chemical nature of the adsorbent solid surface and on the nature of the adsorbing molecules, the $\Gamma(p)$, $\Gamma(c)$, and the interfacial free energy isotherms may have significantly different shapes. For example, the $\Gamma(p)$ isotherm may contain a region corresponding to the adsorption of vapors in narrow capillaries.

The adsorption at the liquid/gas interface can be determined by using a microtome or molecular tracers; however, the most precise and universal methods of determining $\Gamma(c)$ at mobile interfaces, liquid–gas and liquid–liquid, are indirect methods based on the simultaneous use of the Gibbs equation and the surface (or interfacial) tension isotherm, $\sigma(c)$ (Figure 2.8).

A series of $\sigma(c)$ isotherms of aqueous solutions of fatty acids $C_nH_{2n+1}COOH$ is used as a common example. Formic and acetic acids, which do not contain a long-enough hydrocarbon chain, do not

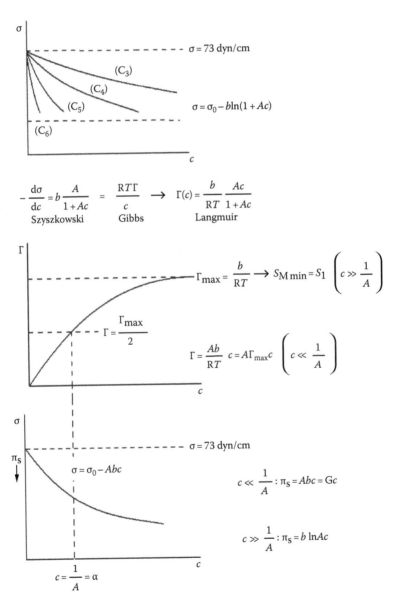

FIGURE 2.8 The Szyszkowski isotherms, $\sigma(c)$, in conjunction with the Gibbs equation, allow for the transition to the adsorption isotherms, $\Gamma(c)$. The figure also shows a comparison of characteristic regions of these isotherms at low and high concentrations of adsorbents.

reveal any surface activity. Higher acid homologues are water-insoluble. The data for intermediate members of the acid series, that is, C_3–C_6, shown in Figure 2.8, were originally obtained by Szyszkowski, who has generalized them empirically by introducing the following expression, now known as Szyszkowski isotherm: $\sigma_0 - \sigma(c) = B \ln(1 + Ac)$. While parameter B is a constant for the entire homologous series, the parameter A increases by a factor of 3–3.5 upon the transition from one homologue to the next one in order. This means that the concentration needed to reach equal lowering of the surface tension is decreased by a factor of 3–3.5 upon the transition to the next member of the homologous series (the Ducleaux–Traube rule). At low concentrations $c \ll 1/A$, the isotherm $\sigma(c)$ displays a linear decay, while at high concentrations $c \gg 1/A$, the isotherm decays logarithmically (within the solubility limits). Substitution of Szyszkowski's isotherm into the Gibbs equation yields

$$\Gamma(c) = -\left(\frac{c}{RT}\right)\frac{d\sigma}{dc} = \left(\frac{c}{RT}\right)\frac{BA}{(1+Ac)} = \Gamma_{max}Ac(1+Ac) \tag{2.13}$$

Equation 2.13 is the *Langmuir adsorption isotherm*, originally derived by Irving Langmuir. As seen in Figure 2.8, the initial portion of the $\Gamma(c)$ Langmuir isotherm is linear, and at high concentrations, the value of $\Gamma(c)$ approaches the limiting value, $\Gamma_{max} = B/RT$. This limiting value can be viewed as representing the dense packing of surfactant molecules in the adsorption layer, thus $1/\Gamma_{max} = RT/B$ is the area occupied by a mole of surfactant molecules. Furthermore, the value of $1/N_A\Gamma_{max}$ yields the area s_1 occupied by a single molecule in the close packed layer, that is, it is the surfactant molecule cross-sectional area. Indeed, such calculations for water-soluble surfactants coincide with the values of the cross-sectional area obtained from measurements with a Langmuir balance using insoluble surfactants.

Figure 2.9 briefly summarizes the main steps in the analysis of the experimental adsorption data using the Langmuir isotherm. It also schematically shows the structure of the solid surface assumed by Langmuir and the summary of the derivation of the isotherm equations from the kinetics of adsorption

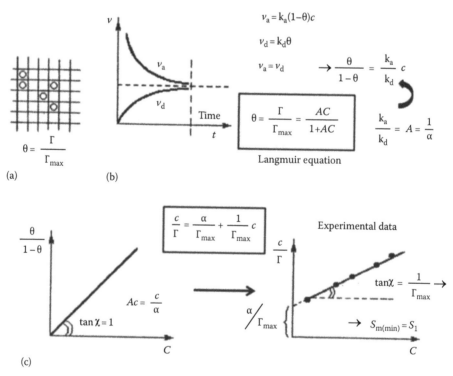

FIGURE 2.9 To the derivation of Langmuir adsorption isotherm (a,b). Application of the Langmuir isotherm to the analysis of the experimental $\Gamma(c)$ (c). (Data from Shchukin, E.D. et al., *Colloid and Surface Chemistry*, Elsevier, Amsterdam, the Netherlands, 2001.)

and desorption. In Langmuir's approach, the surface is modeled as a chessboard, each site of which can host only one adsorbing molecule (Figure 2.9a). The treatment is restricted to the case of localized adsorption, that is, the possibility of migration of molecules from one site to another is not considered. The adsorption and desorption rates, v_a and v_d, are proportional only to the fractional surface coverage, θ, with the proportionality rate constants k_a and k_d, respectively, and an equilibrium between adsorption and desorption, that is, $v_a = v_d$, readily yields Langmuir's isotherm (Figure 2.9b). The ratio of the rate constants, k_a/k_d yields $A = 1/\alpha$, which is the constant of the Langmuir isotherm. The plot of experimental data in the $c/\Gamma - c$ coordinates yields a straight line when the adsorption follows Langmuir's isotherm and allows one to obtain the values of Γ_{max} and $s_{m(min)}$ by linear regression (Figure 2.9c).

2.2 MECHANISMS OF SURFACTANT BEHAVIOR AT DIFFERENT INTERFACES

The adsorption phenomena discussed in the previous section utilized amphiphilic molecules of organic surfactants (predominantly synthetic ones). Here, it would be worthwhile to provide a brief description and classification of the most common types of organic surfactants [11–24].

Synthetic surfactants represent important commodity products, with an annual production estimated at tens of millions of tons. While most of commercially available surfactants have been in production for years, new surfactant molecules, frequently targeting specialized applications, continue to appear.

According to their physical-chemical nature, the surfactants can be subdivided into ionic (about 70% of all surfactants manufactured) and *nonionic*. In nonionic surfactants, the polar part mainly consists of multiple ethylene oxide units, $-(CH_2CH_2O)_n-$, while ionic surfactants can be subdivided into three large classes based on the charge of their polar group, that is, *anionic, cationic,* and *zwitterionic* (amphoteric) surfactants.

In anionic surfactants, the "carrier of surface activity," that is, the hydrocarbon chain, is a part of an anion having a negative charge. A well-known example of such surfactants are *soaps*, which are the alkali or ammonium salts of carbonic acids, for example, $C_nH_{2n+1}COO^-Na^+$. The hydrocarbon chains of such soaps typically contain 12–18 carbon atoms.

The excellent environmental properties of natural soaps come with a serious drawback: weakly ionizable carboxylic groups do not show good solubility in hard water and in acidic medium, while the main commercial source for manufacturing these surfactants is livestock fat.

The most common and inexpensive anionic surfactants are the aryl benzenesulfonates, $R-C_6H_4-SO_3^-Na^+$, containing a strong sulfonic group. Such surfactants are used as wetting agents, dispersion stabilizers, dispersants, and foamers. The drawback of these surfactants is their poor biodegradability. Nonbranched alkylsulfates have better biodegradability characteristics. The example of a common surfactant of such a type is sodium dodecyl sulfate (SDS), $C_{12}H_{25}-OSO_3^-Na^+$. Other common anionic surfactants are alkenesulfonates, containing double bonds.

In cationic surfactants, the carriers of the surface activity are positively charged cations. Along with weakly dissociating alkylamines (e.g., octadecylamine $C_{18}H_{37}NH_2$), these include quaternary ammonium salts, such as cetyltrimethylammonium bromide (CTAB), $C_{16}H_{33}N^+(CH_3)_3Br^-$, and cetylpyridinium chloride $C_{16}H_{33}N^+C_5H_5$ Cl$^-$. The strong ionic group ensures high surface activity and sufficient solubility over a broad pH range and in hard water. Cationic surfactants have numerous industrial applications, such as in corrosion inhibitor formulations and in specialty compositions used to stimulate oil and gas wells.

The main representatives of the amphoteric surfactants class are amino acids, including α-amino acids, $RCH(NH_2)COOH$. These molecules are surface active if they contain a substantially developed hydrocarbon chain. In the acidic medium, such molecules can act as bases by accepting a proton, thus forming NH_3^+ ions, while in the basic medium, they dissociate as acids forming COO^- anions. Amino acids play a significant role in nature. For example, α-aminoacetic acid (glycine) is 1 of the 20 amino acids that take part in the protein synthesis occurring in the human body. Other synthetic amphoteric surfactants may simultaneously contain anions and cations on the same chain. Such surfactants are referred to as zwitterionic—they include betaines and hydroxysultanes, often used as foamers.

In common nonionic surfactants, the polar group consists of chains of polyethylene oxide, $-(CH_2CH_2O)_n-$. These surfactants reveal surface activity at all pH values and in hard water. Some of the oldest and cheapest surfactants of this kind are octyl and nonyl phenol ethoxylates, $C_8H_{17}C_6H_4(OC_2H_4)_nOH$, and $C_9H_{19}C_6H_4(OC_2H_4)_nOH$, respectively. In these surfactants, the values of n are typically in the range of 4–20. Low n values typically correspond to oil-soluble surfactants, while high n values correspond to water-soluble surfactants. The largest drawback of these surfactants is their poor biodegradability and their hazard to the environment. In many countries, these surfactants have already been banned for use in some applications (e.g., in hydraulic fracturing). For these reasons, these materials are being displaced by polyethoxylated alcohols and ethers containing no aromatic rings. Along with alkylsulfates, these surfactants are the main components of synthetic detergents.

An important class of nonionic surfactants is the so-called *poloxamers* or *Pluronics*. They represent block copolymers of ethylene oxide (hydrophilic portion of the molecule) and propylene oxide (hydrophobic portion of the molecule): $H(OC_2H_4)_m(OC_3H_6)_p(OC_2H_4)_nOH$, which are commonly used in oil production and in other areas.

The surfactant market is highly versatile, with numerous brands and compositions available to the end user. Among the major consumers of surfactants are such industries as mining, metalworking, petroleum (demulsifiers, wetting modifiers, foamers, etc.), construction (additives to cement and asphalt), transportation (lubricating oils and greases, lubricating cooling liquids, etc.), pharmaceutical, food, cosmetic, paper, and water treatment. The use of surfactants in household and industrial detergents still remains a major area, consuming significant volumes of manufactured surfactants.

Along with the classification by chemical nature, one can also classify surfactants with respect to the mechanism of their behavior at interfaces. Two main mechanisms of surfactant action at S/L interfaces are of importance in their relevance to contact interactions between particles that we will address broadly throughout this book. These are (1) the control of the wetting of a solid surface by a liquid (see Figures 2.10 and 2.11) and (2) the dispersion action, which facilitates the fracture of solids, which we will discuss in detail in the second part of this book.

The main characteristics and principal concepts of wetting were described in Section 1.1. Here, we will address some specific examples of how surfactants can be used to control wetting in different processes encountered in various areas of technology.

1. Improving surface wetting by water, that is, the hydrophilization of a hydrophobic surface, can be achieved by the adsorption of *water-soluble surfactants* from an aqueous phase. The molecules of such surfactants typically contain a strong polar group promoting good water solubility and a developed hydrophobic chain, which serves as a carrier of surface activity. In accordance with the polarity equalization rule, hydrophobic tails in the adsorption layer are oriented toward the solid surface, while the polar groups are "floating" (hydrated) in the aqueous phase. The most common example of such action is detergency, in which the first step is the act of the wetting of soot or fat particles, fabric fibers, or even the surface of a skin in the course of handwashing. Hydrophilization of plant leaves, chitin shells of insects, and animal furs is necessary for successful application of pesticides, herbicides, and other agricultural chemicals. Surfactants promote the wetting of coal dust in mines and of surfaces in fire extinguishing, etc. In all these applications, which involve making hydrophobic surfaces hydrophilic, the adsorption mechanism is "nonspecific." This means that at the interface between nonpolar surfaces in contact with the nonpolar segments of the surfactant molecules, the compensation of nondispersion interactions takes place, while the interaction between relatively distant polar groups and the surface is of no significance (Section 1.3). As evident from the examples that follow, this is true for both ionic and nonionic surfactants.

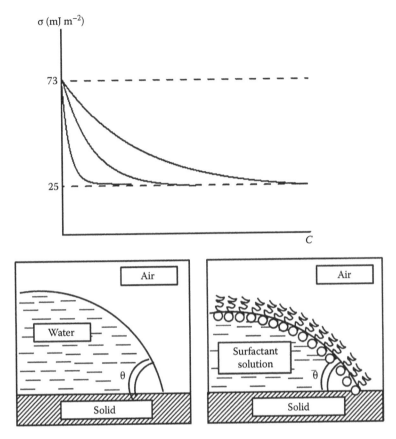

FIGURE 2.10 Moderate improvement in wetting upon the adsorption at the interface between surfactant aqueous solution and air.

2. The hydrophobization of hydrophilic solid surface involves principally different mechanism. In this case, the adsorption layer of *oil-soluble surfactants* with a strongly developed nonpolar part is formed at the S/L interface. Polar group may not necessarily be strongly ionizable but should have a specific affinity to a given solid surface. That is, anionic surfactants with a negatively charged polar group are most effective in the hydrophobization of surfaces bearing a positive charge, such as carbonates and other basic minerals, while cationic surfactants are effective at surfaces bearing a negative charge, such as acidic minerals, clays, oxides and hydroxides, quartz, and glass. Nonionic surfactants can be effective in both of these cases.

It is worth noting that in nature, the surfaces of living organisms are mostly hydrophobic, while those of inanimate matter are mostly hydrophilic.

Traditionally, one of the first and still predominate areas in which hydrophobization has been sought is the textile industry. Natural fibers, such as cotton, flax, silk, or synthetic fibers require "oiling" so that their mutual cohesion can be minimized (see Section 1.3 on the interactions between nonpolar and polar particles), and weaving and spinning processes are facilitated. Another technological example is the modification of polymer filler particles, such as chalk or clay, so that they can be introduced into liquid nonpolar media without undergoing aggregation and caking.

Similar issues are encountered in road construction, where gravel surfaces must be modified for an effective contact with hydrophobic asphalt, or in printing, where selective wetting by a hydrophobic ink is desired on particular sections of the printing plates. All of these examples represent direct applications of physical-chemical mechanics.

A standalone and particularly important example of rendering particles hydrophobic is encountered in ore enrichment. A crushed ore containing a mixture of useful minerals and barren rock undergoes the process of *froth flotation*, in which it is exposed to an aqueous surfactant solution. The surfactants used in this process are capable of selectively "oiling" only one of these fractions (typically desired mineral particles). Upon bubbling the suspension with air, hydrophobized particles, such as particles of metal sulfides, are attached to the hydrophobic air bubbles and are carried to the surface with the froth, while the particles of barren rock (such as sulfates or quartz) undergo settling at the bottom.

In froth flotation, the adsorption takes place from the polar aqueous phase, but the hydrophobic portions of the surfactant molecules in the adsorbed layer must be facing *the polar phase*, which appears to contradict the thermodynamically based polarity equalization rule. This is possible only if surfactant molecules are *chemisorbed* at the mineral surface. Chemisorption involves the formation of a very strong "chemisorption bond" between the polar groups and the particle surface. The energy of such a bond needs to well exceed 50 mJ/m^2 in order to compensate for the free energy increase associated with the formation of the hydrocarbon/water interface (Figure 2.11c).

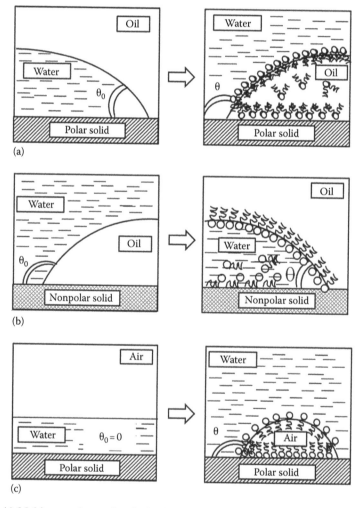

FIGURE 2.11 (a) Making a polar surface hydrophobic and (b) making a nonpolar surface hydrophilic by adsorbing a surfactant at the solid/liquid interface; (c) hydrophobization of a polar surface due to the chemisorption of surfactants from aqueous solutions "against the polarity equalization rule."

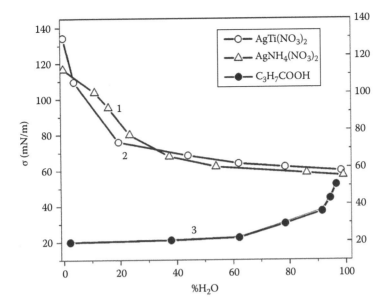

FIGURE 2.12 Rehbinder's illustration of the surface activity of water: the $\sigma(c)$ isotherms for aqueous solutions of $AgTi(NO_3)_2$ (1), $AgNH_4(NO_3)_2$ (2) in comparison with the isotherm of aqueous solution of C_3H_7COOH (3). (Redrawn from Rehbinder, P.A., *Selected Works. Surface Phenomena in Disperse Systems: Colloid Chemistry*, Nauka, Moscow, Russia, 1978.)

Therefore, the use of surfactants for the modification of interfaces is very versatile, both with respect to the nature of the interfaces (between solid and liquid, polar and nonpolar), as well with respect to the assortment of the available surfactants. Up to this point, we have been talking about amphiphilic synthetic organic surfactants. However, the adsorption phenomenon is universal in nature and industry and takes place at all interfaces without any exceptions. It is worth emphasizing one more time that the general reason for the accumulation of surface-active substances at interfaces is the lowering of free energy as a result of the partial compensation of the disrupted bonds between interfacial atoms.

It is noteworthy and rather remarkable that we can distinguish *water* as a surface-active substance (Figure 2.12). Such a distinction was first made by Rehbinder in his classic work [9,25]. While a molecule of water is not amphiphilic, that is, it does not contain a hydrocarbon tail like a surfactant, it does have a large dipole moment and is capable of forming hydrogen bonding. These two features stipulate the capability of water to form strong bonding with the surface of various ionic substances, for example, salts, oxides, and hydroxides. In this situation, the lowering of surface energy of various ionic substances from hundreds of mJ/m^2 to the values close to that of the surface tension of water takes place. One can talk about the universal nature of the adsorption of substances with a low work of cohesion, low melting point, and low value of σ at the surfaces characterized by a high work of cohesion and high values of σ. An example would be the adsorption of hydrocarbons, alcohols, ethers, and volatile substances on the surfaces of substances with ionic or covalent bonding and metals. This is especially well seen at low temperatures, as illustrated by the well-known example of using nitrogen adsorption for the determination of the surface free area of various disperse materials (BET isotherm measurements).

Good wetting of solid metals by metallic melts of certain fusible metals with a low value of σ (such as wetting of iron by zinc, or of copper by bismuth) indicates high values of the work of adhesion, W_a, and the possibility of effective adsorption from either vapor or melt in another low-melting-point metal. This will be the subject of a more detailed discussion on the interactions between solid and liquid metals later in this book.

Adsorption phenomena may also take place at the interfaces between solid phases, such as different phases in multiphase compositions, as well as at grain boundaries in single-phase systems (granulated materials, metals, salts, etc.). All grain boundaries carry uncompensated interactions between the surface atoms and thus have a certain excess of free surface energy associated with them. One typical example of adsorption in such systems is that of sulfur and phosphorus at the grain boundaries in steel. Such adsorption results in a significant change in the mechanical properties of steel and in particular in the lowering of strength. Conversely, the adsorption of carbon or boron results in a significant improvement in the steel quality.

It is worth pointing out here some limitations on the use of the polarity equalization rule. This rule can be directly applied to systems in which one can distinguish the presence and significant role of the dispersion component of interactions in comparison with nondispersion interactions, which are mainly molecular. Obviously, in the case of a contact between a solid and a liquid metal or of a contact between an ionic or covalent body with a fusible salt, or in the case of the adsorption of a third component at such interfaces from a liquid phase, the polarity equalization concept has a more general meaning. Indeed, the rule emphasizes the physical-chemical similarity or dissimilarity between contacting phases. In this sense, the adsorption of water at the surface of ionic compounds "equalizes" the interface with air by replacing a surface having a surface energy on the order of hundreds of mJ/m^2 with the one that has a surface energy on the order of the surface tension of water (~ 72 mJ/m^2).

These basic concepts of surfactant adsorption that we have reviewed should be sufficient for the discussion of the influence of surfactant adsorption on the interactions between disperse phase particles, which we will address in the next section.

2.3   CONTACT INTERACTIONS IN THE PRESENCE OF SURFACTANTS: THE ROLE OF SURFACTANT ADSORPTION

In this section, we will continue the discussion started in previous chapter and address the role of surfactant adsorption in contact interactions. In Section 1.3, we described the concept and experimental method for measuring contact adhesive forces between molecularly smooth particles of different nature in different media. In detail, we explored two limiting cases: that of very hydrophilic glass surfaces and that of hydrophobic surfaces of methylated glass. Along with the air, which is the "most nonpolar" medium, we also compared interactions in liquid media, which can be subdivided into three characteristic groups: (1) adhesive interactions in water and hydrocarbon; (2) adhesive interactions in a series of liquids, including alcohols of different polarities; and (3) interactions in liquids comprising a "continuous spectrum" of polarities, that is, in aqueous solutions of alcohols. Calculations based on Derjaguin's theorem have shown that the free energy of interaction covers a very broad range of four orders of magnitude.

Now, we will turn our attention to experimental studies on contact interactions in the presence of surfactants. Here, we will also devote special attention to the studies involving molecularly smooth surfaces, which allow one to quantitatively estimate the free energy of interaction as the main invariant parameter characterizing adhesion. We will also look at polar and nonpolar particles in various aqueous and nonaqueous media but focus on the role of the presence of small amounts of different surfactants.

Figures 2.13 and 2.14 show the isotherms of the free energy of contact interaction ($\frac{1}{2} F$) between methylated glass spheres as a function of surfactant concentration in aqueous solution for two model ionic surfactants: SDS and CTAB, respectively.

In the solutions of both the anionic SDS surfactant (Figure 2.13) and the cationic CTAB surfactant (Figure 2.14), the systems are in equilibrium and display full reversibility. To some extent, this is also true for a nonionic surfactant, oxyethylated ether (Figure 2.15), but in this case, equilibrium is not reached instantaneously, as the value of p_1 shows strong time dependence (Figure 2.16). In the

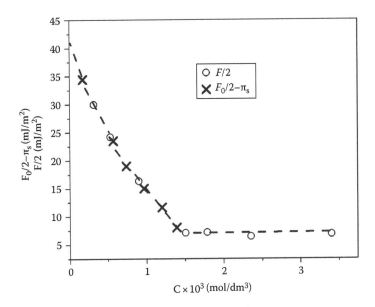

FIGURE 2.13 Free energy of interaction ($\frac{1}{2}$ F) between two methylated glass beads (1); and the value of $\frac{1}{2}$ $F - \pi_s$ (2), erg/cm² in aqueous solutions of sodium dodecyl sulfate as a function of surfactant concentration, $c = 10^{-3}$ mol/L. (Data from Shchukin, E.D. and Amelina, E.A., *J. Dispers. Sci. Technol.*, 24, 377, 2003.)

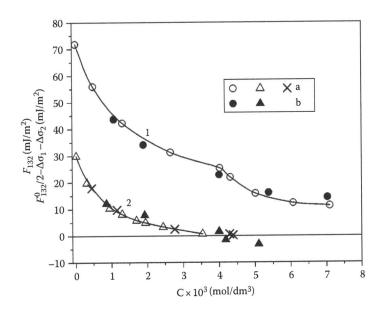

FIGURE 2.14 Free energy of interaction between two methylated surfaces (1) and methylated and acetylated surface (2), erg cm² as a function of the cetyltrimethylammonium bromide concentration determined from measurements of the cohesive force and energy, F_{132} (a) and from the difference between F_{132}^0 in water and the change in the energy of wetting $\Delta\sigma_1$ and $\Delta\sigma_2$ (b). Notations 1, 2, 3 correspond to the methylated surface, the acetylated surface and the medium, respectively. (Data from Shchukin, E.D., *J. Colloid Interface Sci.*, 256, 159, 2002; Shchukin, E.D. and Amelina, E.A., *J. Dispers. Sci. Technol.*, 24, 377, 2003; Yaminskiy, V.V. et al., *Coagulation Contacts in Disperse Systems*, Himiya, Moscow, Russia, 1982; Shchukin, E.D., *Vestnik AN SSSR*, 5, 20, 1976; Shchukin, E.D., Some colloid-chemical aspects of the small particles contact interactions, in: *Fine Particles Science and Technology*, E. Pelizzetti (ed.), Kluwer Academic Publishers, Amsterdam, the Netherlands, 1996, pp. 239–253.)

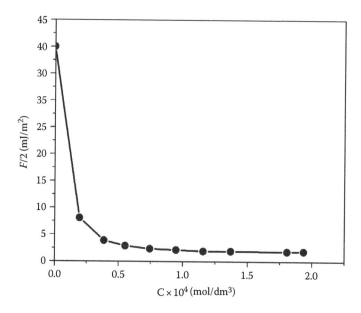

FIGURE 2.15 The equilibrium values of the free energy of interaction between two methylated glass spheres as a function of a nonionic surfactant concentration (c, 10^{-4} mol/L). The nonionic surfactant is oxyethylated ether, $C_{12}E_{20}$. (Data from Shchukin, E.D. and Amelina, E.A., *J. Dispers. Sci. Technol.*, 24, 377, 2003; Shchukin, E.D., *Colloid J.*, 59, 248, 1997.)

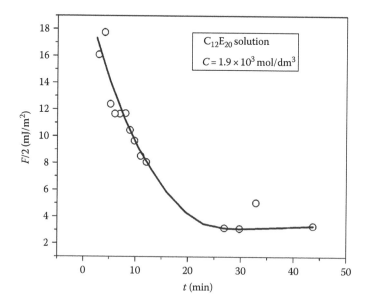

FIGURE 2.16 The free energy of interaction between methylated surfaces as a function of time for surfaces immersed into a 1.9×10^{-5} molar solution of oxyethylated ether, $C_{12}E_{20}$. (Data from Shchukin, E.D. and Amelina, E.A., *J. Dispers. Sci. Technol.*, 24, 377, 2003; Shchukin, E.D., *Colloid J.*, 59, 248, 1997.)

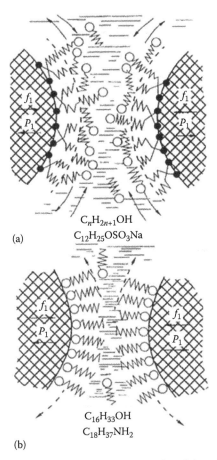

FIGURE 2.17   (a) Schematic illustration of the reversible behavior of the adsorbed layer in the case of non-specific adsorption from an aqueous solution on a methylated surface and (b) in the case of specific adsorption (chemisorption) on polar surface from a nonpolar hydrocarbon. In the second case, a critical compression force needs to be applied in order to cause (with some probability) rupture and the displacement of the adsorption layer. (Redrawn from Shchukin, E.D., *Colloid J.*, 59, 248, 1997.)

latter case, equilibrium is achieved only after the adsorption layer of oxyethylated ether has already formed. Time dependence indicates that for nonionic surfactant, this process is much slower than for a cationic or anionic surfactant.

Figure 2.15 shows a comparison between the interaction of two nonpolar particles and the interaction of a nonpolar particle and a partially lyophilized particle (hydrophilized with acetyl cellulose). In the first case, ½ *F* reaches a limiting value corresponding to low but constant attraction, while in the second case, ½ *F* drops below zero, which is the real, experimentally observed change in the sign of the disjoining pressure, that is, a transition from attraction to repulsion.

All of the illustrated examples correspond to equilibrium and reversible conditions. The response of the adsorption layer formed at the hydrophobic methylated surfaces to the applied compression is schematically illustrated in Figure 2.17. The layer is displaced from the contact zone as the particles are compressed against each other and returns when the particles are pulled apart. The situation is principally different in the case of polar particles and adsorption from nonpolar media, such as in the case of amines on glass spheres. The chemisorption that takes place leads to the formation of an adsorbed layer that has its own mechanical strength, in which case a critical compressive force, f_c, needs to be applied for the adsorption layer to rupture and to be displaced from the contact zone (Figures 2.17b and 2.18).

In retrospectively looking at the data presented in this chapter and the data reported earlier in Section 1.3, we again need to emphasize here that the spectrum of the values of the free energy of

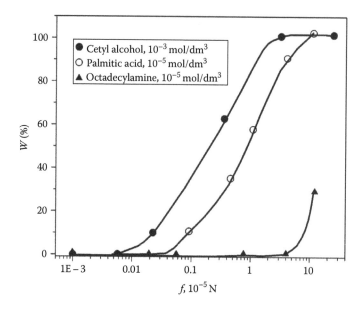

FIGURE 2.18   The probability (W, %) of the rupture of the adsorption layer between glass beads as a function of the compressive force, f, for (a) 10^{-3} mol/dm^3 of cetyl alcohol, (b) 10^{-5} mol/dm^3 palmitic acid , and (c) 10^{-5} mol/dm^3 octadecylamine solutions in heptane. (Redrawn from Shchukin, E.D., *Colloid J.*, 59, 248, 1997.)

interaction is very broad and covers four orders of magnitude. If we were to consider particles with sizes ranging from tens of nanometers to millimeters, the spectrum of the adhesive contact forces, p_1, would cover nine orders of magnitude, from 10^{-8} dynes to tens of dynes, and the corresponding spectrum of the interaction energy values would range from 10^{-15} to 10^{-6} ergs. This can be directly related to the subject of colloid stability, which will be addressed in detail in Chapter 4. A characteristic critical value $w \sim 10kT$ falls in the middle of this range, and thus lyophilicity and lyophobicity are manifested as *properties of the system.*

The described experimental method for measuring the contact interactions between solid particles influenced by surfactant adsorption from various electrolyte solutions allows one to observe transition from lyophilicity to lyophobicity and to study the role of the electrolyte in this transition.

While we will continue to address the subject of using surfactants to control contact interactions in Chapters 3 and 4, we would like to devote the remainder of this chapter to a discussion of the cohesive forces between anisometric particles—specifically, between the cellulose fibers. This study conducted by the authors and their colleagues is of importance in papermaking applications.

2.3.1   Contacts between Cellulose Fibers: The Evaluation of Friction Forces and Contact Strength; The Effect of Common Papermaking Chemicals

The papermaking process is based on the dewatering of a suspension of cellulosic fibers mixed with mineral fillers (e.g., calcium carbonate, clay, titanium dioxide) on a moving wire. In the papermaking process, a suspension of cellulosic fibers undergoes a significant change in rheology dictated by the fiber concentration and fiber–fiber or fiber–filler interactions at different stages in the papermaking process. First, a rather thick and concentrated pulp stock with a consistency of around 4% is diluted with water and pumped at a high rate into the head box of the paper machine. Various papermaking chemicals, such as starches, polymeric coagulants, polymeric flocculants, and colloid particulate suspensions (the so-called "microparticles," most commonly colloidal silica sols) are added to the flowing pulp suspension. From the head box, the pulp is delivered to a moving wire on which water is removed by gravity drainage or vacuum drainage. The paper web is then transferred to the pressing section and the drying section to make a finished product. Contact interactions play an essential role at various stages of this process: upon the dilution of the pulp, they are responsible for controlling rheology; on the moving wire, they control the rate of dewatering and eventually are responsible for good *sheet formation*, which determines the end quality paper; and in the drying stage, contact interactions determine the strength of the paper web, which makes it possible to operate modern paper machines at high speeds [32]. All of these complex phenomena, occurring at various stages of the papermaking process, provide one with a good reason to investigate fiber–fiber contact interactions and to study the impact of common papermaking chemicals on them.

Certain peculiarities of fiber systems are determined, on the one hand, by the inhomogeneous, mosaic-like nature of the fiber surface, and, on the other hand, by their anisometric shape, which allows for the formation of a number of local bonds. The rheological properties of the fiber structures control the resistance of interfiber bonds, both to their rupture under the effect of normal stresses and to shear under tangential ones. This means that friction forces play a significant role in the contact interactions of fibers.

At the same time, the normal loads on the fibers may be small, or absent at the stage of the homogenization of the paper pulp, but significant during pressing. In the case of pulp homogenization, the friction force is caused by the *molecular attraction* between the fibers and depends essentially on their surface properties. Changing the surface properties of the fibers in a desired way can be achieved with the use of surfactants or polymers. This allows one to control the rheological behavior of the pulp and obtain effective adhesion between the paper fibers and the particles of filler, and also to achieve optimal conditions for textile fiber spinning, etc. [33,34].

A small radius of curvature in the fiber–fiber contact areas results in cohesive forces of a very small magnitude. This necessitates the use of highly precise methods and devices for measuring these small forces in contacts between individual fibers, under both normal rupture and shear. Such measurements provide one with the means for the evaluation of a contribution of molecular attraction to friction forces and allow one to obtain a thermodynamic value of the free energy of interaction. The latter is a general characteristic of interactions in a given medium, invariant with respect to the specimen geometry.

The technique and suitable instrumentation for determining the cohesive forces under rupture caused by normal force was reviewed in Section 1.3, and the suitable experimental device was shown in Figure 1.28. The measurement of the cohesion under shear also requires a specialized experimental setup, developed for this purpose, and the scheme of such an apparatus is presented in Figure 2.19. This technique was originally proposed by Yakhnin [35] and subsequently modified and used extensively at the Chemistry Department of Moscow State University [36–40].

The main element of the apparatus for measuring the friction forces between individual fibers is the tilting Π-shaped frame (1). The frame is connected to an electric motor (2) via a reducing gear box (3) to which a rigid holder (4) is attached. The tilt angle of the frame, α, is determined by the position of the pointer (5) on the scale (6). Two threadlike samples (i.e., fibers) (7) and (8) are attached to the frame at end points A and B, as well as C and D, respectively. The fibers contact each other at point O at a right angle. The free end D is connected to the weight P, which exerts a compressive force, N, between the fibers at the point of contact. The crossing fibers can be immersed into a cuvette (10) for measurements in a liquid medium. In the original prototype of this measuring device, the fibers were attached to the frame with glue, which is undesirable since the glue alters the surface properties of the fibers. To alleviate this problem, the fibers were tied with small knots to the extending thread lines, which were mounted in a device with specially designed microclamps. The thread connected to the ends of fiber CD is connected to the frame at point C. The thread (11) connected to the end D is placed over the pulley (12) with a very small friction torque, and a weight (13) is connected to its free end. For more precise measurements, a torsion scale in place of a weight can be used, which allows one to extend the range of the loads P, down to ~10 μN. All of the parts of this setup are mounted on a stand (14).

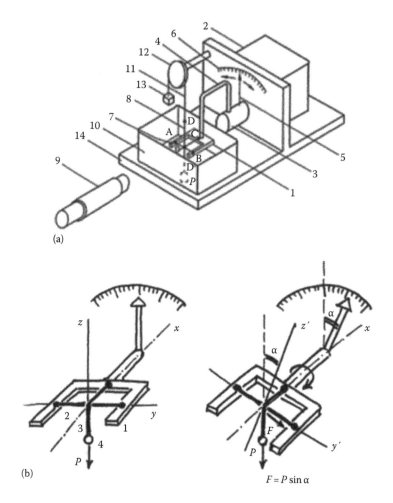

FIGURE 2.19 (a) General schematic illustration of an experimental setup suitable for measuring the friction forces between individual fibers; (b) tilting frame measuring scheme. (Redrawn from Amelina, E.A. et al., *Kolloidnyi Zh.*, 60, 583, 1998.)

At the initial, horizontal position of the frame, $N = P$. Tilting the frame by an angle α exerts a shear force $F = P \sin \alpha$; the corresponding normal force equals $N = P \cos \alpha$. The critical value of the angle α corresponding to a tilt at which fiber CD starts to move along fiber AB is recorded in the course of the measurement with the help of a microscope (9).

The described method was used to measure cohesive force under applied shear between individual cotton fibers in aqueous solution and in solutions of polyethyleneimine (PEI), which is a common papermaking coagulant with a molecular weight of around 60,000 Da. The fiber thickness as determined by microscopy was ~10 μm, and the compression load was varied between 0.057 and 1 g, with a minimum of 30 measurements conducted at each load.

The resulting measurements of the friction force, F, between cotton fibers in water as a function of the applied load, N, is shown in Figure 2.20. As expected, the values of angle α obtained with the same pair of fibers at similar loads, P, show a rather broad scatter. Correspondingly, there is a large amount of scatter in the values $F = P \sin \alpha$ and $N = P \cos \alpha$. Each set of normal and tangential force data is represented by a histogram showing the distribution of these values. The maxima of these distributions are shown as circles, while the distributions themselves are represented by inclined lines for each load P.

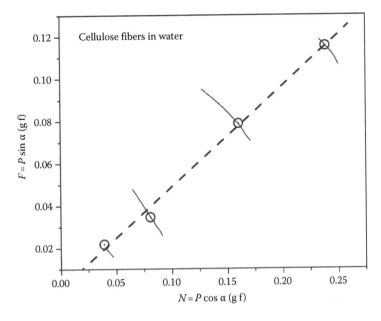

FIGURE 2.20 Friction force between fibers, F (grams of force), as a function of normal load, N (grams of force), in water. (Data from Amelina, E.A. et al., *Colloids Surf.*, 167, 215, 2000.)

Figure 2.20 shows that overall the $F(N)$ data fall on a straight line passing through the origin. Similar trends were also observed for the measurements in air and in PEI solutions. The slope of this line reveals the friction coefficient, μ, which is an invariant characteristic describing fiber–fiber interaction. The described measurements allow one to measure μ with a precision of 5%–10%. Friction coefficients measured between cotton fibers are rather high, $\mu \sim 0.4$–0.5, which may be attributed to the surface roughness of the fibers. This is also supported by experimental observations, which revealed that the highest values of μ are typically obtained in the first few measurements carried out with a given fiber pair. Values of the friction coefficient tend to decrease with consecutive measurements, perhaps due to some "smoothening" of the fiber surface in repeated shearing.

An interesting finding is that similar values of the friction coefficient were obtained in the course of the measurements carried out in air and in water, $\mu = 0.41 \pm 0.03$ and $\mu = 0.47 \pm 0.03$, respectively. One possible reason for this similarity may be related to the hygroscopic nature of cotton fibers, that is, due to the uptake of moisture in air.

The friction coefficients between cotton fibers measured in solutions of PEI of different concentrations are shown in Figure 2.21. The concentration dependence of the friction coefficient reveals a maximum at low concentrations, essentially constant μ in the range of concentrations between 4×10^{-7} and 4×10^{-5} mol/dm^3, and a sharp drop at high concentrations of the polyelectrolyte. These findings agree qualitatively with the known mechanism by which PEI flocculates cellulosic pulp. It is well known that the adsorbed PEI forms positively charged patches on the negatively charged fiber surface, and the flocculation is driven by the attraction of the "positive patch" of one fiber to the bare surface of the other fiber [32,41,42]. Optimum flocculation efficiency is typically observed at about or less than 50% surface coverage and does not require complete surface charge neutralization. In the case of two fibers, this "optimum flocculation" condition logically corresponds to the observed maximum in the friction coefficient. Further increases in PEI concentration would result in increased coverage of the fiber surface by PEI and eventually into a complete charge reversal resulting in electrostatic repulsion between two similarly charged fibers. The trends in the friction

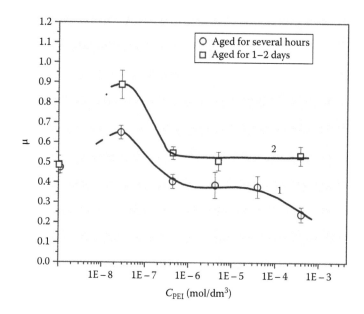

FIGURE 2.21 The friction coefficient between cotton fibers as a function of the concentration of polyethyleneimine: aged for several hours (1); aged for 1–2 days (2). (Data from Amelina, E.A. et al., *Colloids Surf.*, 167, 215, 2000.)

coefficient observed in Figure 2.21 appear to support this argument. These results also agree qualitatively with the trends in the electrokinetic data for cellulosic fibers in the presence of PEI with a lower molecular weight of 30,000 Da [42]. With the addition of PEI zeta potential changes from the original negative values to positive ones, passing through the isoelectric point at $c \sim 10^{-8}$ M of PEI, reaching maximum of +40 mV at $c \sim 10^{-6}$ M and then decaying to 15–18 mV at $c \geq 10^{-4}$ M. At concentrations $c \sim 4 \times 10^{-7}$ to 4×10^{-4} M, the values of the friction coefficient are comparable to the values determined in water without any PEI added. This value probably corresponds to the value of μ measured under conditions corresponding to the absence of electrostatic attraction. One has to keep in mind that despite a general similarity in the trends for the electrokinetic and friction coefficient measurements, it does not seem feasible to completely transfer the results of one set of studies to the other, as the properties of PEI used in the friction coefficient measurements and electrokinetic measurements may be substantially different.

The results presented in Figure 2.21 correspond to the measurements of friction coefficients conducted after the fibers were in contact with PEI for several hours. Data shown in Figure 2.21 also indicate that after ageing the fibers left in PEI solutions for 1–2 days, significantly higher values of the friction coefficient were observed at the corresponding PEI concentrations. Possible reasons for this may include fiber swelling leading to an increase in surface roughness, as well as possible adjustment in the conformation of the adsorbed polymer with time [42,43].

In agreement with Derjaguin's molecular theory of friction [44,45], the friction force is defined as $F = \mu(N + N_0)$, where N is the normal component of the applied load and N_0 is the sum of the molecular attractive forces. In the absence of external loads, the friction force results solely from the action of the forces of molecular attraction, that is, N_0: $F_m = \mu N_0$. Under realistic conditions, that is, in the course of pulp mixing in the chests, the normal loads N are not high, and the contributions of N_0 to the net friction force between the fibers can be quite substantial. However, in the experiments involving the measurement of friction coefficients where there is a small curvature radii of the fibers, the values of N_0 are much smaller than the applied normal loads N. This does not allow one to isolate the portion of the friction force responsible solely to the molecular attraction, that is, $F = \mu N_0$, by a simple extrapolation of $F = F(N)$ data to yield the

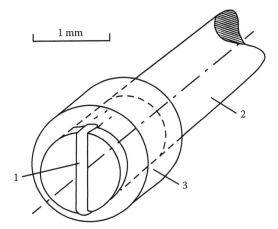

FIGURE 2.22 Schematic illustration of how the fiber is mounted on the L-shaped holders for contact force measurements.

$N = 0$ value. This is an unfortunate drawback of the experimental technique: the estimate for the N_0 is essential for an understanding of the mechanisms of the influence of surfactants and polymers on the friction between fibers and on the pulp rheology. This estimate can be obtained from direct measurements of cohesive forces in contacts between crossed fibers. The molecular component of the cohesion, p, in the contact between two fibers can be obtained as the force necessary to rupture the fiber–fiber contact. The shear friction force can then be determined as the product of the previously determined friction coefficient, μ, and the normal force N_0. In the absence of an external load, the value of N_0 is solely the result of the molecular attraction forces, that is, $F = \mu N_0 = \mu p$.

The cohesive forces between 10 μm fibers are very small but can be measured using the setup described in Section 1.3 and shown in Figure 1.28. The individual fibers can be mounted on the L-shaped holders by using the specialized clamp illustrated in Figure 2.22. The fiber 1 is placed across the end of a hollow polystyrene cylinder 2 and is fixed with a tight-fitting ring 3. In order to achieve better fixation of the fiber, small grooves were made on the outer edge of the cylinder.

The results of the contact force measurements, p, between cellulosic fibers in the presence of PEI are shown in Figure 2.23. The observed trends are similar to those shown in Figure 2.21 for the friction coefficient measurements and are also in good agreement with the "patch model" of flocculation by the PEI, and the previously mentioned electrokinetic studies, as reflected by the maximum in the cohesive force at low concentrations of PEI. As expected, after reaching the maximum, the cohesive forces decrease with a further increase in the PEI concentration. This corresponds to the decrease in the fraction of available negatively charged patches for interaction with PEI and the increase in the electrostatic repulsion due to a continued adsorption of PEI.

The component of the shear friction force in contact, F, due solely to molecular attraction, can be obtained from the measured values of the cohesive forces, p, using the values of the friction coefficients, μ, shown in Figure 2.21. These data are illustrated in Figure 2.24, which shows the friction force, $F = \mu p$ as a function of the PEI concentration. In the concentration interval $c \sim 10^{-9}$–10^{-7} M, the addition of PEI results in an increase in the friction force up to a maximal value observed in the concentration interval between $c \sim 10^{-9}$–10^{-8} M. At PEI concentrations greater than $\sim 10^{-6}$ M, the friction force decreases and reaches values nearly two times less than in water.

The obtained values of the friction force are related to a single "point-like" contact between two fibers crossed at a 90° angle. In the pulp suspension of papermaking consistency, the fibers form numerous such contacts. The summed molecular attractive forces are responsible for the friction forces that determine the rheological properties of the pulp at different stages of the papermaking

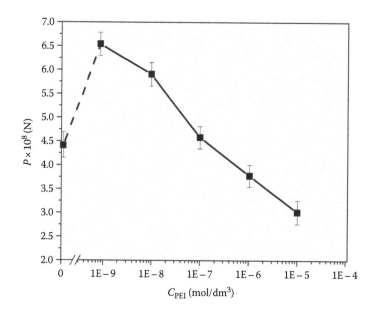

FIGURE 2.23 Cohesive force, p, for a contact between crossed fibers as a function of the PEI concentration. (Redrawn from Amelina, E.A. et al., *Colloids Surf.*, 167, 215, 2000.)

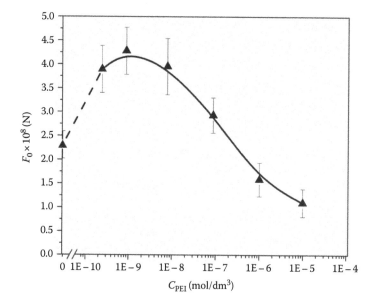

FIGURE 2.24 The molecular component of the shear friction force between cellulosic fibers in water as a function of the PEI concentration. (Redrawn from Amelina, E.A. et al., *Kolloidnyi Zh.*, 63, 132, 2001.)

process. The interactions between fibers in the presence of added papermaking chemicals influence the dewatering of the pulp on the wire and the end sheet formation. The control of the rheological properties is thus dependent on the ability to influence the molecular component of the friction force.

The cohesive force between cellulosic fibers from Kraft bleached pulp was also measured in solutions of a cationic surfactant, tetrabutylammonium iodide (TBAI). The cohesive force, p, as a function of the concentration of TBAI is shown in Figure 2.25. As seen in this figure, the cohesive force undergoes a sharp decrease as the TBAI concentration increases and then passes through a

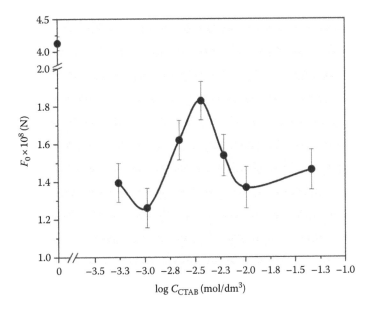

FIGURE 2.25 The cohesive force, p, for a contact between crossed fibers as a function of the TBAI concentration. (Redrawn from Amelina, E.A. et al., *Kolloidnyi Zh.*, 63, 581, 2001.)

minimum at $c \sim 10^{-3}$ M. Further increases in the TBAI concentration result in an increase in the cohesive force up to a maximum value at $c = 3.2 \times 10^{-3}$ M. Measurements of the ζ-potential in the same system revealed that, at concentrations of TBAI of less than 10^{-3} M, the ζ potential value decreases from -2 to -10 mV, most likely due to the specific adsorption of I^- ions. Such an increase in the negative ζ potential results in an increase in electrostatic repulsion, which in turn results in a weakening of the cohesive force. However, the observed decrease in the cohesive force is too large to be explained solely by the modest change in the ζ potential. Most likely, the main contribution to the contact strength decrease is caused by the formation of a hydrophilic adsorption layer. Independent adsorption measurements indicated that at TBAI concentration, $c = 3.2 \times 10^{-3}$ M, the adsorption is 3.7×10^{-4} mol/cm^2, which corresponds to the formation of a polymolecular adsorption layer.

Further increases in the TBAI concentration result in an increase in the ζ-potential due to the preferential adsorption of the surfactant cation. The isoelectric point is observed at $c \sim 3.2 \times 10^{-3}$ M, and the ζ-potential further increases to $+10$ mV at $c \sim 2.5 \times 10^{-2}$ M. Around the isoelectric point, the maximum in the cohesive force has been observed. Further increases in the TBAI concentration result in an increase in electrostatic repulsion, which is the likely reason for the cohesive force increase.

Another interesting study on the cohesive force between bleached pulp fibers was carried out in a solution of papermaking high-molecular-weight cationic polyacrylamide (CPAM) flocculant [46]. In contrast to PEI, the molecular weight of this polyacrylamide flocculant was on the order of tens of millions of daltons. The results of this study are shown in Figure 2.26 in the form of a histogram of the repeated force measurements between two fibers immersed in a 0.001% solution of CPAM. Initially, the cohesive force between the fibers was so high that the contact could not be ruptured by the applied current and had to be ruptured manually. When the fibers were brought together again, the cohesive force was not nearly as high: the contact could be ruptured with a current of sufficient magnitude, and the measured cohesive force, p, was about 4.5 μN. The cohesive force upon the subsequent contact between the same fibers gradually decreased until it reached the minimum value of ~ 0.5 to 0.5 μN. These observations agree very well with the bridging mechanism of pulp using high-molecular-weight flocculants [32]. When the flocculant is added to the flowing pulp slurry,

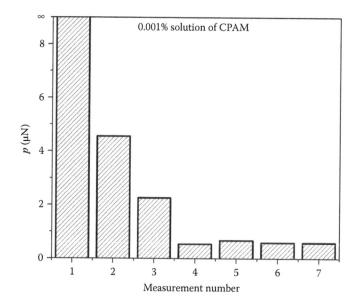

FIGURE 2.26 Cohesive force measured in the contact between two bleached cellulose fibers in a solution of papermaking cationic polyacrylamide. The effect of sequential rupturing and reformation of contacts between the same fibers is shown. (Zelenev, A.S. and Shchukin, E.D., unpublished results.)

very large and strong flocs are initially formed by means of polymer "bridges" between neighboring fibers. These flocs are then broken down by shear stresses, for example, in paper machine screens. Once the flocculated pulp passes the shear point and the relief of shear stress occurs, a reflocculation takes place, but the newly formed flocs are smaller and weaker. The rupture of flocs under papermaking conditions results in the physical rupture of the flocculant molecules with the subsequent reconformation of the residual flocculant on the fiber surface. The degradation in the polymer molecular weight and the reduction in the number of "loops" and "tails" extending into the pulp slurry leads to the formation of smaller and weaker flocs, that is, to the degradation of flocculation effectiveness. This process was to some extent simulated by the repeated rupture of cohesive contacts between the fibers.

REFERENCES

1. Rusanov, A. I. 1967. *Phase Equilibria and Surface Phenomena*. Leningrad, Russia: Himiya.
2. Kerker, M. 1975. *Surface Chemistry and Colloids*. London, U.K.: Butterworth.
3. Adamson, A. 1979. *Physical Chemistry of Surfaces*. Moscow, Russia: Mir.
4. Gibbs, J. W. 1982. *Thermodynamics: Statistical Mechanics*. Moscow, Russia: Nauka.
5. Shchukin, E. D., Pertsov, A. V., Amelina, E. A., and A. S. Zelenev. 2001. *Colloid and Surface Chemistry*. Amsterdam, the Netherlands: Elsevier.
6. Frolov, Y. G. 1982. *Colloid Chemistry: Surface Phenomena in Disperse Systems*. Moscow, Russia: Himiya.
7. Ross, S. and I. D. Morrison. 1988. *Colloidal Systems and Interfaces*. New York: Wiley-Interscience.
8. Hunter, R. J. 1991. *Foundations of Colloid Science*, Vols. 1–2. Oxford, U.K.: Clarendon Press.
9. Malta, V., et al. 1971. Crystal structure of the C form of stearic acid. *J. Chem. Soc.* (B) 548–553.
10. Rehbinder, P. A. 1978. *Selected Works: Surface Phenomena in Disperse Systems: Colloid Chemistry*. Moscow, Russia: Nauka.
11. Shinoda, K., Nakagava, T., Tamamushi, B., and T. Isemura. 1966. *Colloidal Surfactants*. Moscow, Russia: Mir.
12. Schwartz, A., Perry, D., and J. Birch. 1966. *Surfactants and Detergents*. Moscow, Russia: IL.
13. Joensson, B., Lindman, B., Kronberg, B., and K. Holmberg. 1998. *Surfactants and Polymers in Aqueous Solution*. New York: Wiley.

14. Kwak, J. C. T. (Ed.). 1998. *Polymer-Surfactant Systems*. New York: Marcel Dekker.
15. Papirer, E. (Ed.). 2000. *Adsorption on Silica Surfaces*. New York: Marcel Dekker.
16. Mittal, K. L. and D. O. Shah. (Eds.). 2002. *Adsorption and Aggregation of Surfactants in Solutions*. New York: Marcel Dekker.
17. Holmberg, K. (Ed.). 2003. *Novel Surfactants*. New York: Marcel Dekker.
18. Totten, G. E. and Y. Liang. (Eds.). 2004. *Surface Modification and Mechanisms: Friction, Stress, and Reaction Engineering*. Marcel Dekker, New York.
19. Lai, K. Y. (Ed.). 2006. *Liquid Detergents*, 2nd edn. CRC Press/Taylor & Francis Group.
20. Biresaw, G. and K. L. Mittal (Eds.). 2008. *Surfactants in Tribology*. CRC Press, Boca Raton, FL.
21. Rhein, L. D., Schlossman, M., O'Lenick, A., and P. Somasundaran. (Eds.). 2007. *Surfactants in Personal Care Products and Decorative Cosmetics*. CRC Press, Boca Raton, FL.
22. Zoller, U. (Ed.). 2009. *Handbook of Detergents*, Vols. 1–6. CRC Press, Taylor & Francis Group.
23. Ruiz, C. C. (Ed.). 2009. *Sugar Based Surfactants: Fundamentals and Applications*. CRC Press/Taylor & Francis Group.
24. Pillon, L. Z. 2010. *Surface Activity of Petroleum Derived Lubricants*. CRC Press/Taylor & Francis Group, 1–6.
25. Rehbinder, P. A. 1926. Water as a surfactant. *Z. Phys. Chem.* 121: 103.
26. Shchukin, E. D. 2002. Surfactant effects on the cohesive strength of particle contacts: Measurements by the cohesive force apparatus. *J. Colloid Interface Sci.* 256: 159–167.
27. Shchukin, E. D. and E. A. Amelina. 2003. Surface modification and contact interaction of particles. *J. Dispers. Sci. Technol.* 24: 377–395.
28. Yaminskiy, V. V., Pchelin, V. A., Amelina, E. A., and E. D. Shchukin. 1982. *Coagulation Contacts in Disperse Systems*. Moscow, Russia: Himiya.
29. Shchukin, E. D. 1976. Surfactants and their effect on the properties of disperse systems. *Vestnik AN SSSR*. 5: 20–25.
30. Shchukin, E. D. 1996. Some colloid-chemical aspects of the small particles contact interactions. In *Fine Particles Science and Technology*, E. Pelizzetti (Ed.), pp. 239–253. Amsterdam, the Netherlands: Kluwer Academic Publishers.
31. Shchukin, E. D. 1997. Development of Rehbinder's doctrine of strong stabilization factors in disperse systems. *Colloid J.* 59: 248–261.
32. Gullichsen, J. and H. Paulapuro (Eds.). 1999. *Papermaking Science and Technology. Book 4: Papermaking Chemistry*. Helsinki, Finland: Fapet Oy.
33. Reizinsh, R. E. 1987. *Pattern Formation in Suspensions of Cellulose Fibers*. Riga, Latvia: Zinatne.
34. Kerekes, R. J. and C. J. J. Shell. 1992. Characterization of fibre flocculation regimes by a crowding factor. *J. Pulp Paper Sci.* 18: J32.
35. Yakhnin, E. D. 1968. On the relationship between the strength of disperse structure and the interaction forces existing between its elements. *Doklady AN SSSR*. 178: 152.
36. Amelina, E. A., Shchukin, E. D., Parfenova, A. M. et al. 2000. Effect of cationic polyelectrolyte and surfactant on cohesion and friction in contacts between cellulose fibers. *Colloids Surf.* 167: 215–227.
37. Amelina, E. A., Shchukin, E. D., Parfenova, A. M., Bessonov, A. I., and I. V. Videnskiy. 1998. Adhesion of cellulose fibers in liquid media. 1. Measurement of friction forces in contacts. *Kolloidnyi Zh.* 60: 583–587.
38. Shchukin, E. D., Videnskiy, I. V., Amelina, E. A. et al. 1998. Adhesion of cellulose fibers in liquid media. 2. Measurement of attraction forces in contacts. *Kolloidnyi Zh.* 60: 588–591.
39. Amelina, E. A., Videnskiy, I. V., Ivanova, N. I. et al. 2001. Evaluation of the molecular component of friction of fibers in surfactant solutions. *Kolloidnyi Zh.* 63: 132–134.
40. Amelina, E. A., Videnskiy, I. V., Ivanova, N. I. et al. 2001. Contact interactions between the individual fibers of cellulose and its derivatives: Mechanism of action of the cationic surfactants. *Kolloidnyi Zh.* 63: 581–585.
41. Hubbe, M. A. 2007. Flocculation of cellulose fibers. *BioResources*. 2: 296–331.
42. Pfau, A., Schrepp, W., and D. Horn. 1999. Detection of a single molecule adsorption structure of poly(ethylenimine) macromolecules by AFM. *Langmuir*. 15: 3219–3225.
43. Shchukin, E. D. 2001. The role of contact interactions in the rheological behavior of a fibrous suspension. *Colloid J.* 63: 784–787.
44. Derjaguin, B. V. 1963. *What is Friction*, 2nd edn. Moscow, Russia: Izd. AN SSSR.
45. Derjaguin, B. V. 1935. Studies on the external friction and adhesion. V. Theory of adhesion. *Zh. Fizicheskoy Himii*. 6: 1306–1319.
46. Zelenev, A. S. and E. D. Shchukin, unpublished results.

3 Coagulation Structures
Rheological Properties of Disperse Systems

3.1 PRINCIPLES OF RHEOLOGY

Studies on the mechanical properties of solids and liquids indicate that there are many generalities governing the mechanical behavior of bodies of various natures. One can distinguish several basic types of mechanical behavior, a combination of which yields an approximate description of the more complex mechanical properties of real bodies. The science describing the laws that govern the mechanical behavior of solid- and liquid-like bodies is referred to as *rheology* (from the Greek words meaning "the study or theory of flow").

Modern rheology includes rheology itself as an independent fundamental discipline and *rheometry*, which addresses rheological measurements. The rheological behavior of various materials and systems is of importance in a great variety of industries, for example, food, oil and gas production, mining, mineral ore processing, and pharmaceutical. There are numerous publications, monographs, and books dedicated to rheological studies, such as the *Journal of Rheology, Applied Rheology,* and *the Rheological Bulletin* [1–11].

Rheology in general addresses the response of materials to stresses applied in various ways. The main principle of rheology is the description of the mechanical properties of systems using simple idealized models containing a relatively small number of parameters. The simplest approach is the so-called "quasi-steady-state regime," which involves a restriction on uniform shear and low deformation rates.

Inside a physical body, let us consider a cube with an edge of unit length. Let the tangential force, F, (designation used in this chapter) be applied to the opposite faces of the cube to cause a *shear stress*, τ, numerically equal to the applied force (Figure 3.1).

The applied shear stress causes the deformation, or *strain*, of the cube, that is, a shift of its upper face with respect to the lower face by an amount γ. This shift is numerically equal to the tangent of the tilt experienced by the side face, that is, to the *relative shear strain, γ.* When the strains are low, one can write that $\tan(\gamma) \approx \gamma$. The relationship between the stress τ, the strain γ, and their change as a function of time represents the *mechanical behavior*, which is the main subject of rheology. Let us start by reviewing three basic models of mechanical behavior: elastic, viscous, and plastic.

3.1.1 MAIN RHEOLOGICAL MODELS

3.1.1.1 Elastic Behavior

Fully reversible *elastic behavior* is characterized by a direct proportionality between stresses and deformations and is described by Hook's law:

$$\tau = G\gamma$$

where
the proportionality constant G is referred to as the elastic shear modulus [N/m^2]
τ is the shear stress [N/m^2]

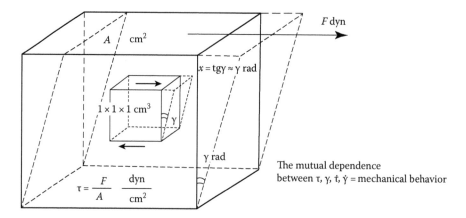

FIGURE 3.1 Deformation (strain), γ, resulting from the applied shear stress τ.

In the rheology of condensed phases, the elastic modulus G is often used as the sole characteristic of elastic behavior. In isotropic media mechanics, it has been established that for solid-like bodies, the modulus G ~ 2/5 of the Young's modulus [10].

Graphically, Hook's law corresponds to a straight line passing through the origin (Figure 3.2), with an inverse slope equal to G. As briefly stated earlier, the characteristic feature of elastic behavior is its complete mechanical and thermodynamic reversibility. This means that the original shape of the body is completely restored upon the removal of the stress, and there is no energy dissipation associated with the application and removal of the stress-causing load. The energy stored by a unit volume of an elastically deformed body is given by the expression

$$W_{el} = \int_0^\gamma \tau(\gamma)\,d\gamma = \frac{G\gamma^2}{2} = \frac{\tau^2}{2G} \tag{3.1}$$

A spring with a spring constant $k = G$ (spring constant is the ratio of the force, F, to the elongation, Δl, caused by that force) can be used as a model of elastic behavior (Figure 3.3).

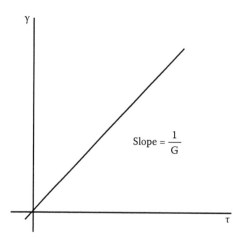

FIGURE 3.2 Elastic deformation.

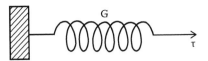

FIGURE 3.3 The model of elastic behavior.

Elastic behavior is typical of solid bodies. The nature of elasticity is related to the reversibility of small deformations of interatomic bonds. Within the limits of small deformations, the potential energy curve can be approximated with a quadratic parabola, which conforms to a linear relationship between γ and τ. The magnitudes of the elastic modulus G depend on the nature of the interactions in the solid body. For molecular crystals, G is $\sim 10^9$ N/m^2, while for metals and covalent crystals, the value of G is on the order of $\sim 10^{11}$ N/m^2. The elastic modulus G reveals either a weak dependence from temperature or no temperature dependence at all.

There are also cases in which the elasticity may have a completely different, entropic nature. This is the case in systems consisting of macromolecules or in clay suspensions. In such systems, the applied shear stress results in a change in the chaotic orientation of the segments of macromolecules or of the clay platelet-like particulates, which causes an ordering and hence a decrease in entropy. The return of the system to the original (disordered) state is associated with thermal motion. The modulus associated with entropic elasticity is small and exhibits strong temperature dependence [10].

As has already been stated, the elastic modulus has the units of pressure, Pa, or N/m^2, which is also equivalent to J/m. The latter means that, in agreement with Equation 3.1, it is possible to formally view the elasticity modulus as twice the elastic energy stored by the unit volume subjected to the unit strain. At a given shear stress, in agreement with Equation 3.1, the smaller the modulus G, the higher the elastic energy density stored by the body.

In reality, the elastic deformation of solid bodies is observed only up to a certain limiting value of shear stress τ, above which the body either undergoes destruction (for brittle objects) or shows residual deformation (i.e., reveals plasticity).

3.1.1.2 Viscous Flow

Viscous flow is characterized by proportionality between the stress and the rate of deformation, that is, between τ and $d\gamma/dt$, and is described by Newton's viscosity law:

$$\tau = \frac{\eta \, d\gamma}{dt}$$

where the coefficient η is dynamic viscosity with the units of Pa s or N s/m^2 (1 Pa s = 10 Poise = 1 cP). Graphically, Newton's viscosity law can be represented by a straight line passing through the origin in the $d\gamma/dt - \tau$ coordinates (Figure 3.4). The inverse slope of this line yields the viscosity η.

In contrast to elastic behavior, this idealized viscous behavior is completely irreversible, both mechanically and thermodynamically. Irreversibility implies that the initial shape of the body is not restored after the shear stress has been relieved. Viscous flow is accompanied by the dissipation of energy, that is, by the conversion of all work into heat. The rate of energy dissipation, that is, the power dissipated per unit volume, is given by

$$\frac{dW_d}{dt} = \frac{\tau \, d\gamma}{dt} = \eta \left(\frac{d\gamma}{dt} \right)^2$$

This quadratic dependence is characteristic of viscous friction. The viscous behavior of a body can be modeled by a dashpot (Figure 3.5). In this model, it is assumed that the ratio of the acting force F to the velocity of a piston, $F/(dl/dt)$, is equal to the viscosity of fluid.

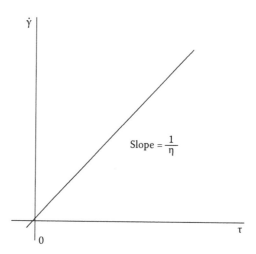

FIGURE 3.4   Viscous flow.

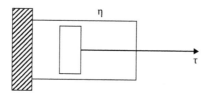

FIGURE 3.5   The model of viscous flow.

The nature of viscous flow is related to self-diffusion, that is, to the mass transfer due to atoms sequentially changing their positions in the course of thermal motion. The applied stress lowers the potential barrier to this mass transfer in one direction and increases mass transfer in the other direction, which results in the gradual development of a macroscopic deformation. Consequently, viscous flow is a thermally activated process, and the viscosity η shows a characteristic exponential dependence on temperature. For different materials, the values of viscosity can cover a very broad range. For low-viscous fluids, such as water or metal melts, $\eta \sim 10^{-3}$ Pa s, while for high-viscous Newtonian fluids, η can be by higher by three to six orders of magnitude and even higher for structured systems. The probability of thermally activated acts (i.e., of diffusion) increases with time, even in the case where the height of the potential energy barrier is significant. For this reason, even solid bodies may reveal liquid-like behavior with viscosities around 10^{15}–10^{20} Pa s or higher. Such flows take place in geological processes.

3.1.1.3 Plastic Flow (Plasticity)

In contrast to the two previous cases, *plastic flow* (*plasticity*) is characterized by the absence of proportionality between the stress and the strain, that is, plasticity represents a case of nonlinear behavior. For plastic bodies subjected to stresses below the critical value, $\tau < \tau^*$ (the so-called *shear yield point*), the rate of strain is zero ($d\gamma/dt = 0$). Plastic flow starts at the yield stress, $\tau = \tau^*$, and does not require further increase in stress (Figure 3.6). Similar to viscous flow, plastic flow is thermodynamically and mechanically irreversible. However, in contrast to the prior case, the rate of energy dissipation in plastic flow is proportional to the rate of strain:

$$\frac{dW_{\mathrm{d}}}{dt} = \tau^* \frac{d\gamma}{dt}$$

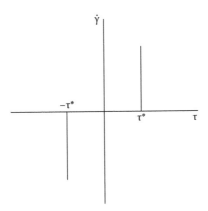

FIGURE 3.6 Plastic flow.

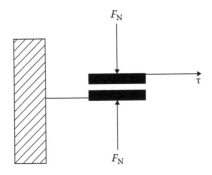

FIGURE 3.7 The model of plastic flow: the "friction element."

This type of dependence is a characteristic of *dry friction*, that is, corresponds to Coulomb's dry friction law, $F_{fr} = \mu F_N$. Plastic behavior can be modeled by a "friction element," that is, two flat plates with a friction coefficient, μ, with respect to each other and compressed with a normal force F_N in such a way that the applied tangential force F_{fr} would correspond to the yield stress of a given material (Figure 3.7).

The nature of plastic flow is a complex of processes involving the rupture and rearrangement of interatomic bonds, which in crystalline solids usually involves the participation of linear defects (dislocations). The temperature dependence of plastic flow may be considerably different from that of a Newtonian fluid. Under certain conditions, various molecular and ionic crystals (e.g., naphthalene, AgCl, and NaCl) reveal behavior that is close to plastic. The values of τ^* typically fall into the range of 10^5–10^9 N/m². Plastic flow is typical for a variety of disperse structures: powders (including snow and sand) and pastes. In the latter case, the origin of the plastic flow is a sequence of acts involving the rupture and reformation of contacts between the particles of disperse phase. In contrast to a fluid, a plastic body maintains its acquired shape after the stresses have been removed. It is noteworthy that the ancient craft of pottery had originated from the plasticity (Greek—πλαστωσ) of wet clay.

These are the three simplest cases of mechanical behavior and the corresponding rheological models. By combining them as elements, one can obtain more complex models describing the rheological properties of various systems. Every such combination is typically viewed within a framework of a specific deformation regime in which one seeks to reveal the qualitatively new properties of a given model as compared to the properties of its elements.

3.1.2 Combination of Rheological Models

a. A combination of elastic models and viscous flow models in series yields the *Maxwell model* (Figure 3.8). Newton's third law, applied to such a combination, dictates that equal forces (shear stresses τ) act on each of the constituent elements, while the deformations are added together

$$\gamma = \gamma_G + \gamma_\eta = \frac{\tau}{G} + \int_0^t \left(\frac{\tau}{\eta}\right) dt$$

where γ is the net strain. Consequently, the rates of strain are additive too:

$$\frac{d\gamma}{dt} = \frac{d\gamma_G}{dt} + \frac{d\gamma_\eta}{dt} \tag{3.2}$$

A characteristic regime that manifests the specifics of the mechanical behavior of this model is the regime in which the strain first instantaneously reaches the value of γ_0 and then stays at that level, that is, $\gamma = \gamma_0 = \text{const}$. At the initial moment, $t = 0$, the strain in the viscous element is zero, so the entire deformation (and the entire work) is concentrated within the elastic element. Consequently, the initial stress is $\tau_0 = G\gamma_0$. This stress further causes the deformation of the viscous element. Since the total deformation is constant, the deformation of the elastic element decreases, resulting in a decrease in the stress. Under conditions of constant net strain, $\gamma = \text{const}$, Equation 3.2 can be written as

$$\left(\frac{1}{G}\right)\frac{d\tau}{dt} + \frac{\tau}{\eta} = 0$$

Integrating this expression, using the initial condition $\tau(t = 0) = \tau_0 = G\gamma_0$, yields

$$\tau = \tau_0 \exp\left(\frac{-t}{t_r}\right)$$

The value $t_r = \eta/G$ has the units of time and is referred to as the *relaxation period*. This value graphically corresponds to the point at which the line tangent to the $\tau(t)$ curve at point $t = 0$ intersects with the x axis (Figure 3.9). This gradual decrease of the stress (*stress relaxation*) is typical in viscoelastic systems. The energy that was previously stored in the elastic element dissipates in the viscous element. As a result, the behavior of such a system is mechanically and thermodynamically irreversible.

The relaxation period defines the behavior of the system, in accordance with the Maxwell model with respect to the timescale of the applied stress. If the time t during which stress is applied is greater than the relaxation period, that is, $t > t_r$, the system has properties similar to those of a viscous liquid, while at $t \ll t_r$, the system behaves like an elastic solid. The flow of glaciers and other processes of strain development in mountains and cliffs are representative examples of such behavior. In rheology, the ratio of a material's characteristic relaxation time to the characteristic flow time is referred to as the *Deborah number*. This parameter plays an important role in describing the response of various materials to different stresses.

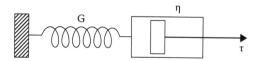

FIGURE 3.8 Maxwell model.

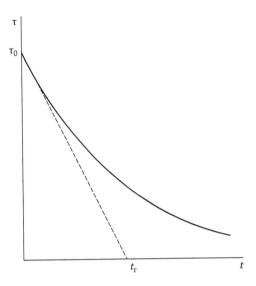

FIGURE 3.9   Stress relaxation.

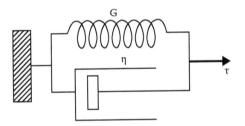

FIGURE 3.10   The Kelvin model.

b. Connecting the elastic and viscous elements to each other in parallel yields the *Kelvin model*, schematically shown in Figure 3.10. In this case, the strain in both elements is the same, while the net stress is the sum of the individual stresses, $\tau = \tau_G + \tau_\eta$. A strain regime corresponding to a constant shear stress, $\tau = \tau_0 = \text{const}$, is of the most interest here. In contrast to the Maxwell model, the presence of the viscous element in the Kelvin model does not allow for an immediate deformation of the elastic element. As a result, the deformation gradually develops over time at a rate given by

$$\frac{d\gamma}{dt} = \frac{\tau_\eta}{\eta} = \frac{(\tau_0 - \tau_G)}{\eta} = \frac{(\tau_0 - G\gamma)}{\eta}$$

The integration of the aforementioned expression yields deformation as a function of time, namely,

$$\gamma = \frac{(\tau_0/G)}{\left[1 - \exp(-t/t_r)\right]}$$

This type of dependence of strain on time is referred to as the *elastic aftereffect*. It is schematically shown in Figure 3.11. The deformation increases at a declining rate to the limit of $\gamma_{max} = \tau_0/G$, which is determined by the elasticity modulus of Hook's element.

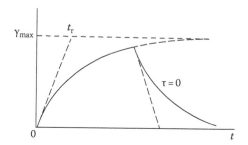

FIGURE 3.11 The elastic aftereffect.

The elastic aftereffect is encountered in solid-like systems with an elastic behavior. The elastic behavior is reversible: when the stress is removed, the strain drops gradually to zero, that is, the initial shape of the body is restored, using the energy stored by the elastic element. However, in contrast to true elastic behavior, the elastic aftereffect is thermodynamically irreversible: the dissipation of energy takes place in the viscous element. The damping of mechanical oscillations in rubber, caused by harmonic stresses, is the example of a process conforming to the Kelvin model.

c. A rheological model that describes the *internal stresses* can be obtained by connecting the elastic element and the nonlinear dry friction element in parallel, as shown in Figure 3.12.

Under conditions when the applied stress τ exceeds the yield stress ($\tau > \tau^*$), the strain $\gamma = (\tau - \tau^*)/G$ appears in the body. This deformation results in the accumulation of energy in the elastic element. If at the same time, $\tau < 2\tau^*$, then the "frozen" residual stresses remain in the body after the stress has been removed due to the action of the dry friction element. This "frozen" stress equals $\tau_0 - \tau^*$ and has the sign opposite to that of the initial stress and can't exceed τ^*.

d. The *Bingham model* represents the situation when the viscous Newtonian element and Coulomb's dry friction element are connected in parallel (Figure 3.13). This model is commonly used for the description of colloidal structures, such as aqueous suspensions of clay-like minerals. Since the elements in the Bingham model are connected in parallel, their deformations are the same, while the stresses are additive. The stress at the Coulomb element can't exceed the yield stress, τ^*. Consequently, the rate of deformation in the viscous element should be proportional to the difference between the acting stress and the yield stress, namely,

$$\frac{d\gamma}{dt} = \frac{(\tau - \tau^*)}{\eta_B}$$

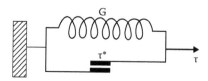

FIGURE 3.12 Model representing internal stresses.

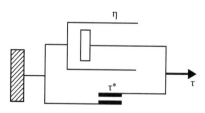

FIGURE 3.13 Bingham model.

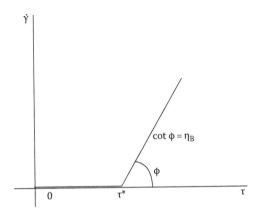

FIGURE 3.14 Viscoplastic behavior.

No flow takes place at $\tau < \tau^*$ (Figure 3.14). Since the η_B parameter defines the derivative, $\eta_B = d\tau/(d\gamma/dt)$, this constant parameter is referred to as the *differential viscosity*, which differs from a variable *effective viscosity*, $\tau/(d\gamma/dt) = \eta_{eff} (d\gamma/dt)$. In rheology literature, $d\gamma/dt$ is typically denoted as $\dot{\gamma}$.

e. *Complex combination models* are often used to describe the behavior of real systems, especially in cases with broadly varying conditions (time, stress, etc.) In these complex models, the described simpler models are present as individual elements. For example, the system may reveal more than one characteristic relaxation time (or a spectrum of relaxation times), as shown in Figure 3.15. As the resulting rheological models become more complex, the mathematical description also becomes more complex.

Employing the so-called *electromechanical analogies* can allow one to substantially simplify the use of rheological models. This method is based on modeling rheological properties using electric circuits. This is possible because the laws describing electric circuits are mathematically identical to the laws describing the deformation of solid and liquid bodies. To illustrate this, let us turn to Figure 3.16. The energy stored by the spring, $G\gamma^2/2$, has exactly the same mathematical form as the expression for the energy of a charged capacitor, $q^2/2C$, where q is the electric charge and C is the capacitance. The energy dissipation in the viscous element, $\eta(d\gamma/dt)^2$, is equivalent to the Joule's heat dissipated by the

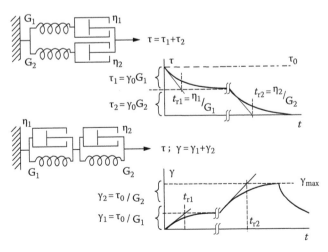

FIGURE 3.15 Spectrum of the stress relaxation and the elastic aftereffect with two components: fast and slow.

$$\eta\left(\frac{d\gamma_\eta}{dt}\right)^2 \propto RI_R^2$$

$$\frac{G}{2}\gamma_G^2 \propto \frac{Q_c^2}{2C} = \frac{1}{2C}\left(\int I_c\,dt\right)^2$$

$$\eta \propto R$$

$$\frac{1}{G} \propto C$$

$$t_r = \frac{\eta}{G} \propto RC = t_r$$

$$\tau_{mech} = \eta\dot\gamma_\eta = G\gamma_G \;\propto\; \frac{1}{C}\int I_c\,dt = RI_R = U_{electr.}$$

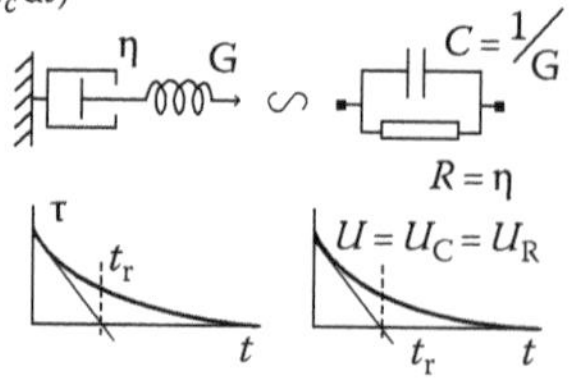

FIGURE 3.16 Schematic illustration of the electromechanical analogy of stress relaxation.

resistance R through which the current I flows, that is, to RI^2. For instance, using these formally identical relationships allows one to model the relaxation of the mechanical stresses in the Maxwell model with a voltage drop occurring during the discharge of a capacitor connected to a resistance in the circuit with the time constant $t_r = RC = \eta/G$. However, at the same time, it is not always possible to describe the behavior of real systems, even with complex models containing only the parameters that do not change in the process of strain: G, η, τ^*. In such cases, one must switch to models that contain variable parameters, for example, the elements of nonlinear elasticity, $G = G(\gamma)$, nonlinear viscosity $\eta = \eta(d\gamma/dt)$, and variable yield stress (hardening) $\tau^* = \tau^*(\gamma)$.

3.2 RHEOLOGICAL PROPERTIES OF COAGULATION STRUCTURES

For a large number of disperse systems with a globular structure mechanical characteristics, such as strength, P_c (N/m^2), and the ability to resist the action of external stresses in general are manifested by the cohesive forces between the particles at points of their contact. Experiments have confirmed [11,12] that the strength P_c can be estimated using the additivity approximation, $P_c = \chi p_1$, where p is the strength (the force measured in N) of individual contacts between particles, and χ (m^{-2}) is the number of such contacts per unit area of the "fracture surface." Estimates for both p and χ can be obtained on the basis of both theory and experimental data.

The value of χ is determined by the geometry of the system, primarily by the particle size (radius, r, for spherical particles) and by the packing density of particles described by porosity, Π. The porosity is a dimensionless characteristic defined as the ratio of the volume of pores, V_p, to the total volume of the porous structure, V, that is, $\Pi = V_p/V$. The $\chi = \chi(r, \Pi)$ dependence can be estimated from data on the degree of dispersion of the particles and the porosity of samples by employing the specific models for disperse structures. For example, in the case of loose monodisperse structures with spherical particles connected into crossing chains with n particles per chain between the nodes (Figure 3.17), the χ function for the case when the porosity Π does not exceed 48% can be described as

$$\chi = \frac{1}{[(2r^2)/n_a^2]}$$

$$\Pi = 1/[\pi/(6n_a^3)]/(3n_a - 2)$$

where n_a is the average number of particles between the nodes in the chains (see Section 7.3 for more details).

For moderately dense structures (such as those described by a primitive cubic lattice with a coordination number of 6, employing a first and most crude approximation yields an estimate of $\chi \approx 1/(2r)^2$. This provides one with the means to estimate the order of magnitude of χ in real systems, namely, for particles having a diameter $2r \sim 100$ µm, we get a value of $\chi \sim 10^3$–10^4 contacts per cm^2;

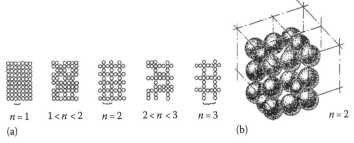

$n = 1$ $1 < n < 2$ $n = 2$ $2 < n < 3$ $n = 3$

(a)

(b)

$n = 2$

FIGURE 3.17 Models describing 2D (a) and 3D (b) globular disperse structures. (From Shchukin, E.D., Physical-chemical theory of the strength of disperse structures and materials, in: *Physical-Chemical Mechanics of Natural Disperse Systems*, E.D. Shchukin, N.V. Pertsov, V.I. Osipov, and R.I. Zlochevskaya (eds.), Izd. MGU, Moscow, Russia, 1985, pp. 72–90.)

for particles with $2r \sim 1$ µm, $\chi \sim 10^7–10^8$ contacts per cm²; while for particles with $2r \sim 10$ nm, $\chi \sim 10^{11}–10^{12}$ contacts per cm². These ranges of values will be different for systems with polydisperse or anisometric particles.

While in the previous chapters we have already addressed particle–particle interactions in some detail, it would be worthwhile to recall the principal concepts here. In *coagulation contacts*, the particle–particle interactions are restricted to contact between the particles via either the remaining equilibrium narrow gaps filled with dispersion medium or directly with each other. Contacts of this type form either in the absence of the Derjaguin–Landau–Verwey–Overbeek potential energy barrier or when the interaction energy is sufficient to overcome the potential energy barrier. Coagulation contacts correspond to particle interactions at the primary potential energy minimum. The *adhesive energy* between the particles in coagulation contact is determined by the depth of the potential energy minimum, namely,

$$u_c \approx \frac{A^* r}{12 h_0},$$

where
A^* is the complex Hamaker constant
h_0 is the equilibrium particle–particle distance
r is the curvature radius of particle surface at the point of contact

For typical lyophobic colloidal systems with a complex Hamaker constant $\sim 10^{-19}$ J and a particle–particle separation of 0.2–1 nm, the adhesive energy in the contact is significantly larger than kT, which indicates that thermodynamics favors the formation of coagulation contacts. The primary potential energy minimum is even deeper in systems composed of coarser particles. At the same time, for the case of low values of the complex Hamaker constant, $10^{-21}–10^{-22}$ J, the adhesion between the particles in systems that are not too coarse (particles with a diameter up to a micron) is overcome by the Brownian motion, and the formation of structures with coagulation contacts is impossible.

The strength of a coagulation contact, that is, the *adhesive force between particles*, is determined primarily by the molecular forces. For spherical particles, this force is given by

$$p_1 \approx \frac{u_c}{h_0} \approx \frac{A^* r}{12 h_0^2}$$

The corresponding estimates of adhesive force, p_1, with a Hamaker constant A^* value on the order of $\sim 10^{-19}$ J and $h_0 \sim 0.2–1$ nm (i.e., $A^*/12 h_0^2 \sim 10^{-1}–10^{-2}$ J/m²) for particles with radii of 10 nm, 1 µm, and 1 mm, are, respectively, $10^{-9}–10^{-10}$, $10^{-7}–10^{-8}$, and $10^{-4}–10^{-5}$ N. Forces that are on the order of 10^{-9} N or higher can be measured experimentally.

In cases when a complete displacement of the dispersion medium from the gap between particles takes place, that is, upon the rupture of the adsorption–solvation layer (or in a vacuum) a direct point-like contact between the particles may be established. Such contact may be formed by one or several atomic cells. In this case, along with the van der Waals forces, the short-range (valent) forces acting over the area of contact may also play some role in particle adhesion. The contribution of such forces to the strength of the contact can be estimated as $p_1 \approx N\,e^2/(b^2 4\pi\varepsilon_0)$, where N is the number of valent bonds, e is the elementary charge, ε_0 is the electric constant, and b is the typical characteristic interatomic distance. In this case where there is a minimum number of valent bonds, $N \sim 1$–10, one finds that $p_1 \sim 10^{-8}$ N. Under these conditions, one does not yet reach the conditions necessary for establishing the bridge-like *phase contacts*, which will be discussed in detail in Chapter 6. Consequently, in lyophobic systems for particles with $r \sim 1$ μm, the contribution of short-range forces to the strength of the contact may be of the same order (or less) as the contribution of the van der Waals forces.

Structures with coagulation contacts typically show low strength and mechanical reversibility, that is, these structures are capable of spontaneous restoration after mechanical destruction (i.e., *thixotropy*). As an example, let us estimate the macroscopic mechanical strength, P_c, of a coagulation structure formed by densely packed particles having radius $r \sim 1$ μm. For a lyophobic system with $A^* \sim 10^{-19}$ J, one gets $P_c \approx \chi p_1 \approx (1/2r)^2 A^*/12 h_0^2 \sim 10^4$ N/m². For a powder or suspension, this value of P_c should be interpreted as the yield stress, τ^*. In systems with coarse particles, for example, those having $r \sim 100$ μm, the estimate for the P_c value is only $\sim 10^2$ N/m². This value is characteristic of systems with high mobility, such as sand in an hourglass. In contrast to these cases, a system with a high degree of dispersion, for example, with particles having $r \sim 10$ nm, is characterized by a mechanical strength, P_c, on the order of $\sim 10^2$ N/m² or higher, which indicates that the system is capable of displaying a substantial resistance to deformation.

It is worth recalling here that a dispersion medium akin to the particles, as well as surfactant adsorption, can lower both the interfacial energy, σ, and the complex Hamaker constant, A^*, by two to three orders of magnitude. In such a lyophilized system, the adhesive energy and force are also lowered by several orders of magnitude. In a concentrated disperse system in which the dispersed particles are mechanically forced to come into contact with each other, the lyophilization manifests itself as a decrease in the resistance to strain τ^*. This means that in concentrated colloidal systems, plasticizing takes place, while in systems with a low concentration of dispersed particles, the lyophilization results in enhanced colloid stability of the free-disperse system (see Chapter 4).

In generally encountered disperse systems, the particle sizes and characteristics of the particle–particle interaction can cover a very broad spectrum of values, which explains the large variation in the rheological properties of different colloidal systems utilized in different areas of technology. At the same time, disperse systems are the main carriers of mechanical properties in both inanimate and live nature [10–17].

Large variety of rheological properties of disperse systems is manifested via a broad range of values of the elasticity modulus, G; Young's modulus, E; viscosity, η; and the yield stress τ^* (the yield point, the ultimate strength).

In continuous systems with solid phases (mineral rocks, construction materials), the parameter G is the elasticity modulus of a solid body, that is, it falls in the range between 10^9 and 10^{11} N/m². It is noteworthy that the elasticity modulus of common liquids under uniform compression conditions is of the same order of magnitude. However, due to low viscosity, the elasticity in liquids can be experimentally determined only by very fast measurements in which the impact time is very close to the relaxation time. For this reason, under common conditions, liquids with low values of η behave like viscous media.

In systems with solid and liquid phases, the elasticity modulus is determined by the interactions between the particles of the dispersed phase. For porous disperse structures of globular type with phase contacts between the particles, the elasticity modulus of the system is determined by the elasticity modulus of the substance making up the solid phase and by the number and area of the contacts between the particles. This is typically so, regardless of whether the other phase is liquid or gaseous. For example, the elasticity modulus of porous crystallization structures may fall in the range between 10^8 and 10^{10} N/m^2. These types of structures are brittle and reveal a tendency toward irreversible erosion without any noticeable preceding residual deformation. The disintegration occurs at a yield stress value that is below the stress level at which plastic flow can occur.

The elasticity modulus of coagulation colloidal structures with solid and liquid phases can have a different nature if the formation of such structures takes place under conditions of a relatively low volume fraction of the dispersed phase, when the degree of dispersion is high or in the case when the particles are anisometric. Examples of such systems include hydrogels of vanadium pentoxide and structured colloidal suspensions of bentonite clay in water. As will be discussed in detail in the following text, the shear elasticity in such systems may be stipulated by a higher or lower degree of coorientation between the particles in the course of the strain. Such coorientation increases the ordering in the system and hence decreases entropy. When the load is removed, the Brownian (rotational) motion of the particles restores their chaotic orientation and thus restores the original shape of the body due to the change in the configuration entropy [18]. Shear elasticity has, therefore, an entropic origin, similar to the entropic origin of the elasticity of gas pressure or osmotic pressure. The elasticity modulus, G_{el}, is on the order of nkT, where n is the number of particles per unit volume that participate in the Brownian motion, that is, the number of kinetically independent units. For instance, in a dilute suspension of finely disperse clay particles having $n \sim (3–5) \times 10^{23}$ particles per m^3, the value $G_{el} \sim 10^3$ N/m^2.

Elastic deformation can also be observed in foams and concentrated emulsions. In such cases, the yield stress is determined by the increase in the interfacial area upon the deformation of the particles. The mechanical properties of solidified foams and other solid-like materials with a cellular structure are defined by the degree of their dispersion, their backbone structure, and the combination of the mechanical properties of dispersion medium and dispersed phase.

The viscosity of dilute *free-disperse systems* is mainly determined by the viscosity of the dispersion medium, which, in principle, may vary over many orders of magnitude. Gases, for instance, have a viscosity on the order of 10^{-5} Pa s, while liquid-like materials have viscosities ranging from 10^{-2} to 10^{10} Pa s and for glasses and solids viscosity values fall in the range between 10^{15} and 10^{20} Pa s or higher.

Einstein showed that the viscosity of dilute suspensions in the absence of interactions between the particles is proportional to the volume fraction of the dispersed phase, ϕ, that is, the addition of particles to the dispersion medium results in energy dissipation due to the rotation of the particles in the shear force field:

$$\frac{(\eta - \eta_0)}{\eta_0} = k\phi$$

where η_0 is the viscosity of dispersion medium. For spherical particles, $k = 2.5$. Thus, in the absence of interactions between the particles, the system behaves as a Newtonian fluid but with a viscosity slightly higher than that of a dispersion medium.

In a suspension containing anisometric particles (ellipsoids, platelets, rods) or "soft" deformable particles (droplets or macromolecules), various tendencies may be revealed. Shear stresses accompanied by particle rotation tend to deform the particles and orient them in the flow in a particular way (Figure 3.18). The rotational diffusion of the particles in this case tends to oppose the orienting action. As a result, the extent of the particle orientation strongly depends on the rate of the strain. That is, at a low flow rate, the particles may be completely disoriented, while at a high flow rate, they may exhibit a high degree of orientation, which can be registered by optical methods. This in turn leads to the dependence of the viscosity on the flow rate (or the shear stress). In this case, a single variable of the Newtonian viscosity, $\eta = d\tau/d(d\gamma/dt)$, is no longer sufficient for describing the system and the shear-rate-dependent *effective viscosity*, $\eta_{eff} = \tau/(d\gamma/dt)$. The effective viscosity is maximal at a low strain rate, and then it gradually increases with the increase in the shear rate to

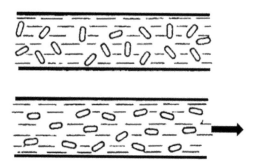

FIGURE 3.18   The orientation of anisometric particles in a flowing suspension.

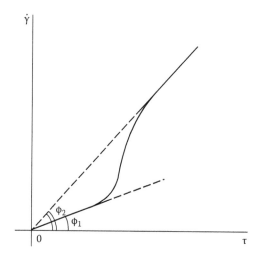

FIGURE 3.19   Rheological curve of a free-disperse system with anisometric particles: $\tan^{-1} \phi_1 = \eta_{max}$, $\tan^{-1} \phi_2 = \eta_{min}$.

some minimum value corresponding to the state in which the particles are completely oriented in the flow (Figure 3.19). Once that minimum value has been reached, the effective viscosity does not undergo any further change.

When certain hydrophilic polymers are dissolved in water (e.g., polyacrylamide), the flow rate may increase significantly due to the suppression of the turbulence (the Toms effect of *friction reduction*) resulting in an increase in the efficiency of pumping. This phenomenon is commonly used in the petroleum industry to fracture subterranean formations and in firefighting to increase the power of pressure pumping. Oppositely, under certain conditions, some systems may undergo an increase in the effective viscosity as the flow rate increases. Such systems are referred to as *dilatant* or shear thickening. This may be caused, for instance, by a significant deformation of the macromolecules in the flow due to conformation changes. In a concentrated suspension, a significant increase in viscosity with increases in flow rate is referred to as *rheopecty (or rheopexy)*. This phenomenon can be observed during the initial stages of the starch dissolution process when solid starch is suspended in water.

These types of phenomena can't be described in terms of simple rheological models with constant parameters. Systems that reveal the dependence of the viscosity on the flow rate are referred to as *anomalous* or *non-Newtonian*. In dilute suspensions, changes in the viscosity associated with the orientation and deformation of the particles in the absence of particle–particle interactions are typically not too large.

The viscosity in *connected-disperse systems* with coagulation structures changes more abruptly than the viscosity in free-disperse systems. In this case, one can encounter an entire spectrum of states between two limiting cases: that of a completely intact structure and one corresponding to the

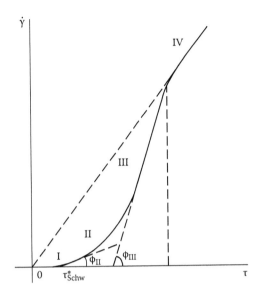

FIGURE 3.20 A complete rheological curve of structured disperse system: $\tan^{-1} \phi_{II} = \eta_{Schw}$, $\tan^{1} \phi_{III} = \eta_{B}$.

fully destroyed state. Depending on the magnitude of the applied stress (flow rate), the rheological properties of structured systems can vary over a broad range: from those typical of solid-like materials to those characteristic of Newtonian fluids. This rheological behavior diversity of a real disperse system with coagulation structure is described by a *complete rheological curve*. An example of such a curve in the form of $d\gamma/dt = f(\tau)$ for aqueous suspension of finely dispersed bentonite is shown in Figure 3.20. A complete rheological curve may also be presented as a dependence of the effective viscosity, $\eta_{eff} = \tau/(d\gamma/dt)$, on the shear stress, τ (Figure 3.21).

As seen in Figure 3.21, a complete rheological curve contains four characteristic regions. Region I corresponds to low stresses under which the system may demonstrate a solid-like behavior with high viscosity (Kelvin model). This case is characteristic of the already mentioned bentonite clays. The studies of relaxation structures in moderately concentrated suspensions of bentonite clays indicated the appearance of elastic aftereffect at low shear stresses. This effect has an entropic nature, as it is associated with the

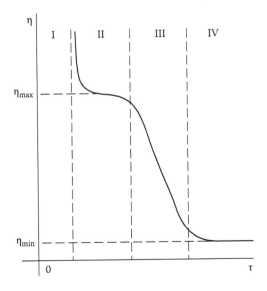

FIGURE 3.21 A complete rheological curve in η–τ coordinates.

mutual orientation of the anisometric particles that are capable of participating in the thermal motion. High values of viscosity are associated with the flow of the dispersion medium from the cells that are shrinking in size to the neighboring ones and by the occurring sliding of particles against each other.

When the shear stress reaches a particular value, τ_{Schw}^*, the system reveals a viscoplastic flow with an essentially preserved structure and enters the so-called Schwedow creep region (region II in Figures 3.20 and 3.21). In this region, shear is caused by the fluctuation process of the destruction and subsequent restoration of the coagulation contacts. Due to external stresses, this process becomes directional. This mechanism of creep may be considered by the analogy with the mechanism of flow of liquids developed by Frenkel [19].

As a result of Brownian motion, the particles agglomerated into a single coagulation structure undergo oscillation around their position in the contacts. Because of thermal fluctuations, some contacts are ruptured, but new contacts between particles in other locations are formed. On average, the number of contacts in the formed structure does not change over time and remains close to the maximum possible number of contacts. In the absence of shear stresses, the process of the rupture and reformation of the contacts in any section occurs with equal probability in all directions. Upon the application of an external force field, these processes become directional, and slow macroscopic shear motion (creep) is observed. Creep is possible over some particular range of τ values, at which a balance between the small number of rupturing and reforming contacts is reached. Creep region II, as well as the subsequent region III (Figure 3.21), can be described in terms of a viscoplastic flow model with a low yield stress, τ_{Schw}^*, and a very high differential viscosity, η_{Schw}:

$$\tau - \tau_{Schw}^* = \eta_{Schw} \frac{d\gamma}{dt}$$

where η_{Schw} is given by the inverse slope ϕ_{II} of the curve in region II, that is, $\tan^{-1} \phi_{II} = \eta_{Schw}$. Consequently, the effective (variable) viscosity in this region is also high:

$$\eta_{eff} = \frac{\tau}{(d\gamma/dt)} = \frac{\eta_{Schw}}{[1 - (\tau_{Schw}^*/\tau)]}$$

Overall, strains γ on the order of a few percent are typical for low shear stress regions I and II in Figure 3.21. Upon the action of long-lasting shear stresses, large deformation can develop. This is the case in geological processes, such as during the movement of glaciers.

Once a certain shear stress, τ_B^*, is reached, the equilibrium between the formation and rupture of the contacts is shifted toward rupturing. The higher the shear stress τ, the larger the shift. This flow regime, in which the structure undergoes intensive destruction, corresponds to the viscoplastic flow region III in Figure 3.20. This region can be described in terms of Bingham's model with a relatively large yield stress τ_B^* and a low differential Bingham viscosity, η_B:

$$\tau - \tau_B^* = \eta_B \frac{d\gamma}{dt}$$

Bingham's yield stress, τ_B^*, corresponding to the onset of an intensive structure degradation, can be viewed as a system's strength characteristic.

The shift in the equilibrium toward the rupture of the contacts results in a drop of the effective viscosity, sometimes by several orders of magnitude:

$$\eta_{eff} = \frac{\tau}{(d\gamma/dt)} = \frac{\eta_B}{\left[1 - (\tau_B^*/\tau)\right]}$$

For suspensions of bentonite clays, the values of the effective viscosity may vary over several orders of magnitude, that is, from 10^6 to $\sim 10^{-2}$ Pa s.

After the structure has been completely destroyed, under the conditions of a laminar flow, the disperse system behaves as a Newtonian fluid (region IV in Figure 3.20) with a constant and lowest viscosity, η_{min} (Figure 3.21). The viscosity η_{min} of such a system is higher than that of the dispersion medium to a degree that is greater than predicted by Einstein's viscosity law. The reason for such a larger increase in viscosity is the interactions between particles at the sufficiently high concentration of a suspension. Further increases in shear stress result in a deviation from the Newtonian behavior due to the transition into a turbulent flow regime. Region IV (Figure 3.21) may sometimes not be observed due to the early appearance of turbulent flow.

The rheological properties of structured disperse systems may undergo a substantial variation under the action of vibration. Vibration favors the rupture of the contacts between particles and results in a liquefying of the system at lower shear stresses. As a result, the $d\gamma/d\tau - \tau$ curve can shift to the left, as illustrated in Figure 3.22. Vibration is commonly used in various applications for controlling the rheological properties of various disperse systems, such as concentrated suspensions, powders, and pastes.

When the contacts between the particles and the entire structure of the solid phase dispersion are completely destroyed, the rheological behavior of the system is similar to that of a liquid medium. In the opposite case, there is a critical shear stress that is the source of the internal friction, preventing compact packing. Liquids, including those with high viscosity, form a horizontal surface upon spilling. At the same time, dry powders, when poured, form a cone-shaped "mountain" with an angle ϕ (Figure 3.23).

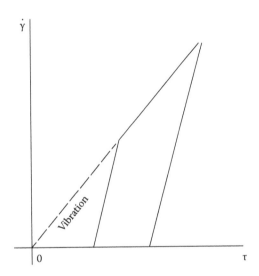

FIGURE 3.22   The impact of vibration on the rheological properties of a structured disperse system.

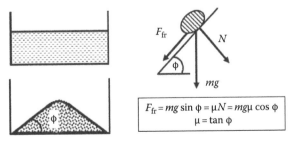

FIGURE 3.23   A scheme illustrating the difference between the rheology of powders and the rheology of fluid-like media: the resistance to shear, F_{fr}, increases with an increase in the normal load P_0.

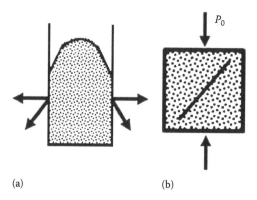

(a)   (b)

FIGURE 3.24 The tangential pressure acting on vessel walls (a), the resistance to shear increases with increases in the normal load, P_0 (b).

A single particle stays on a ramp if the component of its weight, $mg \sin \phi$, is balanced by the friction force $mg\mu \cos \phi$, where μ is the friction coefficient. This establishes a critical tilt angle of the ramp, that is, $mg \sin \phi = mg\mu \cos \phi$ or $\mu = \tan \phi$. For many systems, this tilt angle is around 30°. The surface of a powder that has been loosely poured into a vessel may also form a conical tilt. In contrast to a liquid medium, the pressure on the vessel walls may also have a tangential (vertical) component in addition to the normal component. This tangential component corresponds to the Coulomb friction acting along the surface of the vessel (Figure 3.24). This friction prevents a complete filling of the vessel with a powder.

An important subject that one needs to address here is the role of moisture. Typically, in the case of a complete flooding of powder with a wetting liquid, the internal friction and critical tilt angle are significantly lower than in a dry system. However, at a certain "average" content of the liquid phase, the cohesion between particles and the resistance to shear significantly increase due to the appearance of capillary attractive force (see Chapter 1).

The specifics of the rheology of powders are completely revealed in the process of their compaction, that is, pressing in a mold in a cylindrical matrix. The axial pressure, P_0, acting in a given compact structure results in the appearance of the tangential components of a stressed state, which manifest shear deformation and compaction. At the same time, the compression stresses are acting in directions normal to the possible directions of sliding. According to Coulomb's law, these stresses oppose shear deformations and compaction (Figure 3.24). When there are no factors limiting the expansion of a body in the direction perpendicular to the applied uniaxial compression, the tangential stresses are maximal in planes oriented at a 45° angle to the axis. The values of these stresses reach $\frac{1}{2}P_0$. Similar normal compressive stresses also act along these planes. These are the conditions that one encounters in the course of pressing a disc or a coin. If there is a resistance from the wall of cylindrical matrix, the stress state also includes uniform compression, and the shear deformations become "jammed."

In the case of plastic particles, the residual deformations in the structure may occur due to the deformation of the particles themselves and due to the growth in the contact area between the particles. For solid particles, the compaction of powders becomes possible only by the comminution of particles and requires very high stresses. The latter constitutes one of the major problems of powder technology, in particular, that of carbide materials [20]. The problem can be solved by applying vibration.

Let us now provide a quantitative description of the compaction of a powdered material on the basis of the work by Spasskiy et al. [21–25]. These studies deal with tungsten powders compacted in a cylindrical "floating" matrix with two mobile punches with diameters of 2 cm and a cross-sectional area, $S = 3.14$ cm². Punch (*1*) with a mass $m_1 \sim 400$ g is compressed against a powder column with force, F, from 3 to 30 kg by means of a spring (Figure 3.25). The compaction is determined by the displacement of the punch from the value h_0 based on the initial height of the powder column to the current one, $h(t)$,

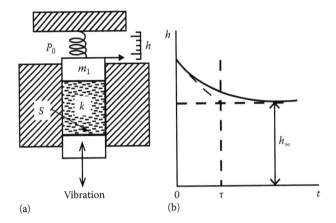

(a)   (b)

FIGURE 3.25 Schematic illustration of the vibration compacting of the powder (a); the kinetics of the compaction process (b).

and further to the final value of h_4. The compaction is then calculated as $\gamma = (m/S\,h_\infty \rho)$, where m is the powder mass and ρ is the powder density (19.3 g/cm^3 for tungsten). The second punch (2) is connected to a vibrator. The experiments were conducted with vibration frequencies, f, in the range between 100 and 1000 Hz. The acceleration, a, of punch (2) was measured with a special sensor. The amplitude of the acceleration was chosen between 10 and 100 m/s^2. The tungsten powders contained particles between 3 and 10 μm. The mass of the specimen was only 10 g, that is, the tungsten columns were essentially 0.5 cm tall tablets. This allowed one to neglect matrix wall effects.

The parameters of interest were the static compression force F, applied by a soft spring to punch (1), vibration frequency f, the acceleration a, the power of vibrations W, and the compaction achieved γ, as a function of these parameters. The combined data are shown in Figure 3.26.

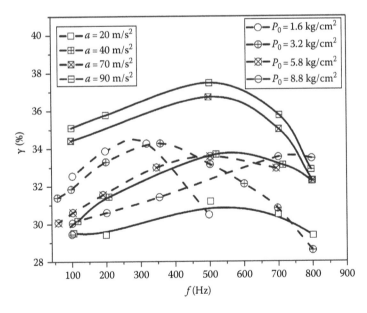

FIGURE 3.26 The vibration compaction, γ, (%), as a function of the vibration frequency, f, (Hz), at constant stress, $P_0 = 5.8$ kg/cm^2, for different values of acceleration amplitude, $a = 20, 40, 70,$ and 90 m/s^2 (solid lines); the compaction, γ, (%), as a function of the vibration frequency, f, (Hz), at constant acceleration, $a = 40$ m/s^2, for different values of $P_0 = 1.6, 3.2, 5.8,$ and 8.8 kg/cm^2 (dashed lines). (Data from Dobrushina, I.A. et al., *Doklady AN SSSR*, 189, 525, 1969; Spasskiy, M.R. et al., *Doklady AN SSSR*, 189, 767, 1969.)

The experiments conducted allowed one to draw the following conclusions, which define one's possibility to approach optimal compaction conditions.

First, there is a clear and reproducible *nonmonotonous* dependence between the compaction achieved, γ, and the vibration frequency, f. It is *a priori* clear that the optimum vibration frequency, the frequency at which maximum energy absorption takes place, corresponds to a certain resonance frequency in the system, f_{res}. Intuitively, one could assume that this is related to the mass of the punch (*1*) (400 g) and the stiffness of the compressing spring, k_s (reasonably low, around a few kg/mm). However, direct observations of the oscillations of a weight of a given mass on this spring point to a frequency $f \sim 1$ Hz, which is far from the realistically optimal frequencies of 100–1000 Hz. Our experiments have shown that, with a high probability, the resonance frequency is determined by the mass m_1 and the effective elasticity of the powder structure, $k_p : f_{res} = (1/2\pi)(k_p/m_1)^{1/2}$. Using the solution of the Hertz problem and employing dimensional analysis, one finds that the effective elasticity modulus of the pack of spheres with Young's modulus E under these special conditions (i.e., when the area of the contacts is a nonlinear function of the load) is $E_{eff} \propto (E^2 P_0)^{1/3}$. For tungsten, using $E = 4 \times 10^{12}$ dyn/cm^2 and for $P_0 = 5.8 \times 10^6$ dyn/cm^2, one finds that $E_{eff} = 4.5 \times 10^{10}$ dyn/cm^2, which is much smaller than E. For a given volume of powder in a matrix with the area $S = 3.14$ cm^2 and a column height, $h = 0.5$ cm, one gets $k_p = bE_{eff}S/h \approx b \cdot 30 \times 10^{10}$ dyn/cm. Depending on the *structure model* utilized, the dimensionless coefficient b, describing the degree of structure damageability, may be much smaller than 1. Assuming that b is on the order of 1%, one can estimate the elasticity of the powder structure, $k_p \sim 3 \times 10^9$ dyn/cm. In this case, for $P_0 = 5.8 \times 10^6$ dyn/cm^2, one finds $f_{res} \sim 430$ Hz, which is close to the experimental observations. One can thereby see significant differences in the application of the elasticity of a continuous medium to a partially damageable structure. This difference manifests itself in the parameter b.

Second, the main stage of the compaction in the $h(t)$ curve can be approximated using an exponent (Figure 3.25). On the one hand, at the extreme compression of the punch (*1*), the powder is jammed and does not undergo compaction. On the other hand, when F is too small, the powder "swells."

Comparing the optimal values of the axial compression stresses, P_0, with the corresponding resonance frequencies, f_{res}, at a punch acceleration, $a = 40$ m/s^2, one gets a peculiar linear dependence between P_0 and f_{res} with a slope of 1.6×10^4 dyn/cm s (Figure 3.27). The values of P_0 in the vibration compaction regime are on the order of a few kg/cm^2 instead of hundreds of kilograms or tons per cm^2 for static powder pressing.

Third, let γ_{max} be the compaction corresponding to the resonance frequency, f_{res}. With respect to achieving maximum compaction, the dependence of γ_{max} on the amplitude of the acceleration, a, is important. The amplitude of acceleration is indicative of the intensity of the vibration action at a given P_0 and corresponding f_{res}. The variation of the punch acceleration at optimal values, $P_0 = F/S$, corresponds to the monotonous increase of γ_{max} (Figure 3.28). The dynamic component of the pressure can be estimated as $P_d \sim am_1/S \sim (10–100) \times 10^2$ cm/s$^2 \sim (0.1–1) \times 10^6$ g/cm s$^2 \sim 0.1–1$ kg/cm^2. This value is by an order of magnitude lower than P_0, which also reflects the specifics of the rheology of the partially damaged structure.

Fourth, it is also of interest to obtain estimates of the power used, W. To obtain this estimate, one can turn to the role of the contact interactions and specifically to the dependence of the frequency of the contact rupture on the vibration frequency. For particles with a diameter of 10 µm, one gets the number concentration of about 10^9 particles per cm^2. The number of contacts ruptured per cycle is of the same order of magnitude. Since the work of rupture of an individual contact in a given system is on the average estimated by u_1, that is, it is approximately 10^{-8} erg (see Sections 1.3 and 4.4), for 10^3 cycles the power is no more than $W \sim 10^3$ s$^{-1} \times 10^9$ cm$^{-3} \times 10^{-8}$ erg $\times 4 \times 3$ cm$^3 \sim 10^5$ erg/s $\sim 10^{-2}$ W. In reality, the power consumed under the optimal conditions is around 0.1 W.

A discrepancy of one or two orders of magnitude would be indicative of the ineffectiveness of the method, that is, of the dissipation of 99% of the power into heat inside the machine and especially into the friction of the particles, that is, dissipation due to the entire spectrum of contact interactions

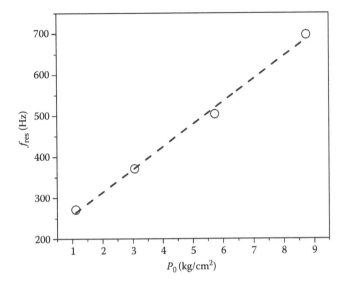

FIGURE 3.27 The resonance frequency, f_{res}, as a function of the stress, P_0, for an acceleration, $a = 40$ m/s^2. (Data from Dobrushina, I.A. et al., *Doklady AN SSSR*, 189, 525, 1969; Spasskiy, M.R. et al., *Doklady AN SSSR*, 189, 767, 1969.)

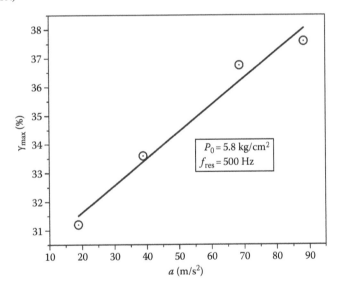

FIGURE 3.28 The compaction γ_{max} as a function of the acceleration, a, at $P_0 = 5.8$ kg/cm^2 and $f_{res} = 500$ Hz. (Data from Dobrushina, I.A. et al., *Doklady AN SSSR*, 189, 525, 1969; Spasskiy, M.R. et al., *Doklady AN SSSR*, 189, 767, 1969.)

in the structure. However, in actual applications of vibration technology (e.g., in filling hoppers, transport in pipes, liquifaction of cement slurries or pulp), even these excessive power amount constitute a rather small contribution to the total balance of power consumption; hence, vibration technology can be regarded as a rather economical one [26,27].

The crude estimate provided here indicates that the problem of reaching optimal parameters in vibration technology is many sided and needs to be taken into account in solving various engineering problems.

The rheological behavior of a structured thixotropic disperse system depends to a great extent on to which side the equilibrium between the formation and the rupture of contacts is shifted. Since the

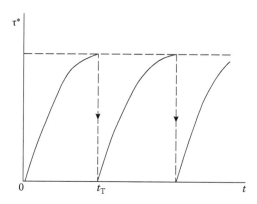

FIGURE 3.29 The illustration of a schematic yield stress, τ^*, as a function of the time, t, necessary to restore the thixotropic structure.

rate of contact restoration, which is associated with the Brownian motion of the particles, is finite, a certain time is required for the equilibrium to establish. For this reason, a spontaneous thixotropic restoration of the structure takes place in real time. Due to a complete destruction of the structure, which takes place in region IV, the strength of the system, that is, the yield stress τ^*, sharply drops (all the way to zero in the limiting case), and the system exhibits a liquid-like behavior.

When at rest, the system gradually restores its strength, that is, again acquires solid-like behavior. The strength (i.e., the yield stress, τ^*) of a fully restored structure does not depend on the number of structure destruction cycles (Figure 3.29). The time needed for a complete thixotropic restoration of the structure is referred to as the *period of thixotropy*, t_T. If the rheological measurements are conducted before the dynamic equilibrium between the rupture and the formation of the contacts has been reached, the measured mechanical properties of the system will strongly depend on the structure of the system at the time of measurement, that is, on the extent of the structure destruction at the time of measurement.

In practice, when working with thixotropic systems, one often uses some apparent values of the rheological characteristics, rather than the equilibrium ("true") ones. The apparent characteristics can be evaluated at a particular time, for instance, after a complete degradation of the thixotropic-reversible structure. In some instances, a prolonged constant-rate deformation of the system is required for the equilibrium between the contact rupture and restoration to be established. Such conditions may not be always achievable in the laboratory.

Let us list some examples illustrating the role that the thixotropy of disperse systems plays in nature and technology. The thixotropic properties of bentonite clays are the main reason for the use of bentonite clay suspensions in drilling muds in oil industry. These suspensions behave as ordinary fluids under the normal operation of a bore. One of the functions of the drilling mud is to carry the drilled-out rock to the surface. When there is a need to stop the bore, as in the case when the casing needs to be extended, there is a danger that coarse rock particles will sediment and cause jamming of the bore. The thixotropic properties of the fine disperse clay suspension will result in the formation of a coagulation structure network that will trap the suspending particles and prevent sedimentation. Once the operation of the bore is restored, the coagulation structure formed is rapidly degraded, and the system again acquires liquid-like properties.

The thixotropic properties of flooded clay-based grounds also need to be taken into account in civil construction projects, such as in planning the construction of buildings, bridges, and roads.

The thixotropic properties of pigment structures in oil-based paints provide the paint with the necessary rheological properties. Mixing results in the destruction of the coagulation structure and allows one to apply the paint as a thin layer on the surface. Quick restoration of the coagulation structure prevents the paint from gravity-caused draining downward.

Another area in which the structuring of colloidal systems needs to be taken into account is in the introduction of fillers into various types of polymeric materials. If the task is to obtain a material with high strength and hardness (at the expense of elasticity), it is beneficial to reach a more complete particle packing and use as much filler as possible. To achieve this, one has to prevent the formation of a loose spatial network of particles, that is, it is necessary to weaken interparticle cohesion while maintaining a good cohesion between the particles and the matrix. Since fillers typically consist of polar particles, while the matrix is typically nonpolar, good cohesion is achieved by using chemisorbing surfactants that make the surface of the filler particles hydrophobic. In the case of aluminosilicates, this can be achieved by using sufficient amounts of cationic surfactants.

If the task is to preserve a high degree of elasticity, then one must cut back on the use of filler, so that a loose network of filler particles can be created. This task also requires fine-tuning of the cohesive forces between the filler particles. Indeed, on the one hand, if the degree of lyophilization of the system is too high (weak cohesion between the particles), the filler particles will sediment and a nonuniform material will be obtained as a result. On the other hand, excessive lyophobization (too strong cohesion) will also result in a nonuniform material because of filler caking due to aggregation. It has been demonstrated by Taubman and Nikitina [28] that the ideal structuring of the system takes place under the conditions when the adsorption layer is approximately half of the monolayer. This illustrates the universal role of surfactants in fine-tuning the cohesion between the particles of the disperse phase, as a result of the structural and rheological (mechanical) properties of disperse systems and materials.

The analysis of a full rheological curve indicates how the complex mechanical behavior of the system can be decomposed into a combination of a number of regions, where each region can be described in terms of a simple model utilizing one or two constant parameters, as shown in Figure 3.30. This allows one to use the same model with very different parameters to describe such complex and molecularly different phenomena as Schwedow's creep and Bingham's viscoplastic flow. The decomposition of a complex behavior into a limited series of regions described by simpler models is the main technique used in *macrorheology* for solving complex engineering problems.

At the same time, the understanding of the mechanism of each of these elementary behaviors requires the application of the framework of molecular-kinetic theory and can be characterized as the *microrheological approach*. To further illustrate the microrheological approach, let us address in detail two particular studies that reveal the role of contact interactions in physical-chemical mechanics.

One such study is the analysis of the nature of elastic deformation, that is, the mechanically reversible behavior of dilute clay suspensions, which find broad use in drilling muds. We will address this subject by presenting a theoretical description of the microrheology of the behavior of a coagulation structure with anisometric particles resulting from the action of small stresses. Under these conditions, the system reveals elastic aftereffect without contact rupture [18]. This phenomenon underlies the ability of clay suspensions to prevent the sedimentation of mineral particles in drilling operation when the bore is stopped.

The second study focuses on the summation of the contact interactions using the estimation of the rheological parameters of a coagulation structure formed with substantially anisometric particles (fibers) [29].

3.2.1 Mechanism of Elastic Aftereffect in Structured Dilute Suspensions of Bentonite

The rheology of bentonite suspensions has been studied by numerous authors. A peculiar high elasticity (the so-called elastic aftereffect) in the coagulation thixotropic structures that develop in aqueous bentonite suspensions was demonstrated in early studies by Serb-Serbina et al. [30–33]. Under constant shear stress, the strain develops in bentonite suspensions. The strain gradually increases with time and gradually decays, once the shear stress has been removed. Earlier in this chapter,

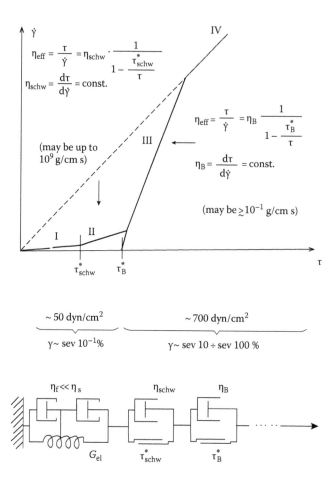

FIGURE 3.30 Description of a full rheological curve by the combination of quantitative two-parameter models. (From Shchukin, E.D. et al., *Colloid and Surface Chemistry*, Elsevier, Amsterdam, the Netherlands, 2001.)

we explained the entropic nature of these phenomena [18]. The original work by Fedotova et al. [34], focusing on the rheology of dilute (<3% v/v) aqueous colloidal dispersions of bentonite, has shown that the coagulation structures that develop in such suspensions can be spontaneously restored in an isothermal process after degradation, that is, these suspensions reveal thixotropy.

The dependence of the strain rate on the applied uniform shear stress in a steady-state flow reveals several characteristic regions that correspond to different physical phenomena.

In the stress interval between 50 and 500 dyn/cm^2, there is a slow creep regime with a constant Schwedow's viscosity, $\eta_0 \sim 10^9$ poise. This corresponds to a solid-like structure in which the fraction of the ruptured contacts is not yet very large. Under these conditions, the structure is restored in the flow by thixotropy. At higher stresses, a viscoplastic regime with significant rates of deformation is observed (Bingham's region). With the increase in shear stress, the effective viscosity decreases to the minimal constant value, $\eta_{min} \sim 0.1$ poise, which is a characteristic of the viscous flow of a system with a totally destroyed structure. At the same time, at small shear stresses, lower than about 40–50 dyn/cm^2, there is practically no creep, and the deformations observed in this regime are completely reversible (by magnitude). These deformations develop until they reach an equilibrium value corresponding to the magnitude of the applied stress, that is, these deformations are the elastic deformations of elastic aftereffect. Figure 3.31 shows shear deformation, ε, as a function of time for shear stress levels slightly lower than 50 dyn/cm^2. The deformation characteristic, ε, corresponds to the "technical deformation," γ, defined in continuum mechanics, that is, it corresponds to twice the pure

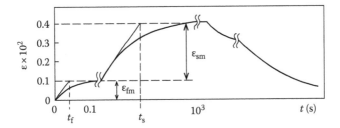

FIGURE 3.31 Shear deformation, ε, as a function of time during the elastic aftereffect at the stages of fast and slow elastic deformations.

shear value (see Section 5.1 for more details). This region corresponds to the onset of a rheological curve and is characteristic of a system without any residual strain.

The dependence of the shear strain on time, shown in Figure 3.31, clearly reveals two characteristic stages of the elastic strain: "fast" and "slow." Both of these strain regimes can be approximated using exponential functions with the corresponding time constants. These exponents show an increase in the strain with time during the action of shear and a time-dependent decay in the strain upon removing the load. During the first "fast" stage the time constant, t_f, is on the order of ~ 10^{-2} s, while during the second, "slow" stage the time constant, t_s, is on the order of ~ 10^2 s. The corresponding limiting equilibrium deformations at a shear stress ~50 dyn/cm² are ε_{fm} ~ 0.1% and ε_{sm} ~ 0.3%, respectively. The elasticity modulus value is thus ~10^4–10^5 dyn/cm².

A qualitative explanation of this peculiar behavior of bentonite suspensions observed even at very low volume fractions of the disperse phase is as follows. The reversibility of the strain under conditions when there is none of the "normal" elasticity typical for a solid (elasticity modulus ~ 10^{10}–10^{12} dyn/cm²) can be explained by the changes in the configuration entropy of the system due to changes in the mutual orientation of the particles. This description of the elasticity is similar to the description of the elasticity in polymer systems, in which it is explained by the statistics of the polymer chain configuration [35]. The fast stage of the aftereffect may be due to the particles turning and rolling without any translational motion of the particles with respect to each other. The slow stage may be related to the further coorientation of particles with some sliding motion, resulting in a translational motion of the points of the coagulation contacts along the particle surface.

The processes involved in the elastic aftereffect can be described by a simple rheological model consisting of two Kelvin's elements connected in series, as shown in Figure 3.32a. It is worth emphasizing here that this model is applicable only in the region of low shear stresses, below the onset of Schwedow's creep. From this model, one gets the following values for the slow and fast elastic strain moduli:

$$G_f = \frac{\tau}{\varepsilon_{fm}} = \frac{50}{10^{-3}} = 5 \times 10^4 \text{ dyn/cm}^2$$

$$G_s = \frac{\tau}{\varepsilon_{sm}} = \frac{50}{3 \times 10^{-3}} = 1.7 \times 10^4 \text{ dyn/cm}^2$$

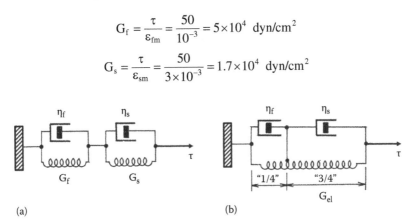

FIGURE 3.32 The corresponding rheological models describing two stages of the elastic aftereffect. The two Kelvin elements connected in series (a); G_f and G_s strains are replaced with a single elasticity modulus, G_{el} (b).

The superposition of Kelvin models is suitable here only for the analysis of the equilibrium states and for the estimation of the aftereffect time constants, while the kinetics of the development of the strain and the kinetics of the relaxation are considerably more complex and can't be described in terms of this model.

Both "fast" and "slow" strains have the same entropic nature, for which reason it is possible to introduce a single "common" elasticity modulus, G_{el}, corresponding to the net equilibrium deformation (Figure 3.32b):

$$\varepsilon_m = \varepsilon_{fm} + \varepsilon_{sm} \approx 0.1 + 0.3 \approx 0.4\%$$

mainly

$$G_s = \frac{\tau}{\varepsilon_m} \approx \frac{50}{4 \times 10^{-3}} \approx 1.25 \times 10^4 \ \text{dyn/cm}^2$$

The deformabilities at these two stages, which corresponding to the different viscous elements, η_f and η_s, are different and amount to ~¼ and ~¾, respectively (Figure 3.32b). Since the time constant of the elastic aftereffect is defined as the ratio of the viscosity to the corresponding elasticity modulus, the two corresponding viscosities are

$$\eta_f = G_f \times t_f \approx 5 \times 10^4 \times 10^{-2} \approx 5 \times 10^2 \ \text{poise}$$
$$\eta_s = G_s \times t_s \approx 1.7 \times 10^4 \times 10^2 \approx 1.7 \times 10^6 \ \text{poise}$$

Consequently, at the stage characterized by "fast" elasticity, the effective viscosity of the structure formed exceeds the viscosity of the dispersion medium (water) by a factor of 10^4–10^5. Where the elastic deformation is "slow," the ratio of the structured system viscosity to the viscosity of the dispersion medium is ~10^8.

The estimated microrheological parameters corresponding to the behavior of disperse systems subjected to small shear stresses may be described quantitatively utilizing the concepts of the microprocesses taking place in the thixotropic coagulation structures. In the section that follows, we will provide such a quantitative analysis.

3.2.1.1   Elasticity Modulus

Let us examine a structured system under the condition of uniform shear at constant temperature, T, and constant shear stress, τ, which, in this case, acts as pressure, p. The behavior of such a system can be described using the thermodynamic potential (per unit volume), $\Phi(T, \tau) = U - ST - \tau\varepsilon$, where the deformation $\varepsilon = \varepsilon(T, \tau)$, and the $\tau\varepsilon$ term is positive when the work performed over the system is positive. Neglecting the true elastic deformations, which are very small at these low stresses, and focusing only on the elastic deformations associated with the change in particle configuration, we may state that $U = \text{const}$, $S = S_1 = S_2(\varepsilon)$, where S_1 is independent of the deformation ε, and $S_2 = S_2(\varepsilon)$ is the "configuration component" of the entropy. Furthermore, $\varepsilon = 0$ and $S_2 = 0$ at $\tau = 0$.

At constant values of T and τ, the equilibrium in a given system corresponds to the minimum of the potential Φ. The strain ε is then the only variable parameter, that is,

$$\left(\frac{\partial \Phi}{\partial \varepsilon} \right)_{T,\tau} = -T \left(\frac{\partial S_2}{\partial \varepsilon} \right) - \tau = 0$$

Since ε and τ are of the same sign, the entropy should decrease with an increase in the strain. The required condition that an extremum is a minimum is that

$$\left(\frac{\partial^2 \Phi}{\partial \varepsilon^2} \right)_{T,\tau} = -T \left(\frac{\partial^2 S_2}{\partial \varepsilon^2} \right) > 0$$

which indicates that the entropy decrease with increases in strain must be at least as steep as $-\varepsilon^2$.

Let us model a coagulation structure by using a system of identical anisometric particles with a linear size δ. For strongly anisometric particles (platelets or rods), let $\delta = l$, where l is the largest size. For particles that are not strongly anisometric (ellipsoids), let $\delta = l - d$, where l and d are the largest and the smallest linear dimensions, respectively.

The equilibrium state of the system in the absence of shear stresses is characterized by a chaotic arrangement of the particles, and the mean projection of the particle dimension, δ, on any direction has the same value:

$$z_{av} = \left(\frac{1}{N}\right)3z_i = \frac{\delta}{2}$$

where N is the number of disperse phase particles in a unit volume of the system.

Indeed, in the $\theta - \phi$ coordinate system, the projection of particle dimension onto the polar axis is $z = \delta \sin \theta$, as evident from Figure 3.33. Within the element of the solid angle $d\omega = \cos\theta \, d\theta \, d\phi$, the number of particles, dN, equals $(N/2\pi) \cos\theta \, d\theta \, d\phi$. We restrict this consideration to a single hemisphere, that is, $0 < \phi < 2\pi$ and $0 < \theta < \pi/2$.

The mean projection of the sizes of all the particles on the selected axis is

$$z_{av} = \left(\frac{1}{N}\right)\int\delta\sin\theta\, dN = \left(\frac{1}{N}\right)\left(\frac{N\delta}{2\pi}\right)\int\int\sin\theta\,\cos\theta\, d\theta\, d\phi = \frac{\delta}{2}$$

The fraction of the particles in the interval $dz = \delta \cos\theta\, d\theta$, that is, within the "belt" $d\omega' = 2\pi \cos\theta\, d\theta$, is

$$\frac{dN}{N} = \left(\frac{1}{2\pi}\right)2\pi\cos\theta\, d\theta = f(z)dz = f(z)\delta\cos\theta\, d\theta$$

Consequently, the distribution function can be written as $f(z) = \text{const} = 1/\delta$, and the dispersion of z can be written as

$$\sigma_z^2 = \int(z - z_{av})^2 f(z)dz = \left(\frac{1}{\delta}\right)\int_0^\delta\left(\frac{z-\delta}{2}\right)^2 dz = \left(\frac{1}{12}\right)\delta^2$$

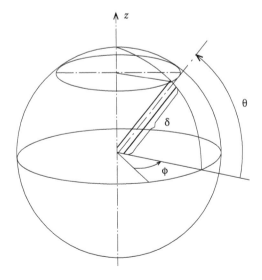

FIGURE 3.33 Evaluation of particle projection on a given axis.

Let us now assume that, due to the stress applied along the selected polar axis, a partial coorientation of the particles has taken place (i.e., the system acquired "texture"). As a result, the value of the
particle size projections mean on that axis has increased. To estimate the probability of this state, we
will use the approach developed by Yushenko on the basis of suggestions offered by Kolmogorov.
Since the projection z of each particle is a random number, the Lyapunov central limit theorem can
be applied to the system. Then the distribution around the mean value of the projection, z_{av}, is distributed normally with a mean, $\mu = \delta/2$, and a dispersion, $\sigma^2 = \sigma_z^2/N = \delta^2/12N$. The probability of a
deviation of a given mean of a projection on a given axis from the most probable value of the mean,
$\delta/2$, in the absence of coorientation is given by the function

$$f\left(z_{av}\right) = [1/\left(2\pi\right)^{1/2}\sigma]\exp-\left[\left(z_{av}-\mu\right)^2/2\sigma^2\right] = \left[(6/\pi)^{1/2}N^{1/2}/\delta\right]\exp-\{(3/2)N[\left(z_{av}-\delta/2\right)/(\delta/2)]^2\}$$

At the same time, this deviation has the physical meaning of a quantity close to the value of the
relative strain in the system:

$$\frac{\left(z_{av}-\delta/2\right)}{(\delta/2)} \approx \varepsilon$$

The case of a strong anisometry of the particles is assumed in this case. If the anisometry is weak,
it is possible to assume that the strain, ε, is a function of the characteristic describing the degree
of anisometry, which in the simplest case is a liner function:$\varepsilon \sim (\delta/l)(z_{av} - \delta/2)/(\delta/2)$, where l is the
maximum linear dimension.

The probability of deformation, ε, due to the coorientation of the particles is given by the
distribution function $[(6/\pi)^{1/2}N^{1/2}/\delta]\exp-\{(3/2)N\varepsilon^2\}$, and the change in the natural log of the
probability upon the transition from the state corresponding to $\varepsilon = 0$ to a state with some particular value of ε is given by $\Delta\ln W = -(3/2)N\varepsilon^2$. This corresponds to a change in the system
entropy, $\Delta S_2 = -(3/2)Nk\varepsilon^2$. Along the equilibrium curve $\varepsilon = \varepsilon(\tau)$ corresponding to the minimum
of the thermodynamic potential Φ, one has $\tau = -T(\partial S_2/\partial\varepsilon) = 3NkT\varepsilon$, which yields the equilibrium
elasticity modulus as

$$G_{el} = \frac{\tau}{\varepsilon} = 3NkT$$

The particles that make the colloidal suspension of bentonite are small platelets with a maximum
linear dimension, $l \sim 100-200$ μm, and having thickness, $d \sim 10$ μm. The number of particles per
unit volume in a 3% dispersion is

$$N \approx \frac{C}{(l^2d)} \approx 10^{17} \text{ cm}^{-3}$$

The estimate for the equilibrium elasticity modulus at room temperature is then $G_{el} \approx 3 \times 10^{17}\cdot 4 \times
10^{-14} = 1.2 \times 10^4$ dyn/cm^2, which agrees well with the order of magnitude of the experimentally
determined value of 1.25×10^4 dyn/cm^2. This agreement suggests that the proposed views on the
mechanism of the reversible deformations are valid.

3.2.1.2 Viscosity of the Elastic Aftereffect

The analysis of the elastic aftereffect also requires that a particular model be introduced. Let us first
examine the first stage of the elastic aftereffect: the "fast" elastic strain related to the appearance of
the viscosity, $\eta_f \sim (10^4-10^5)\eta_w$, where η_w is the viscosity of water.

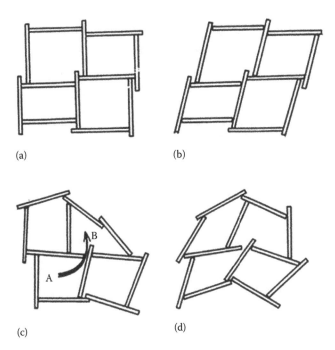

(a) (b)

(c) (d)

FIGURE 3.34 Shear deformations in idealized (a and b) and real (c and d) spatial coagulation structures. (From Shchukin, E.D. and Rehbinder, P.A., *Kolloidnyi Zh.*, 33, 450, 1971.)

Figure 3.34a and b shows the scheme of deformation of coagulation structure cells. This strain is the result of mutual turning of particles (platelets) without the sliding motion of points of contact. It is assumed here that the contacts can be viewed as "ideal joints" and do not require that any work be expended in turning the particles. The mutual turning of the particles results in their coorientation, that is, in an increase in the mean value of the projection on a given axis. It is clear that within the framework of this idealistic model, there is no reason for a substantial increase in the effective viscosity of the system relative to that of the immobilized water forming the dispersion medium. Indeed, the deformation of each cell is caused by the same uniform shear as that which causes the strain of the whole system.

Here, we are not going to consider the possibility of a viscosity increase in small volumes of liquid resulting from the contact of the liquid with the particle surface. While this factor requires a separate discussion and analysis, it still can't explain the high effective viscosity of the system, which exceeds that of water by several orders of magnitude. This becomes clear if one examines the thixotropic reversibility of the coagulation structures. Indeed, a complete destruction of the structures results in a substantial drop in the viscosity. The resulting viscosity is not too different from that of a dispersion medium (e.g., within one order of magnitude of the latter) and agrees with Einstein's viscosity law. The mean size of the dispersion medium "cells" remains approximately the same, that is, on the order of 100 Å. This means that the total volume of the dispersion medium is in fact a thin boundary layer with a thickness of several tens of Å. If the main source of the viscosity increase had been the interactions between the dispersion medium layers and the neighboring solid phase, the destruction of a structure would not have been expected to cause a strong decrease in the viscosity.

If the shear ε occurs over a time t with the average rate of $d\varepsilon/dt = \varepsilon/t$, the estimate for the work of the viscous forces per unit volume is $\eta \, d\varepsilon/dt \, \varepsilon = \eta \varepsilon^2/t$, and for a cell with a volume on the order of about l^3, it is

$$W \sim \frac{\eta \varepsilon^2 l^3}{t}$$

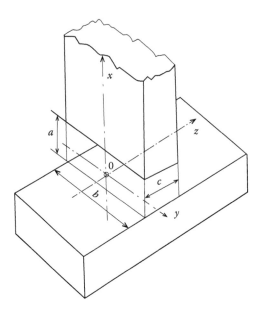

FIGURE 3.35　Schematic illustration of the flow in the gap.

The natural resolution to an effort to understand the issue of increased viscosity is the consideration of a real chaotic structure of particles in which the deformation of the cells is strongly influenced by the flow of the dispersion medium from one cell to another via relatively thin bridges (Figure 3.34c and d). By utilizing this approach, we can state that the shear, ε, causes a change in the cell volume approximately by $l^3\varepsilon$ and consequently results in the flow of water, amounting in volume to $V = \beta l^3\varepsilon$, between the neighboring cells. Here, β is the coefficient on the order of several units, which takes into account the microscopic process of the flow of multiple elementary volumes of water between the neighboring cells in a chaotic, "nonideal" particle structure. If the flow takes place in a channel having thickness a, width b, and length c (Figure 3.35), the average flow rate is $w_{av} = \beta l^3\varepsilon/abt$. The distribution of the flow rates is then given by a parabolic law

$$w(x) = \frac{1}{2}\left(\frac{\Delta p}{c\eta_w}\right)\left(\frac{a}{2}\right)^2\left[1 - \left(\frac{x}{1/2a}\right)^2\right]$$

where Δp is the pressure drop along the channel length, c. It can be shown that

$$\Delta p = 3\eta_w\left[\frac{c}{(1/2a)^2}\right]w_{av} = 3\eta_w\left[\frac{c}{(1/2a)^2}\right]\beta l^3\varepsilon/abt$$

and the dissipation of the energy in the channel is given by

$$W_f = \Delta p\beta l^3\varepsilon = 12\eta_w(c/ba^3)(\beta l^3\varepsilon)^2 1/t$$

Since we have shown that $W_f \approx \eta_f\varepsilon^2 l^3/t$, the effective viscosity of the "fast" elastic deformation is

$$\frac{\eta_f}{\eta_w} = 12\beta^2\left(\frac{c}{ba^3}\right)l^3$$

Assuming that the thickness and length of the channels are approximately the same as the thickness of the particles (platelets), and that the width of channels approximately equals the maximum particle diameter, $a = c \approx d$, $b \approx l$, one finds that

$$\frac{\eta_f}{\eta_w} = 12\beta^2 \left(\frac{l}{d}\right)^2$$

If the coefficient β is on the order of a few units and the ratio $l/d \sim 10$ (or several tens), the aforementioned expression indicates that the viscosity of the fast elastic deformation should exceed the viscosity of water by a factor of 10^4–10^5, which agrees well with experimental observations.

It is necessary to emphasize here that the estimate for η_f is still rather crude. The aforementioned relationship does not include in any direct form the concentration of the dispersed phase. However, this concentration is dictated by the same model and amounts to several particles (n) per volume βl^3, where $\beta \approx n$. Consequently, the concentration of the dispersed phase can be assumed to be equal to $nl^2d/\beta l^3 \approx d/l$, which is on the order of 10^{-1} or somewhat less, which again agrees well with the experimental conditions.

The described method of the viscosity estimation for the "fast" stage of the elastic aftereffect can be further extended to obtain an estimate for the much higher viscosity of the second stage of the elastic aftereffect. In this case, it is assumed that the particle orientation changes owing to a small-scale sliding of the points of contacts along the particle surface without the destruction of the majority of the contacts. To a degree, this sliding process can also be described using the flow of viscous fluid in a gap, but the thickness of the gap needs to be much smaller than that in the case described earlier. While in the previous case, we assumed the gap to be on the order of the thickness of the platelets, $d \sim a \sim 10\text{Å}$ (or slightly larger), in the present case, the gap thickness is on the order of several angstroms, that is, $a = \theta d$, where the coefficient θ is around 0.1. Substituting this value into the equation for the effective viscosity, one obtains the following result for the viscosity of the "slow" elastic aftereffect regime:

$$\frac{\eta_s}{\eta_w} \sim 12\beta^2 \left(\frac{1}{\theta^3}\right)\left(\frac{l}{d}\right)^2$$

which is three orders of magnitude higher than the viscosity of the "fast" stage and is close to the experimentally determined viscosity of $\sim 10^6$ poise. However, the latter estimate requires further accounting for the changes of the fluid properties in the gap, since the gap thickness is comparable to several molecular dimensions. Furthermore, additional analysis is also needed to understand the ratio of the strains at the "fast" and "slow" stages, $\varepsilon_{fm}:\varepsilon_{sm} \propto 1:3$.

In the next section, we will describe the approximate microrheological estimation of the effective viscosity of a fibrous suspension in a steady-state thixotropic regime carried out by Shchukin [29].

3.2.2 The Role of Contact Interactions in the Rheological Behavior of Fibrous Suspensions

The system of interest here is that of a slurry containing fibrous matter with randomly crossing fibers. A common and well-studied example of such a system is aqueous slurry of cellulosic fibers, that is, papermaking pulp. We will present a model that allows one to estimate the number of contacts between the fibers and to provide an estimate for the mechanical properties of the structure. We will also illustrate the impact of common papermaking coagulants (e.g., polyethyleneimine [PEI] and quaternary alkyl ammonium salts) on the macroscopic rheological properties of pulp suspensions. These studies have an important practical application, as they provide an insight into the control of the rheology of cellulosic slurries at different stages of papermaking. The rheological behavior of papermaking furnish undergoes drastic changes throughout the papermaking process. The slurry must behave as a liquid at the point of entry from the paper machine head box to the wire and must display a fast increase in the effective viscosity and the critical yield stress upon dewatering and forming a sheet.

In the description of the simplest models of globular disperse structures, that is, suspensions, porous media and powders consisting of isometric (spherical) particles, one needs in reality only two parameters to describe the system: the average particle radius, r, the dimensionless porosity, Π (cm^3 of pores/cm^3 of porous body) or conversely the continuity of the solid phase, $V_s = 1 - \Pi$, cm^3. Alternatively, one can also operate with the mean coordination number in a chaotic structure or the mean knot-to-knot number of particles in a model of crossing chains [36]. The two parameters allow one to determine the number of contacts between the particles per unit cross-sectional area, χ. Multiplying the parameter χ by the mean contact strength p_{av} allows one to obtain an estimate for the strength (i.e., the critical shear stress), $P_c = \chi p_{av}$. Numerous experimental studies frequently referenced throughout this book have proven that such a summation is indeed a valid approach.

One can utilize various models for modeling fibrous structures. In the case of a system composed of long elastic entangled threads and fibers (fabrics and felts), the own strength of fibers becomes a determining factor in the strength of the entire structure. For hard fibers that are not too anisometric (linear dimensions ratio up to 100), the model based on the summation of the contact interactions may still be valid. However, in describing the structure and calculating the number of contacts, the use of the two-parameter model is no longer possible, and one needs at least three parameters. These parameters may be the volume fraction of the solid phase, V_s, the fiber length, l, and the fiber diameter, $d = 2r$. In the following text, we will present the model for calculating the number of contacts between the fibers and compare those results with the results of experimental studies on the cohesion of cellulosic fibers and the rheology of pulp suspension [37–44].

Let us consider the uniform deformation of a unit volume of a disperse system under the condition of a steady-state viscoplastic flow or the "slow" stage of the elastic aftereffect under the action of a shear stress τ. The system consists of fibers of length l and diameter d occupying the volume fraction V_s. The contact between the crossed fibers can be characterized by the average tangential friction force (which in turn characterizes the mean force of resistance to the shear motion in the contact), p_{tg}. The number of particles, ν, per unit volume of the disperse system is

$$\nu = \frac{4V_s}{\pi d^2 l}$$

The mean distance between the fibers (distance between their centers) in a given direction (x, y, or z) can be estimated as

$$b = \nu^{-1/3}, \quad b \ll l$$

In a cylindrical volume with radius b and length l, the approximate number of neighboring fibers next to a given fiber, ν_c, is given by

$$\nu_c \approx \pi b^2 l \nu = \pi \nu^{1/3} l = \frac{\pi l}{b}$$

If the fraction α of these neighboring fibers is in contact with a given reference fiber, one can write for the number of contacts along the reference fiber, n_c,

$$n_c = \frac{1}{2} \alpha \pi \nu^{1/3} l$$

where the numerical coefficient ½ takes into account the fact that the contact belongs to both fibers in contact. The number of contacts per unit volume of the system, n, can be estimated as

$$n = n_c \nu = \frac{1}{2} \alpha \pi \nu^{4/3} l$$

The most complex step here is the estimation of the *contact cross section layer thickness*, that is, the thickness of the layer in which the destruction of contacts take place upon the destruction of the structure or conversely of the layer in which the contacts together form the basis of the resistance to shear, that is, the net friction force acting under conditions of steady-state flow. In the two limiting cases, this thickness is either l or b, while in the general case, it is a combination of l and b with the units of length, that is, $l^k b^{1-k}$, and $0 \leq k \leq 1$. It is then possible to calculate the number of contacts per unit area in the contact section as

$$\chi = n l^k b^{1-k} = \frac{1}{2} \alpha \pi \nu^{1+k/3} l^{1+k}$$

Assuming that $k = 1$, one can write

$$\chi_l = nl = \frac{1}{2} \alpha \pi \nu^{4/3} l^2$$

for $k = 0$, one has

$$\chi_b = nb = \frac{1}{2} \alpha \pi \nu l$$

The ratio of χ values corresponding to these two limiting cases is $l/b = l\nu^{1/3}$ times.

If direct measurements of the molecular component of the cohesive force between crossed fibers yield an average value of p_n and independent measurements yield the value of μ for the friction coefficient, the contact friction force is

$$p_{tg} = \mu p_n$$

And the macroscopic resistance to shear is given as

$$\tau \approx \chi p_{tg} \sim n l^k b^{1-k} = \frac{1}{2} \alpha \pi \nu^{1+k/3} l^{1+k} p_n$$

The term "molecular component" implies that the normal component of the load (i.e., the compression in the contact) is absent. It may appear strange that in the final expression for χ, there are only two and not three components. The reason for this is that within the framework of the model of rigid thin fibers, the smaller dimension, $d \ll l$, was neglected in the process of counting the number of contacts. This smaller dimension has been used only in the transition from the concentration of the disperse phase, V_s, to the number of particles per unit volume, ν.

Let us further utilize the literature values of V_s, l, d and the contact cohesive forces [39–43]. The length l of softwood and hardwood fibers used in the experimental studies was 1–2 mm and 20 µm, respectively. The consistency of the pulp suspensions was 1% and 2%. The experimental rheological studies conducted in the elastic aftereffect regime were performed with the constant shear stress ranging between 10 and 90 Pa. From the experimental data for 2% softwood pulp ($l = 2$ mm), the shear rate in the uniform shear, $d\gamma/dt \approx 10^{-4}$ cm^{-1} s^{-1} at $\tau = 20$ Pa, which yields a macroscopic viscosity, $\eta \sim 2 \times 10^5$ Pa s. Assuming that the thickness of the contact layer equals l, for $\alpha = 1$, one finds the following values: $\nu \sim 3 \times 10^4$ cm^{-3}, $b \sim 0.03$ cm, and $l/b \sim 7$, $n_c \sim 10$ and $n \sim 3 \times 10^5$ cm^{-3}, $\chi_l \sim 6 \times 10^4$ cm^{-2}.

According to the published data [42], the molecular component of the friction force in the individual contact between two fibers, $F = p_{tg} = \mu p_n$ in the presence of PEI coagulant is on the order of 10^{-3} dyn. For the shear stress, τ, this yields $\tau = \chi_l p_{tg} \sim 6 \times 10^4$ cm$^{-2} \times 3 \times 10^{-3}$ dyn $= 180$ dyn/cm$^2 = 18$ Pa

This value agrees reasonable well with the experimentally determined value of 20 Pa, despite some apparent approximations, such as with regard to the homogeneous flow of the pulp suspension.

Earlier experimental studies [37–39] indicate clear trends in the influence of the surface active agents (PEI, quaternary alkyl ammonium salts, etc.) on the rheological properties of pulp and on the strength of the individual contacts between fibers. This has already been discussed in detail in Chapter 2. In both cases, the dependence of these parameters on the additive concentration passes through maxima, the location and depth of which depend on the type of the additive and the type of the fiber. It has been demonstrated that such dependence is related to the effect of surface active additives on the surface charge of the fibers and the possibility of generating patches of opposite charges on the fiber surface. It has also been suggested [29] that the fiber surface can be made both hydrophilic and hydrophobic, depending on how much of an additive has been adsorbed. The reader is referred to Section 2.3.1 for a detailed discussion on the effect of polyelectrolytes and surfactants on the cohesive forces between the fibers.

A significant increase in the pulp mobility (liquification) and homogenization can be achieved by vibration [27,44]. According to the published data [44,45], in a rotational setup operating at a rate of 3000 rev/min, the viscosity of a 15% pulp slurry was about 10 Pa s with the power consumption on the order of a few watts per cm^3. It is of interest to estimate the fraction of consumed power that has been used in overcoming the friction forces in the contacts between fibers.

To estimate the number of particles and the number of contacts per unit volume, ν, let us employ the approach described earlier. Even if we assume that all the contacts are ruptured with every half-period of vibration (i.e., 100 times/s), we may underestimate the frequency of the fiber collisions. The rate s of the relative fiber motion reported in [44,45] was on the order of 10 cm/c. With the average distance between the particles being $b = \nu^{-1/3}$ (in the present case $b \approx 0.016$ cm), this would correspond to the frequency of the formation and rupture of each contact equal to $\omega = s/b \sim 10$ cm/s 0.016 cm $\sim 10^3$ s^{-1}.

This estimate yields the maximum number of the collisions between the fibers. Then the number of fiber collisions and the number of ruptured contacts per unit volume of the disperse system is $n\omega$ (cm^3/s).

Let us assume that the work, u, needed to rupture the contact is predominantly the work that needs to be spent to overcome friction upon moving the contact point along the fiber by a distance $l/2$ on average. This elementary work is then $u \approx p_{tg} l/2$. The power U dissipated by all contacts in a unit volume can be estimated as $U \sim n\omega u \sim n(s/b) p_{tg} l/2 \sim 10^6$ erg/cm^3/s. This value represents the highest possible estimate, and it is an order of magnitude lower than the actual consumed power, N. This large discrepancy between the estimated and the real power consumption can be explained by the cyclic processes of flow (turbulent) of the dispersion medium between the neighboring small volumes ("flocs") of the pulp slurry. This flow of the dispersion medium is characterized by a high effective viscosity, $\eta_{eff} \sim 10^4 \eta_w$, as is schematically illustrated in Figure 3.36. The flocs with a linear size on the order of $10l$ were indeed observed in the experiments. Contact interactions are responsible for the formation of flocs upon the action of vibration. In order to explain the observed high value of the effective viscosity, we can refer here to the model of the rheological behavior of clay suspensions (see Section 3.2.1): the flow of the dispersion medium between neighboring microscopic volumes yields a very similar estimate for η_{eff}.

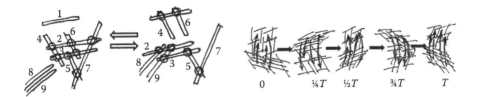

FIGURE 3.36 Periodic structural changes that occur in a fibrous suspension due to vibration. The structural changes are caused by the rupture and restoration of contacts between the fibers upon the localized flow of the dispersion medium caused by the rotation of the fibers and the sliding of the points of contact along the fiber surface in the contacts that remain intact. (From Schchukin, E.D. private communication.)

REFERENCES

1. Reiner, M. 1960. *Deformation, Strain and Flow: An Elementary Introduction to Rheology*. London, U.K.: H. K. Lewis.
2. Macosco, C. W. 1994. *Rheology: Principles, Measurements, and Applications*. New York: Wiley.
3. Tanner, R. I. 2000. *Engineering Rheology*, 2nd edn. Oxford, U.K.: Oxford University Press.
4. Morrison, F. A. 2001. *Understanding of Rheology (Topics in Chemical Engineering)*. Oxford, U.K.: Oxford University Press.
5. Giese, R. F. and C. J. van Oss. 2002. *Colloid and Surface Properties of Clays and Related Minerals*. New York: Marcel Dekker.
6. Shramm, G. 2003. *Principles of Practical Rheology and Rheometry*. Moscow, Russia: KolosS.
7. Mezger, T. G. 2006. *The Rheology Handbook*. Hannover, Germany: Vincentz Network.
8. Pal, R. 2007. *Rheology of Particulate Dispersions and Composites*. Boca Raton: CRC Press.
9. Chhabra, R. P. and J. F. Richardson. 2008. *Non-Newtonian Flow and Applied Rheology: Engineering Applications*, 2nd edn. Amsterdam, the Netherlands: Elsevier.
10. Shchukin, E. D., Pertsov, A. V., Amelina, E. A., and A. S. Zelenev. 2001. *Colloid and Surface Chemistry*. Amsterdam, the Netherlands: Elsevier.
11. Shchukin, E. D., Pertsov, N. V., Osipov, V. I., and R. I. Zlochevskaya (Eds.). 1985. *Physical-Chemical Mechanics of Natural Disperse Systems*. Moscow, Russia: Izd. MGU.
12. Yaminskiy, V. V., Yaminskaya, K. B., and E. D. Shchukin. 1987. Particle adhesion and coagulation structure formation. *Kolloidnyi Zh.* 49: 967–971.
13. Rehbinder, P. A. 1979. *Selected Works: Surface Phenomena in Disperse Systems: Physical-Chemical Mechanics*. Moscow, Russia: Nauka.
14. Shchukin, E. D. 2006. The influence of surface-active media on the mechanical properties of materials. *Adv. Colloid Interface Sci.* 123–126: 33–47.
15. Izmaylova, V. N., Alexeeva, I. G., Shchukin, E. D., and P. A. Rehbinder. 1972. Rheological properties of interfacial adsorption layers of proteins and surfactants. *Doklady AN SSSR*. 206: 1150–1153.
16. Izmaylova, V. N., Alexeeva, I. G., Shchukin, E. D., and P. A. Rehbinder. 1973. Rheological properties of interfacial adsorption layers of gelatin on the boundary with benzene. *Kolloidnyi Zh.* 35: 860–866.
17. Izmaylova, V. N. and P. A. Rehbinder. 1976. *Structure Formation in Protein Systems*. Moscow, Russia: Nauka.
18. Shchukin, E. D. and P. A. Rehbinder. 1971. The mechanism of elastic after effect in structured bentonite suspensions of low concentration. *Kolloidnyi Zh.* 33: 450–458.
19. Frenkel, Y. I. 1975. *Kinetic Theory of Liquids*. Moscow, Russia: Izd. AN SSSR.
20. Balshin, M. Y. 1948. *Powder Metallurgy*. Moscow, Russia: Metallurgizdat.
21. Dobrushina, I. A., Spasskiy, M. R., Shatalova, I. G., and E. D. Shchukin. 1969. The kinetics of the vibrational compaction of tungsten powder. *Doklady AN SSSR*. 189: 525–527.
22. Spasskiy, M. R., Spasskaya, I. A., Shatalova, I. G., and E. D. Shchukin. 1979. The effect of the geometric properties of powders on their compaction. *Fizika i Himiya obrabotki materialov*. 3: 147–151.
23. Polukarova, Z. M., Shatalova, I. G., Yusupov, R. K., and E. D. Shchukin. 1968. Use of vibration compacting for increasing the strength of compacts. *Powder Metall.* 6: 54–56.
24. Spasskiy, M. R., Dobrushina, I. A., Shatalova, I. G., Shchukin, E. D., and P. A. Rehbinder. 1969. The energy parameter of vibrocompaction of powders. *Doklady AN SSSR*. 189: 767–770.
25. Spasskiy, M. R. and E. D. Shchukin. 1973. Model of the vibrational compaction of disperse medium. *Kolloidnyi Zh.* 35: 897–905.

26. Rehbinder, P. A. and N. V. Mikhaylov. 1979. Physical-chemical mechanics—The scientific basis of optimal technology of concrete and reinforced concrete. In *Selected Works: Surface Phenomena in Disperse Systems: Physical-Chemical Mechanics*, Shchukin, E. (Ed.), pp. 324–335. Moscow, Russia: Nauka.
27. Uriev, N. B. and A. A. Potanin. 1992. *Fluidity of Suspensions and Powders*. Moscow, Russia: Himiya.
28. Taubman, A. B. and S. A. Nikitina. 1962. Structural and mechanical properties of the surface layers of the emulsifier and mechanism of stabilization of concentrated emulsions. *Kolloidnyi Zh.* 24: 633–666.
29. Shchukin, E. D. 2001. The role of contact interactions in the rheological behavior of a fibrous suspension. *Colloid J.* 63: 784–787.
30. Serb-Serbina, N. N. and P. A. Rehbinder. 1947. Structure formation in aqueous suspensions of bentonite clays. *Kolloidnyi Zh.* 9: 38–43.
31. Abduragimova, L. A., Rehbinder, P. A., and N. N. Serb-Serbina. 1955. Elastic-viscous properties of thixotropic systems. *Kolloidnyi Zh.* 17: 184–195.
32. Rehbinder, P. A. 1950. *New Methods of Physical-Chemical Studies of Surface Phenomena*. Moscow, Russia: Izd. AN SSSR.
33. Rehbinder, P. A. 1954. Coagulation and thixotropic structures. *Discuss Faraday Soc.* 18: 151–160.
34. Fedotova, V. A., Hodzhaeva, H. A., and P. A. Rehbinder. 1967. Elastic modulus, efficient and extremely plastic viscosity of thixotropic-hardened solid-like coagulation structures. *Doklady AN SSSR*. 177: 155–158.
35. Volkenstein, M. V. 1958. The configurational statistics of polymeric chains. *J. Polym. Sci.* 29: 441–454.
36. Shchukin, E. D. 1985. Physical-chemical theory of the strength of disperse structures and materials. In *Physical-Chemical Mechanics of Natural Disperse Systems*, E. D. Shchukin, N. V. Pertsov, V. I. Osipov, and R. I. Zlochevskaya (Eds.), pp. 72–90. Moscow, Russia: Izd. MGU.
37. Amelina, E. A., Shchukin, E. D., Parfenova, A. M., Bessonov, A. I., and I. V. Videnskiy. 1998. Adhesion of cellulose fibers in liquid media. 1. Measurement of friction forces in contacts. *Kolloidnyi Zh.* 60: 583–587.
38. Shchukin, E. D., Videnskiy I. V., Amelina, E. A. et al. 1998. Adhesion of cellulose fibers in liquid media. 2. Measurement of attraction forces in contacts. *Kolloidnyi Zh.* 60: 588–591.
39. Amelina, E. A., Shchukin, E. D., Parfenova, A. M. et al. 2000. Effect of cationic polyelectrolyte and surfactant on cohesion and friction in contacts between cellulose fibers. *Colloids Surf.* 167: 215–227.
40. Amelina, E. A., Videnskiy, I. V., Ivanova, N. I. et al. 2001. Evaluation of the molecular component of friction of fibers in surfactant solutions. *Kolloidnyi Zh.* 63: 132–134.
41. Amelina, E. A., Videnskiy, I. V., Ivanova, N. I. et al. 2001. The interactions of individual fibers during adsorption modification of their surface. *Vestnik MGU, Ser. 2: Khimia.* 42: 49–54.
42. Reizinsh, R. E. 1987. *Pattern Formation in Suspensions of Cellulose Fibers*. Riga, Latvia: Zinatne.
43. Kerekes, R. J. and C. J. J. Shell. 1992. Characterization of fibre flocculation regimes by a crowding factor. *J. Pulp Paper Sci.* 18: J32.
44. Laine J. and P. Stenius. 1997. The effect of the charge on the fiber and paper properties of bleached industrial kraft pulp. *Pap puu* 79(4): 257–266.
45. Gullichsen, J. and E. Harkonen. 1981. Medium Consistency Technology. *Tappi*, 64(6): 69–72.
46. Schchukin, E. D. and A. S. Zelenev. Private communication.

4 Contact Interactions and the Stability of Free-Disperse Systems

The concept of stability applies to virtually everything that surrounds us: the geographical terrain, snow and mud banks in the mountains, buildings and various other civil structures, currency exchange rates, the political situation in the world, etc. In colloid science and physical-chemical mechanics, we study the *stability of disperse systems*, that is, of heterogeneous systems consisting of two or more phases in which the constituent particles are small (<1 μm), that may participate in Brownian motion but still retain the physicochemical and mechanical properties of the phase and the interface. We distinguish diluted free-disperse systems and concentrated connected-disperse systems (Figure 4.1).

In connected-disperse systems consisting of coagulation structures, gels, and sediments, the cohesive forces between the particles are sufficiently strong to withstand both the slacking effect of the thermal motion and external mechanical impacts. In the latter case, we are interested in the stability of the connected-dispersed state of the system, that is, the stability of the structure and its resistance to transition into a free-disperse state as a result of the peptization process. The response of connected-disperse systems to applied stress is the subject of rheological studies. We have discussed the principal concepts of rheology in Chapter 3 and have described the role of entropy in the change in rheological behavior upon the coorientation of anisometric particles.

Free-disperse systems comprise dilute emulsions, sols, and suspensions in which the participation of particles in thermal Brownian motion plays a dominant role over the cohesive forces between them. In these systems, we are particularly interested in the stability resisting the transition from the free-disperse state to the connected-disperse state via aggregation, flocculation, or sedimentation (Figure 4.2).

It is worth pointing out again that in both free-disperse and connected-disperse systems, the interaction and cohesion between particles play the determining role. The principal difference between the two cases is as follows. Stability and its loss in connected-disperse systems (structures) are governed by the competition between cohesive forces and applied mechanical stresses, while in the free-disperse system, they are governed by the competition between the kinetic energy of particles participating in the Brownian motion and the energy of cohesion. These, however, are the two elementary approaches: the loss of stability in a connected-disperse system may, to a large extent, be driven by thermal fluctuations, while in a free-disperse system, the loss of stability can be initiated by external forces, such as gravity (in sedimentation) or a velocity gradient (in orthokinetic coagulation).

In this chapter, we will address the thermodynamic and kinetic aspects of colloid stability in free-disperse systems. We will discuss the concept of the factors for *weak* and *strong* stabilization, the possibility of spontaneous dispersion, and the conditions necessary to form thermodynamically stable colloidal systems. Furthermore, we will discuss the necessary conditions for the coagulation–peptization (dispersion) transition and the equilibrium between a coagulate comprising the connected-disperse system and the free-dispersed system formed in the course of dispersion. The fundamentals of colloid stability have been partially discussed in Chapters 1 and 2 and are covered to a great detail in textbooks on colloid and surface science [1–29]. We will address here the subject of colloid stability to the extent appropriate to the general scope of this book.

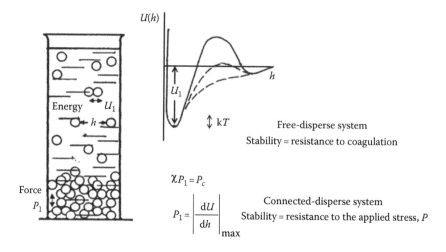

FIGURE 4.1   The concept of stability in dilute and concentrated disperse systems.

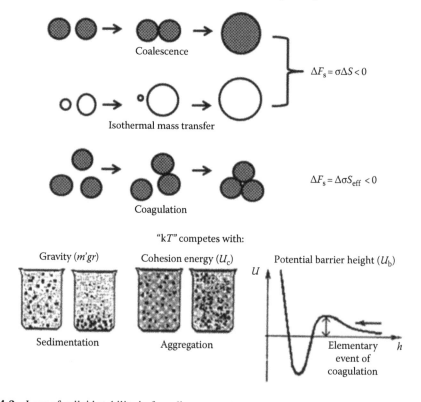

FIGURE 4.2   Loss of colloid stability in free-disperse system.

4.1   KEY FACTORS GOVERNING THE STABILITY OF FREE-DISPERSE SYSTEMS

A general approach to the analysis of the conditions of stability in free-disperse systems is the analysis of the interactions in the contact established between the particles in a given dispersion medium. This is essentially the analysis of the properties of a thin film of dispersion medium in the gap between the particles (see Section 1.2). The quantitative thermodynamic description of the properties of the film is based on the analysis of the change in *free energy of interaction*, $\Delta\sigma_f(h)$, when the surfaces are brought close together from an infinite gap to a given width of the gap, h (Figure 4.3).

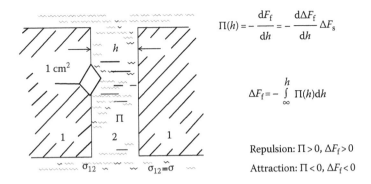

$$\Pi(h) = -\frac{dF_f}{dh} = -\frac{d\Delta F_f}{dh}\Delta F_s$$

$$\Delta F_f = -\int_\infty^h \Pi(h)dh$$

Repulsion: $\Pi > 0$, $\Delta F_f > 0$

Attraction: $\Pi < 0$, $\Delta F_f < 0$

FIGURE 4.3 To the definition of the disjoining pressure, $\Pi(h)$, in a thin flat–parallel plate according to Derjaguin.

The derivative $-d\Delta\sigma(h)/dh = \Pi(h)$ is the force per unit cross-sectional area, also known as Derjaguin's *disjoining pressure* [4,5]. Within the context of this definition, both $\Delta\sigma(h)$ and $\Pi(h)$ are positive in the case of a repulsion and negative in the case of an attraction. Molecular attraction forces prevail at long distances, while repulsive forces prevail at very short distances (the so-called Born repulsion). The principal theory that describes the interactions in a thin film is the well-known Derjaguin–Landau–Verwey–Overbeek (DLVO) theory, which focuses on the analysis of the competitive contribution of molecular (dispersion) attractive forces and electrostatic repulsion to the interaction between surfaces separated by a liquid film.

The dispersion component of the molecular attractive force per unit cross-sectional area in a flat–parallel gap of thickness h is characterized by the free potential energy of interaction, $U_{mol}^d(h) = \Delta\sigma_f^d(h) = -A^*/12\pi h^2$, where the complex Hamaker constant, A^*, takes into account the molecular interactions in the solid phase and the liquid medium and between the two (Figure 4.4). This is the *molecular component of the free energy of interaction*. Its derivative $-\Pi_{mol}^d(h) = -d\Delta\sigma_f^d(h)/dh = -A^*/6\pi h^3$ is the *molecular component of the disjoining pressure*. Due to the natural character of the dispersion forces and the use of the Lennard-Jones "6–12" potential [2,11,26], the attraction at very short distances corresponding to a contact between electron shells is replaced by rapidly increasing repulsion—the so-called Born repulsion.

The value of $|\Delta\sigma_f(h_0)| = \Delta\sigma_f = F$ characterizes the extent of cohesion between the surfaces in direct contact at the *primary* free energy minimum.

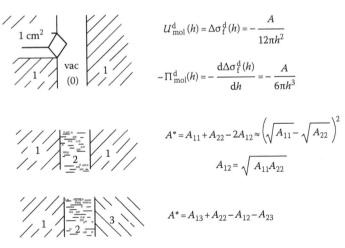

$$U_{mol}^d(h) = \Delta\sigma_f^d(h) = -\frac{A}{12\pi h^2}$$

$$-\Pi_{mol}^d(h) = -\frac{d\Delta\sigma_f^d(h)}{dh} = -\frac{A}{6\pi h^3}$$

$$A^* = A_{11} + A_{22} - 2A_{12} \approx \left(\sqrt{A_{11}} - \sqrt{A_{22}}\right)^2$$

$$A_{12} = \sqrt{A_{11}A_{22}}$$

$$A^* = A_{13} + A_{22} - A_{12} - A_{23}$$

FIGURE 4.4 The molecular (dispersion) component of the free energy of interaction—the *negative component*, $\Delta\sigma_f^d(h) = \Delta F_{f(mol)}(h)$; the definition of the Hamaker constant and the complex Hamaker constant.

The *electrostatic component* of the disjoining pressure is given by the positive term describing the interaction between the electrical double layers of the two surfaces in contact. The detailed description of the electrical double layer can be found in textbooks on colloid science and in specialized textbooks on electrokinetic phenomena at interfaces [5,8–13]. We will restrict ourselves to a very brief revision of the basic concepts associated with the electrical double layer.

The electrical double layer at the interface between the solid and liquid phases can be viewed as a capacitor formed by the combination of a charged surface, which is a carrier of fixed *potential-determining ions*, and the equivalent number of counter ions distributed in a volume of the dispersion medium adjacent to the surface. The distribution of a potential, $U(h)$, within the region of interest (a *diffuse* part of electrical double layer) is described by an exponential function decreasing with distance, $U(h) = \text{const} \times \exp(-h/\delta)$, where $\delta = (\varepsilon\varepsilon_0 kT/2z^2e^2n_0)^{1/2}$. Here, k is the Boltzmann constant, T is the absolute temperature, ε is the dielectric constant, ε_0 is the electric constant, e is the elementary charge, and n_0 is the number concentration of ions. The concentration of ions has a strong influence on the value of δ, which is the effective thickness of the diffuse double layer. The meaning of parameter δ is similar to that of the Gibbs' discontinuity surface. In the Debye–Hückel theory of strong electrolytes, δ is known as the *Debye length*.

The value of the preexponential constant, $\text{const} = (4kT/ze) \tanh (ze\phi_0/4kT)$, includes the dependence of a potential on these parameters and on the surface potential, ϕ_0. At high electrolyte concentrations, $c \sim 0.1–1$ mol/dm³, the double layer is compressed to a thickness on the order of fractions of a nm, while at low electrolyte concentrations, $c \sim 10^{-3}–10^{-4}$ mol/dm³, the diffuse double layer thickness is on the order of tens of nanometers.

According to DLVO theory, the repulsion of the surfaces across the gap is determined by the overlap between the double layers of both surfaces at a distance $h \sim 2\delta$. The essence of the calculations discussed in the traditional literature on colloid science is presented in Figure 4.5. In the middle of the (film filling the) gap between the two identical surfaces, the potential of electric field is $2U(h/2) \times \text{const}_1 \times e^{-h/2\delta}$. The charge density in the middle point, given by the derivative of the potential with respect to distance, is given by $\rho(h/2) \times \text{const}_2 \times e^{-h/2\delta}$. The product of the charge density and the potential given by $\text{const}_3 \times e^{-h/\delta}$ characterizes the work of concentrating the charge (the density of the free energy excess) and thus the pressure force per unit area (J/m³ = N/m²). In this way, we have come up with an exponential function for the dependence of the electrostatic component of the disjoining pressure on the distance in a flat–parallel gap of thickness δ

$$\Pi_{el}(h) = \text{const}_4 \times e^{-h/\delta}$$

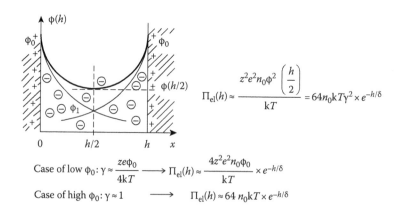

Case of low ϕ_0: $\gamma \approx \dfrac{ze\phi_0}{4kT}$ $\longrightarrow$ $\Pi_{el}(h) \approx \dfrac{4z^2e^2n_0\phi_0}{kT} \times e^{-h/\delta}$

Case of high ϕ_0: $\gamma \approx 1$ $\longrightarrow$ $\Pi_{el}(h) \approx 64\,n_0kT \times e^{-h/\delta}$

FIGURE 4.5 The ion-electrostatic component of the disjoining pressure, $\Pi_{el}(h)$—the positive term.

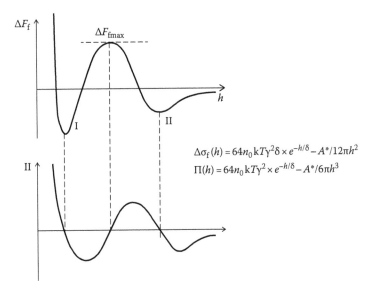

FIGURE 4.6 The free energy of interaction, $\Delta\sigma_f(h) = \Delta F_f(h)$, and the disjoining pressure, $\Pi(h)$, in a thin film calculated by a superposition of the molecular (attractive) and the electrostatic (repulsive) components.

where the constant $\text{const}_4 = 64 n_0 k T \gamma^2$ and $\gamma = \tanh(ze\phi_0/4kT)$. Integrating this expression over distance yields the ion-electrostatic component of the free energy of interaction

$$\Delta\sigma_{fel}(h) = \text{const}_5 \delta \times e^{-h/\delta}$$

The main concept of DLVO theory is the summation of the positive (exponential) and negative (hyperbolic) terms of the free energy of interaction, $\Delta\sigma_f(h)$, and the disjoining pressure, $\Pi(h)$,

$$\Delta\sigma_f(h) = 64 n_0 k T \gamma^2 \delta \times e^{-h/\delta} - \frac{A^*}{12\pi h^2}$$

$$\Pi(h) = 64 n_0 k T \gamma^2 \times e^{-h/\delta} - \frac{A^*}{6\pi h^3}$$

For a given concentration of a monovalent 1:1 electrolyte (e.g., $c \sim 10^{-3}$ mol/dm³), the free energy of interaction and the disjoining pressure are shown in Figure 4.6. The function $\Delta\sigma_f(h)$ reveals two minima separated by a potential barrier. These minima are referred to as the *primary minimum* and the *secondary minimum*. Even a qualitative analysis of the $\Delta\sigma_f(h)$ functions points to the relation between the height of the potential energy barrier and the ability of particles engaged in thermal motion to overcome it and thus coagulate in the primary minimum. In Section 1.2, the description of the interactions between surfaces was related to the thickness of the flat–parallel gap between the particles. This is not sufficient for a quantitative assessment of the comparison of the potential energy barrier to the value of thermal energy, kT. The latter requires knowledge of the interaction energy between particles of a particular size. Within the first approximation, we can multiply $\Delta\sigma_f(h)$ and $\Pi(h)$ by some effective area, S_{eff}, characterizing the zone of particle contact. Alternatively, one can utilize the main concept of DLVO theory, which defines the necessary condition for coagulation as the disappearance of potential energy barrier. Figure 4.7 shows the $\Delta\sigma_f(h)$ curves obtained by varying the electrolyte concentration, n_0 (number of ions per unit volume, or c, mol/dm³), which is the principal factor determining the shape of these curves. At very high electrolyte concentrations, the electrical double layer is compressed, and consequently, the electrostatic repulsion is completely

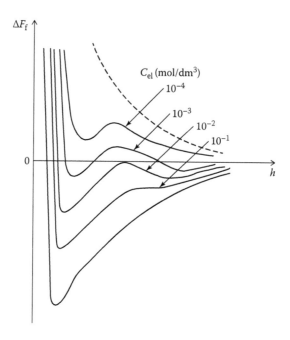

FIGURE 4.7   The curves $\Delta\sigma_f(h) = \Delta F_f(h)$ calculated for various concentrations of the electrolyte.

suppressed, leaving only the molecular attraction described by the hyperbolic term and the primary minimum. At very low n_0, electrostatic repulsion can play a predominant role at substantial distances, yielding a potential barrier and a secondary minimum. At some critical value of concentration, n_c, the positive potential barrier can disappear. At this point, the function $\Delta\sigma_f(h)$ and its derivative $\Pi(h)$ have the same value, which allows one to obtain the value of the *critical coagulation concentration* (c.c.c.), n_c:

$$n_c = \frac{k_1(\varepsilon\varepsilon_0)^2(kT)^5\gamma^4}{A * z^6 e^6}$$

This expression is the principal result of DLVO theory. The steep z^{-6} dependence of the critical coagulation concentration has been observed experimentally for strongly charged surfaces and is commonly known as the Schulze–Hardy rule. In the case of weakly charged surfaces, a less steep dependence of the critical coagulation concentration on the valence (z^{-2}) is observed. In fact, it was the empirical dependence of the c.c.c. on the valence established in the course of experimental studies on coagulation that inspired the development of DLVO theory.

The typical $\Delta\sigma_f(h)$ isotherms (force–distance curves) calculated using the DLVO theory have been experimentally verified in a large number of studies using specialized experimental techniques. In contrast to the measurement of the contact forces discussed in Sections 1.2 and 1.3, the determination of the long-range DLVO forces requires simultaneous measurements of *both* the forces and the distances. The latter is a rather complex experimental task. Modern instrumentation for such studies has been developed and improved over the several past decades [10–13]. The surface force apparatus developed by Israelashvili [10] for measuring the interaction forces between crossed cylindrical surfaces of mica utilized an interferometric technique for measuring the distance. Another technique that is commonly used nowadays is atomic force microscopy (AFM). This technique involves the scanning of the surface with a cantilever tip of diameter of ~20 Å or measuring the interaction force between the cantilever and the surface while keeping the distance between the surface and the cantilever tip constant. Another technique originally developed at Sandia National Laboratory is interfacial force microscopy (IFM). It functions similar to AFM

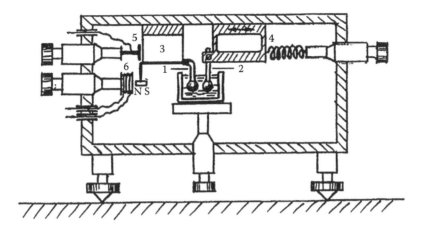

FIGURE 4.8 Schematic illustration of the device for simultaneous measurement of ultralow forces and distances utilizing a capacitor-based ultra-microdynamometer: 1, 2, sample holders; 3, double-console dynamometer spring; 4, double-console spring of a manipulator; 5, capacitor with air-filled gap between the plates; 6, electromagnetic loading system. (From Shchukin, E.D. et al., *J. Disp Sci. Technol.* 24, 377, 2003.)

but has a force sensor of zero compliance. In IFM, the induction system is used to control the distance between the cantilever tip and the surface. In the Russian Institute of Physical Chemistry, the author and his colleagues developed a capacitor-based sensor for measuring the distance between two particles [30]. This device is shown in Figure 4.8.

While describing the basic mechanisms of the coagulation and stabilization of aqueous sols in the presence of electrolytes, DLVO theory has limitations when working with dilute sols of hydrophobic particles. Such sols are weakly stabilized and readily undergo coagulation. DLVO theory is also not applicable to the description of the stability of concentrated colloidal systems that remain stable at high electrolyte concentrations and freezing, that is, *strongly stabilized systems*. In this sense, the electrostatic component of the disjoining pressure can be referred to, following Rehbinder's terminology, as the "weak factor of the colloid stability." This factor has a peculiar dual nature: while the origin of the potential barrier is *thermodynamic*, the ability of particles to overcome it in the course of their thermal motion is a phenomenon of *kinetic* nature.

In addition to the ion-electrostatic factor of colloid stability, which we have discussed, there are also other factors leading to the appearance of a positive disjoining pressure, $\Pi(h) > 0$, that prevents particles from approaching each other under the action of the attractive (negative) component of the disjoining pressure, $\Pi_{mol}^d(h) < 0$. The most essential factors are as follows (Figure 4.9):

1. The stretch of a thin soap film requires that more surfactant molecules enter the surface from the bulk, and consequently, the bulk of the solution become deficient in surfactant. At the same time, with respect to the new equilibrium state, both of these factors lead to a decrease in the surface surfactant concentration, as compared to the initial one, that is, they result in an increase in the surface tension (in Section 1.1, we did not discuss the possibility of this effect). The increase in surface tension as film stretching increases is known as the *characteristic of effective elasticity, the Gibbs effect*. The latter is a purely thermodynamic phenomenon that results in a resistance to film thinning.

2. As the soap films stretch and become thinner, there is also a kinetic phenomenon, referred to as the Marangoni–Gibbs effect. When drainage of liquid from the film into the channels bordering the film takes place under the action of negative capillary pressure (without film stretching), the surfactant concentration in the film decreases, resulting in an increase in the surface tension.

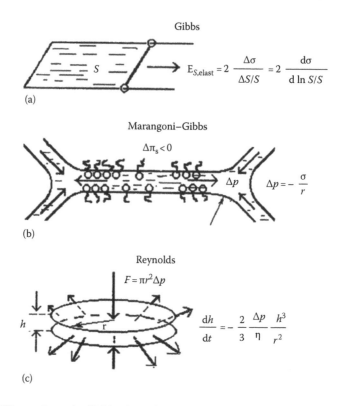

FIGURE 4.9 (a) Thermodynamic, (b) kinetic, and (c) hydrodynamic factors of colloid stability.

3. Hydrodynamic factor: According to Reynolds, at a given compression force, the rate of fluid drainage from the gap between round flat–parallel plates drops as the cube of gap thickness, h. Such a steep drop in the fluid drainage rate makes drainage slow and, when the gaps are very small, prevents a complete approach between the surfaces and limits film thinning.

These three phenomena, which prevent a decrease in film thickness and retard the drainage of fluid from the gap between particles, may also be classified as weak factors of colloid stability. In contrast to these, Rehbinder introduced the concept of a *factor of strong stabilization* of disperse systems. This factor ensures colloid stability in sols and emulsions against coagulation and coalescence at high concentrations of electrolytes, in concentrated disperse systems, under substantial changes in temperature or due to the action of mechanical forces. This factor, referred to as *the structure-mechanical barrier*, arises from a combination of the structural–mechanical (rheological) properties of the interfacial adsorption layer, the ability of such a layer to resist deformation and destruction, and the extreme lyophilization of the outer part of the adsorption layer facing the dispersion medium. This is schematically illustrated in Figure 4.10 and is discussed in detail further in this chapter.

4.2   REHBINDER'S LYOPHILIC STRUCTURAL–MECHANICAL BARRIER AS A FACTOR OF STRONG COLLOID STABILITY

The original teachings of Rehbinder and his school on colloid stability were based on some qualitative concepts that were known with regard to protective colloids. As one of the factors of colloid stability, Rehbinder's teachings included the concept of the so-called steric stabilization, which was fully developed at a later time. The principal achievements and contribution of Rehbinder and

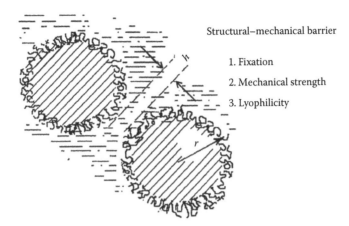

Structural–mechanical barrier

1. Fixation

2. Mechanical strength

3. Lyophilicity

FIGURE 4.10 Schematic illustration of Rehbinder's structural–mechanical barrier as a strong factor of colloid stability.

his coworkers are the development of experimental methods for the investigation of the structural–mechanical barrier, that is, the methods that allow one to study the strength of the adsorption layers and understand their critical role in controlling the coalescence of emulsion droplets [1,31–38]. It has been shown that strong colloid stability is achieved as a result of the high lyophilicity of the adsorption layer facing the dispersion medium. The importance of these teachings for the improvement of various industrial processes and a deep understanding of the stability of living cells and live tissues should not be underestimated.

The concept of the strong stabilization of colloidal systems may be illustrated by a simple experiment—adding a teaspoon of table salt into a glass of milk. If we address colloid stability strictly from the standpoint of DLVO theory, the system should be stable at 10^{-3}–10^{-4} M salinity and should coagulate at ~10^{-1} M of salt. A teaspoon contains 5–7 g of salt, which would produce an electrolyte concentration well exceeding the critical coagulation concentration. Nevertheless, coagulation does not occur—there is no deposit forming on the glass walls. As is also well known, milk is stable at both high and low temperatures: it can be both boiled and frozen. At the same time, squeezing a little bit of lemon juice into the same glass of milk results in immediate coagulation and a drastic loss of colloid stability.

Rehbinder was the first one to formulate the principals of the thermodynamic and "nonthermodynamic" factors of colloid stability. The thermodynamic factors are referred to as the factors of the stability of thin films that can be described in terms of the thermodynamics of reversible equilibrium processes. Such are the factors described by DLVO theory and Gibbs elasticity. The *nonthermodynamic* factors include the kinetic and hydrodynamic factors described in the previous section (e.g., Marangoni effect and Reynolds effect), but also, and even more so, the mechanical resistance of films to rupture and displacement. There are numerous examples of systems in which the factors of both kinds determine stability. Those belonging to the first group include not too thin electrolyte-stabilized wetting films and foam films, as well as dilute sols. Examples of common systems representing the second group include milk, natural latex dispersions, the cell walls in living tissue, and crude petroleum.

Appealing to concepts regarding thermodynamic and *nonthermodynamic* stabilization factors, we can state that strong stabilization against both high electrolyte concentration and high concentration of the disperse phase is due to the mechanical resistance of film to rupture and resistance to a displacement from the gap between droplets, bubbles, and particles. In the case of solid particles, this can be achieved by the firm attachment of the adsorption layer to the surface (i.e., by chemisorption), whereas in the case of emulsions (mobile fluid–fluid interface), it is necessary for the interface itself to have sufficient mechanical stability. The latter include high (nonlinear) viscosity,

enhanced elasticity, and *strength*, which is the main characteristic of the stabilizing layer responsible for strong stabilization. Nevertheless, this is not it. Mechanical strength provides stability
against the coalescence caused by the action of external forces. Resistance to coagulation requires
that the outer part of the stabilizing layer is *lyophilic*, that is, must have a nature close to that of a
dispersion medium. Consequently, one can talk about a *lyophilic structural–mechanical barrier.*
Within the framework of the concepts dealing with *protective colloids*, such a complex approach
can be regarded as quantitative development of these concepts in the light of the physicochemical
mechanics of disperse systems.

There are a number of experimental methods that can be used to evaluate these two factors of
a lyophilic structural–mechanical barrier separately. We will briefly review these methods in the
succeeding text and then focus on specific experimental data that have been collected by employing
these methods.

1. A torsion pendulum device was developed by the scientists at the Department of Colloid
   Chemistry of Moscow State University, that is, by Izmaylova et al. [35–37]. This device,
   shown schematically in Figure 4.11, allows one to evaluate the mechanical characteristic of thin-film behavior at both the liquid–air interface and the liquid–liquid interfaces.
   Nowadays, similar studies can be conducted with commercial high-sensitivity shear rhe-
   ometers using a special bicone tool.

   Figures 4.12 and 4.13 show some results obtained with the torsion pendulum instrument
   by Izmaylova et al. [35–37]. These figures show the typical rheological and deformation
   curves of the 2D interfacial layers formed at the interface between an aqueous gelatin solution and benzene. The deformation curves show the force as a function of time at a constant
   rate of deformation, and the rheological curves show the steady-state rate of the deformation
   as a function of the shear stress. These experimental data allow one to obtain direct quantitative characteristics of the elasticity in the slow and fast regions and, most importantly, the
   critical stress values that correspond to strength (by extrapolation to the *x*-axis, Figure 4.13).

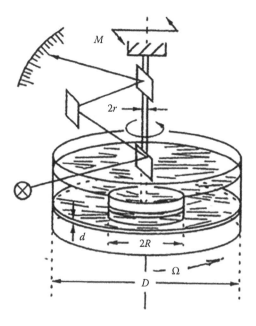

FIGURE 4.11 Schematic illustration of a torsion pendulum device for studying the rheological properties of the interfacial adsorption layer formed at an interface between polar and nonpolar liquids. (From
Izmaylova, V.N. et al., *Doklady AN SSSR*, 206, 1150, 1972; Izmaylova, V.N. et al., *Kolloidnyi Zh.*, 35, 860,
1973; Izmaylova, V.N. et al., *Surface Phenomena in Protein Systems*, Khimiya, Moscow, Russia, 1988.)

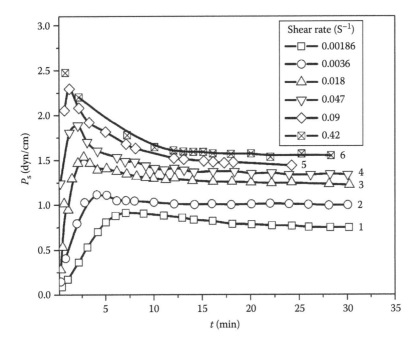

FIGURE 4.12 Shear stress, P_s (dyn/cm), as a function of time, t (min), of the interfacial layer at the interface between an aqueous solution of gelatin (0.5 g/100 mL) and benzene at various strain rates (s^{-1}): 0.00186 (1), 0.0036 (2), 0.018 (3), 0.047 (4), 0.09 (5), 0.42 (6). (Redrawn from Izmaylova, V.N. et al., *Kolloidnyi Zh.*, 35, 860, 1973.)

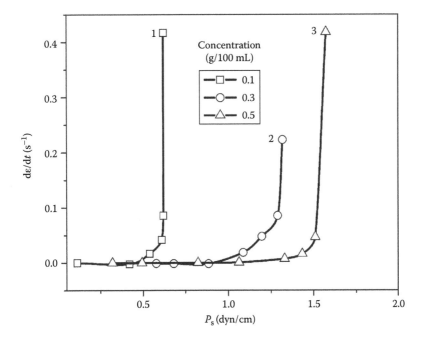

FIGURE 4.13 Steady-state deformation rates, $d\varepsilon/dt$ (s^{-1}), as a function of the shear stress, P_s (dyn/cm), for the interfacial adsorption layers at the interface between benzene and aqueous solutions of gelatin for three different concentrations: 0.1 g/100 mL (1), 0.3 g/100 mL (2), 0.5 g/100 mL (3). (Redrawn from Izmaylova, V.N. et al., *Kolloidnyi Zh.*, 35, 860, 1973.)

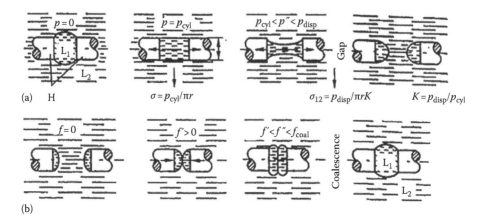

FIGURE 4.14 Rupture of a droplet of liquid L_1 placed between cylindrical holders with a diameter $2r$ immersed into a dispersion medium L_2 resulting from the applied extension stress, p, (a). Coalescence of two newly formed droplets under the action of the compression force, f, (b). The tension strength, p_{cyl}, at which the droplet assumes a cylindrical shape yields the value of the interfacial tension, $\sigma_{12} = p_{cyl}/\pi r$. With an increase in the value of p_{cyl}, the rupture force p_{disp} also increases. This rupture force characterizes the strength of the asymmetric adsorption film between phases L_1 and L_2. The force necessary to cause coalescence, f_{coal}, is determined by the strength of a symmetric film between the droplets of L_1.

Like any other experimental method, the torsion pendulum method has its advantages and disadvantages. While the method has the advantage of making it possible to carry out studies under *pure* stress conditions, it has a limitation due to the nonuniformity of the stress in the gap between the pendulum and the cuvette.

2. The Langmuir trough allows one to study film behavior under conditions of one-sided compression or one-sided tension. In both physical-chemical mechanics and material science, one needs to compare different stressed states. Tens of dynes per cm in compression or tension applied to the adsorption layer determine the critical stress, that is, the conditions corresponding to the mechanical collapse of the adsorption layer in one-sided compression.

3. A neat direct method for studying droplet rupture and coalescence, allowing one actual visual observation of individual droplets, was developed by Parfenova et al. [38,39]. A very small droplet (about 0.3 mm in diameter) is immersed into another liquid phase and stretched under controlled conditions (see Figure 4.14). The interfacial tension and rupture force are measured at the point when the droplet assumes cylindrical shape due to deformation. This corresponds to uniaxial tension resulting from asymmetric uniaxial stretching of the film (membrane) at the interface between the polar and nonpolar phases (which can represent both the dispersed phase as well as dispersion medium). In subsequent compression of two half droplets, one measures the critical force causing coalescence, f_{coal}, that is, the rupture of a double-sided emulsion film. These results correspond to a symmetrical double-sided axisymmetric stretch of the membrane.

The study of droplet rupture and coalescence by direct visual observation has been utilized in numerous essential studies [39–43]. Of principal importance are the experimental studies by Amelina et al. on the analysis of colloid stability in artificial blood substitutes [40–43]. These studies involved the use of various nonpolar phases, including perfluorinated systems, such as perfluorodecalin (PFD), perfluorotributylamine (PFTBA), perfluoromethylcyclohexylpiperidine (PFMCHP), and "conventional" hydrocarbons, such as heptane. Stabilizing agents included Pluronic surfactants (ethylene oxide (EO)/propylene oxide (PO) block copolymers), as well-fluorinated surfactants, such as perfluorodiisononylene with 20 mol of EO (φ-PEG). Tables 4.1 and 4.2 show some very characteristic results.

TABLE 4.1

Stability of Droplets Formed by Various Nonpolar Liquids in Aqueous Solutions of Pluronic F-68 (5×10^{-9} mol/dm³)

Media	Heptane	PFD	PFMCPH	PFTBA
f_{coal} (dyn)	<0.01	2.5	9	>10
$\Delta\sigma$, erg/cm²	15	27	12	10

Note: The critical coalescence force, f_{coal} (in dynes), in comparison with the interfacial tension lowering, $\Delta\sigma$, erg/cm² at the corresponding interfaces.

TABLE 4.2

Stability of Droplets Formed by Various Nonpolar Liquids in Aqueous Solutions of Fluorinated Surfactant Perfluorodiisononylene-Polyethylene Glycol, (ϕ-PEG) (5×10^{-6} wt %)

Media	Heptane	PFD
f_{coal} (dyn)	1.2	0.1
$\Delta\sigma$, erg/cm²	4	14

Note: The critical coalescence force, f_{coal} (in dynes), in comparison with the interfacial tension lowering, $\Delta\sigma$, erg/cm², at the corresponding interfaces.

These tables indicate that in the presence of Pluronic surfactant the strength of the adsorption layer at the interface between fluorinated phases and water was two orders of magnitude higher than the strength of the layer formed by the same surfactant at the water–heptane interface. In the case of an adsorption layer formed with the fluorinated surfactant, the opposite effect was observed: the strength of this adsorption layer at the water/heptane interface was an order of magnitude higher than that of an adsorption layer formed at the fluorinated hydrocarbon/water interface.

When discussing the mechanical properties of interfacial adsorption layers, one should clearly identify the role of the so-called steric factor among other factors responsible for colloid stability. The concept of a "steric factor" was introduced at a much later time than Rehbinder's concept of the lyophilic structural–mechanical barrier. The steric factor predominantly addresses the contribution of the configuration flexibility of the loops and tails of the lyophilic portions of the adsorbed macromolecules. These loops and tails penetrate the dispersion medium as flexible "tentacles." This *steric factor* has an osmotic origin and represents only the entropic contribution to the elastic strength. The magnitude of this contribution is rather small, and hence the *steric factor* alone cannot be responsible for strong colloid stability. This is illustrated by the data shown in Tables 4.1 and 4.2, which can't be explained in terms of the *steric factor* concept alone. It is especially evident from a comparison of the Pluronic adsorption layer strength at the interface with nonpolar PFD and more polar PFTBA. Most likely, due to an unfavorable interaction with the dispersed phase, the "tentacles" are forced into the adsorption layer and participate in the formation of a dense mechanically strong layer. In the general case, within such a layer, both polar and nonpolar groups interact with each other and are responsible for the strength, which is a *nonthermodynamic* characteristic.

The method described represents a rather attractive tool. A number of other known experimental approaches are as follows:

4. Methods based on the analysis of the interaction between a bubble or a drop with a flat surface.
5. Methods based on the analysis of the behavior of foam columns, foam caps, and especially of the flow of foams.
6. Methods based on studies of individual foam and emulsion films, for example, using a Mysels–Sheludko cell.
7. Various optical methods focusing on the study of the surface layer. These include ellipsometry or the observation of capillary waves. Usually, these methods are restricted to linear characteristics, such as constant viscosity (serving as a dissipation factor) or surface elasticity.

One *does* indeed need so many different experimental approaches because each of these tools provides a new equation, which allows one to estimate new parameters. The more independent equations are available, the better. Some of these methods allow one to conduct experiments not only with liquid phases but also with solid particles, and hence new parameters specific to the interactions with solid phases become available. Here, it is worth mentioning an eighth technique that focuses on the analysis of contact interactions between solid particles:

8. The measurement of contact forces between plastic particulates immersed in different media [31,44]. In this method, a particle is compressed against another particle with a given force, f, over a given period of time and then separated, as schematically shown in Figure 4.15. The magnitude of the measured contact strength, p_1, at a given compression, f, serves as an indication of whether or not the adsorption layer has been ruptured.

The results of the cohesive force measurements between two silver chloride crystals compressed against each other and immersed in different media are shown in Figure 4.16. These are histograms reflecting the probability of getting a particular strength of contact, p_1,

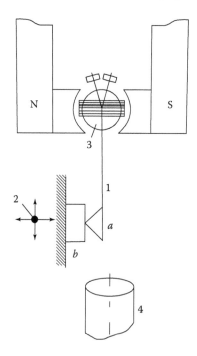

FIGURE 4.15 A schematic diagram of the device used to measure cohesive forces in a contact: particles (*a, b*) are mounted on the arm attached to the galvanometer core (1) and on the manipulator (2). Force is determined by running the current through the galvanometer frame (3), and the contact formation and rupture can be observed visually using a microscope (4).

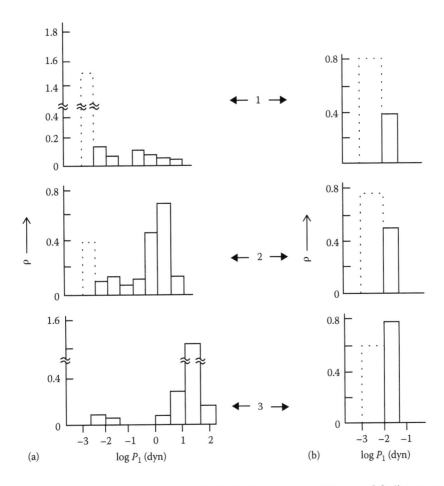

(a)

(b)

$\log P_1$ (dyn)

$\log P_1$ (dyn)

FIGURE 4.16 Histogram of the contact strength distribution: the probability, ρ, of finding contacts with strength p_1 upon compressing two silver chloride crystals with a force f_1: (a) untreated crystals with a virgin AgCl surface; (b) crystals coated with a monolayer of octadecylamine at three levels of compression force, $f_1 = 1$ (1), 10 (2), 50 dyn (3). (Redrawn from Shchukin, E.D. and Amelina, E.A., *J. Dispers. Sci. Technol.,* 24, 377, 2003.)

between crystals. As one can see, there are either very weak coagulation contacts or very strong *phase contacts*. Strong contacts emerge when a particle attaches to another particle as a result of plastic deformation in the zone of contact at sufficiently high contact stresses. The higher the applied force, f_1, the greater the fraction of these strong (phase) contacts reflecting the situation of local coalescence, that is, the crystals become fused together. The left column (a) in Figure 4.16 shows the results of measurements conducted in air with untreated crystals, while the right column (b) shows the results of measurements between crystals coated with a complete monolayer of octadecylamine. In the latter case, even the maximum compression applied was unable to rupture the adsorption layer: only weak contacts were observed. At the microscopic level, this means that the near-surface dislocations in the zone of contact are blocked by the adsorption monolayer of octadecylamine.

Since it is difficult for one to assess the absolute strength values, in reality, one can only talk about the *probability* of rupturing the adsorption layer. This probability varies significantly in cases involving nonspecific (physical) and specific adsorption (chemisorption). Both of these cases are presented in Figure 4.17, which illustrates the fraction of the phase contacts formed in solutions of alcohols in heptane (physical adsorption) and in the

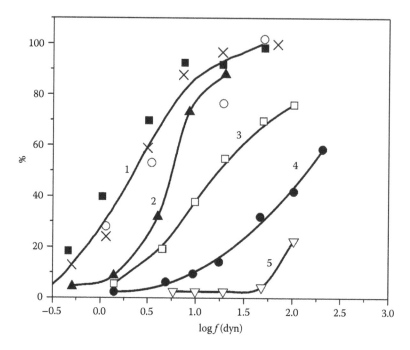

FIGURE 4.17 The fraction of the phase contacts ($p_1 > 0.1$ dyn) formed between the silver chloride particles as a function of the compression force f_1 in air, water, and heptane (1), in heptane solutions of decyl alcohol (2), in heptane solutions of cetyl alcohol (3), and in a heptane solution of octadecylamine (4), and in water with traces of octadecylamine present (5). (Redrawn from Shchukin, E.D. and Amelina, E.A., *J. Dispers. Sci. Technol.*, 24, 377, 2003.)

    presence of octadecylamine (chemisorption). This figure clearly shows the *protective* role of the adsorption layer of octadecylamine. The formation of the phase contacts is significantly influenced by the degree of the completeness of the adsorption layer. Model studies conducted by this method, in addition to being of an academic value, are also significant for understanding such areas of technology as friction and wear and specifically the role of greases with chemisorbing additives.

9. Finally, it is worth mentioning another method that can be used to study the structural–rheological (mechanical) properties of film. Sclera is the opaque fibrous outer layer of the eye, which contains collagen and elastic fiber. Changes in the sclera properties are responsible for a very strong nearsightedness that develops with ageing. In terms of material science, we talk here about the creep of the sclera under the action of internal eye pressure. The creep of the sclera results in an increase in the thickness of the eye, causing a distortion in focus and resulting in myopia. Every extra millimeter results in three extra diopters in the lens strength. Over many years, the action of a pressure of 20 mm of Hg on a relatively thin film that makes up the sclera constitutes a fairly strong mechanical action. It is, indeed, somewhat surprising, in this case, that such strong nearsightedness affects only 10% of people.

There are three basic questions that need to be addressed here: first, what kind of rheological behavior does sclera exhibit and what types of measurements and stress regimes are needed to study it? The second question is related to which factors (substances present in the media or enzymes) catalyze sclera creep? Finally, the third question is how to find a solution that inhibits sclera creep?

    The early seminal studies were carried out by Bessonov et al. [45,46] using the instrument shown in Figure 4.18. His results from the study of bovine sclera creep are shown in Figure 4.19.

    The onset of all three curves corresponds to the tests in physiological solution without any additives. All tests were conducted at the same temperature using a specimen cut out of the same sclera.

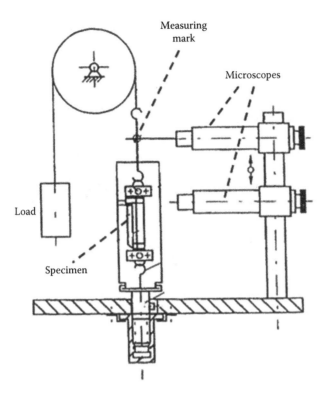

FIGURE 4.18   Bessonov's instrument for studying sclera creep. (From Shchukin, E.D. et al., *Kolloidnyi Zh.*, 56, 463, 1994; Shchukin, E.D., et al., *Vestnik Ophtalmolo.*, 3, 3, 1997.)

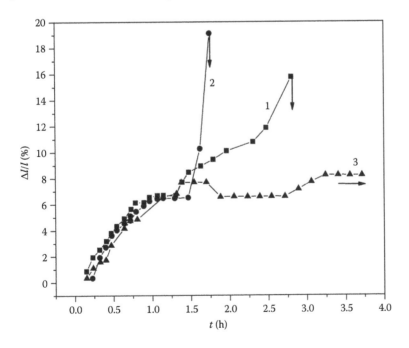

FIGURE 4.19   Creep curves for 0.6 × 2 × 20 mm samples of bovine sclera, showing the relative elongation ($\Delta l/l$, %) as a function of time (t, h) at a constant load of 30 g in physiological solution (1), and after introducing (at 1 h) collagenase (2) and elastase (3). (From Shchukin, E.D. et al., *Kolloidnyi Zh.*, 56, 463, 1994; Shchukin, E.D., et al., *Vestnik Ophtalmolo.*, 3, 3, 1997.)

In the absence of enzyme additives, a creep leading to an eventual mechanical collapse in about 3 h was observed. The addition of collagenase catalyzed the creep and caused failure to occur much sooner. The same figure shows the impact of elastase, which drastically inhibits the creep. It is possible that the enzymes differently affect the proteoglycans and glycoproteins that act as adhesives for the collagens, making a complex network of sclera.

When discussing the factor of the strong stabilization of a thin film, we have primarily focused on a nonthermodynamic factor, that is, the mechanical resistance to media displacement and the mechanical stability of film against coalescence. Now, we will address the subject of colloid stability against *coagulation*, which is due to the competition between particle adhesion and particle involvement in thermal motion. If the latter prevails, the system can be characterized as *lyophilic* or is said to be lyophilized.

Experimentally, it is advantageous to focus on the adsorption of surfactant molecules on nondeformable surfaces, that is, to work with solid particles. The relevant experiments were conducted over a number of years at the Chemistry Department of Moscow State University using the methodology described earlier (see Figure 4.15).

According to Gibbs, the physical discontinuity surface is a layer of finite thickness in which major properties differ from those in bulk phases (Figure 1.2). If two solid surfaces are brought close together (in a given gas or liquid media), their discontinuity surfaces start to overlap, and Derjaguin' disjoining pressure, $\Pi(h)$, develops as a function of the gap thickness h between the surfaces (Figure 1.24). This pressure can be both positive and negative: positive disjoining pressure corresponds to repulsion, while negative disjoining pressure corresponds to attraction. The first integration of disjoining pressure over the distance h yields the free energy of interaction, $\Delta\sigma_f(h)$, in erg/cm^2 in a flat–parallel gap. Since it is difficult to maintain the necessary small gap between the surfaces, direct experimental measurements are possible only in a limited number of cases, for example, for thin wetting films [5,13,16,47]. However, one can utilize the Bradley–Derjaguin theorem [4,5]. It is essential that the second integration of the disjoining pressure yields the *cohesive force*, $p_1(h)$, between two spherical particles (or any other second order surfaces). This force can be compared to the free energy of interaction in a plane–parallel gap given by the first integral of the disjoining pressure (Figure 1.26). The force and free energy of interaction are proportional to each other and differ by a constant factor that includes the curvature radius, $p_1(h) = \pi r \Delta\sigma_f(h)$. This means that, by measuring the force $p_1(h)$ of the interaction between the particles, one obtains the important characteristic describing the interaction between a given phase and a medium, that is, one gets an estimate of the lyophilicity or lyophobicity of the system, $\Delta\sigma_f(h)$.

Surface forces have been the subject of numerous studies by Derjaguin, Churaev, Israelachvili, Ninham, and many other researchers [4,5,10–13,21–23,48,49]. In most cases, the focus has been on long-range interactions. The latter is an important, but not the only, aspect of the problem. In this book, we focus on the interactions occurring directly in the contact between the particles. Derjaguin's theorem is also valid for a direct equilibrium contact between the particles, that is, when $h = h_0$. When the geometry is known, it is sufficient to conduct a *one-time* measurement of the force, $p_1(h) = p_1$, in order to determine the free energy of interaction (cohesion), $\Delta\sigma_f(h_0) = \Delta\sigma_f = F$. If one is certain that the media has been completely displaced from the gap, half of this value, $\sigma_{12} \sim \frac{1}{2}|F|$, gives us the interfacial energy at the solid–liquid interface, which is a fundamental characteristic of the lyophilicity and lyophobicity, similar to the free energy of interaction. In further discussions, we will operate, for simplicity, with absolute values of p_1 and $\Delta\sigma_f = F$. However, one must remember that these variables characterize the cohesion in the contact and are both negative.

The experimental method involves the compression of the particles against each other with a force f, with the subsequent rupture of the contact. The precision of the strength determination in this experiment is 10^{-3} dyn; however, higher precision can be achieved. The main difficulty is associated with the influence of the meniscus formed in a liquid medium. This issue may be resolved by using a special L-shaped holder (Figure 4.20 shows a schematic illustration of how two particles are brought together in a contact). A more serious issue is associated with the integration of the thermodynamic equations: the number of variables exceeds the number of available equations. In this case, the free energy of

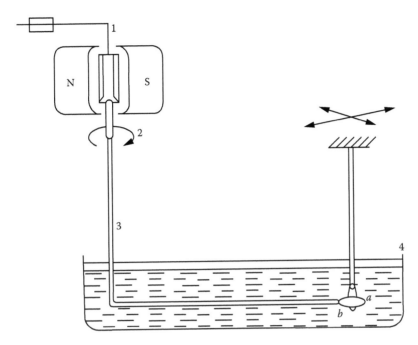

FIGURE 4.20 The scheme of a device for the measurement of the cohesive forces.

interaction $\Delta\sigma_f$, the interfacial energy σ_{12}, and the free energy of the film in the gap, σ_f, are all linked via the same equation, that is, $\sigma_f = 2\sigma_{12} + \Delta\sigma_f$. This means that at least one more independent equation is needed in order to establish pairwise relationships. To achieve this, we can utilize the change in the interfacial energy, $\Delta\sigma_{12}$, as the work of wetting, using the experimental values of the contact angles, or conducting the adsorption measurements and subsequently integrating the Gibbs equation.

Table 4.3 contains the data obtained by Amelina et al. [30,50–53] corresponding to the limiting cases of the interactions. For the cases of hydrophilic surfaces in air and in nonpolar media, as well as hydrophobic surfaces in polar media, the free energy of interaction is on the order of tens of erg/cm². This corresponds to the case of *lyophobicity*. Conversely, for a hydrophilic surface in water or a hydrophobic surface in hydrocarbon, the free energy of interaction does not exceed 0.01 erg/cm², which corresponds to the case of *lyophilicity*. Consequently, this gives one a precise quantitative definition of the lyophilicity and lyophobicity. However, one must note here that what has been described is a characteristic of the interactions in a given medium and does not represent the *true* lyophilicity/lyophobicity of the disperse system, because the thermal motion (kT factor) has not been yet considered.

For hydrophobic particles, the data gap for the interface between water and heptane can be filled with the data for other liquids with a gradual change in polarity (these data are shown in Table 4.4).

TABLE 4.3

Free Energy of Interaction between Different Interfaces ($F/2 = p_1/2\pi r$, erg/cm²)

Surface	Air	Heptane	Water
	Medium		
Glass	~40	25	0.01
Methylated glass	22	0.01	40
Fluorinated glass	28	5	50

Note: The surface free energy of the methylated surface in air, $\sigma^d = \sigma_c = 22$ erg/cm².

TABLE 4.4

Free Energy of Interaction between Hydrophobic Surfaces Immersed in Various Media ($F_{SL}/2$, erg/cm²) and the Corresponding Work of Wetting ($\sigma_L \cos \theta$)

Medium	$F_{SL}/2$	$-\Delta \sigma_f/2 = (F_s - F_{SL})/2$	$\sigma_L \cos \theta$	$\sigma_L = 22 - \sigma_L \cos \theta$
Water	40.6	−18.8	−17	~39
Ethylene glycol	16.1	5.7	0.8	0.14
Ethanol	1.6	20.2	N/A	N/A
Propanol	0.1	21.7	20	<2
Heptane	0.01	21.8	N/A	N/A

Note: In air $F_s/2 \sim 22$ erg/m².

Figure 4.21 shows the values for aqueous solutions of various alcohols—from methanol to butanol—which represent a continuous spectrum of values of the free energy of interaction. Very similar results were observed in the case of an aqueous solution of a *classic* anionic surfactant—sodium dodecyl sulfate (Figure 4.22). In this case, one is dealing with a fully reversible true equilibrium system. The same behavior was also observed for a cationic surfactant, cetylpyridinium bromide (CPB) (Figure 4.23), and to some extent, in a solution of a nonionic surfactant, polyoxyethylated ether (Figure 4.24). In the latter case, however, the system is a nonequilibrium one: there is a dependence of the measured p_1 value on time.

It is also of interest to compare the results for the case when both particles are nonpolar to the case when one particle is partially lyophilized (hydrophilized) with acetylcellulose. These data are shown in Figure 4.25. The first isotherm displays low (but finite) values of attraction. The second

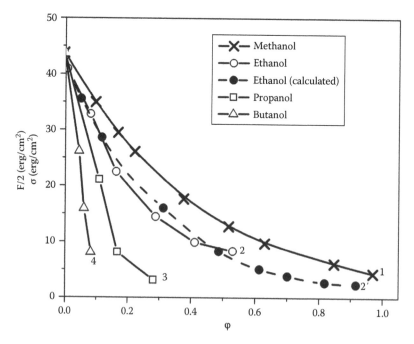

FIGURE 4.21   Absolute value of the free energy of interaction between methylated glass particles as a function of the volume fraction of various alcohols in water: methanol (1), ethanol (2), propanol (3), butanol (4); the interfacial tension isotherm (σ) for the interface of ethanol solutions in water at the interface with solid paraffin (2′). (Redrawn from Shchukin, E.D. and Amelina, E.A., *J. Dispers. Sci. Technol.*, 24, 377, 2003.)

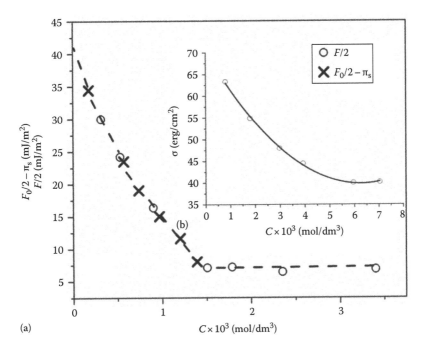

FIGURE 4.22 The energy of interaction, $F/2$, between methylated glass beads (1) and the value of $F_0/2 - \pi_s$ (2) in aqueous solutions of sodium dodecyl sulfate (a); the surface tension isotherm of aqueous sodium dodecyl sulfate solutions (b).

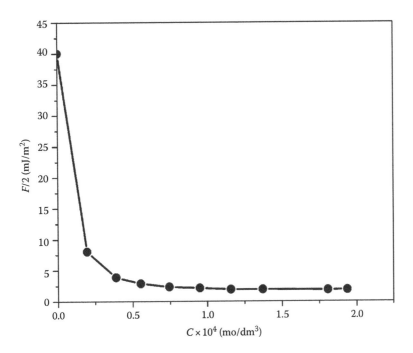

FIGURE 4.23 The equilibrium values of the free energy of interaction between two methylated glass spheres as a function of a nonionic surfactant concentration (c, 10^{-4} mol/L). The nonionic surfactant is oxyethylated ether, $C_{12}E_{20}$.

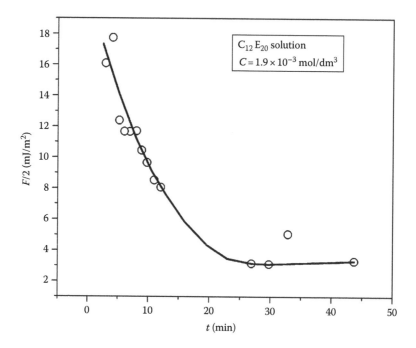

FIGURE 4.24 The free energy of interaction between methylated surfaces as a function of time for surfaces immersed into a 1.9×10^{-5} molar solution of oxyethylated ether, $C_{12}E_{20}$.

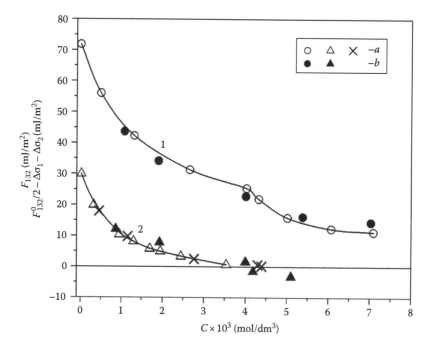

FIGURE 4.25 Free energy of interaction between two methylated surfaces (1) and between a methylated surface and an acetylated surface (2) obtained from the measurements of the cohesive force and cohesive energy F_{132} (a), and by the difference between F_{132}^0 in water and the change in the energies of wetting, $\Delta\sigma_1$ and $\Delta\sigma_1$, in solutions (b). The indices 1, 2, and 3 correspond to the methylated and acetylated surfaces and to the medium, respectively.

isotherm had originally drawn the close attention of Derjaguin, as it shows a change in the sign of the disjoining pressure, that is, a transition from attraction to repulsion.

As one can see, the spectrum of the free energy of interaction covers a very broad range of values—four orders of magnitude. For particles ranging in size from tens of nanometers to millimeters, the spectrum of cohesive forces covers nine orders of magnitude from 10^{-8} dyn to tens of dynes. The spectrum of the interaction energies also covers a very broad range, from 10^{-15} to 10^{-6} erg. This is of direct relevance to the subject of colloid stability, because the typical critical value of w is on the order of $10\,kT$, which falls in the middle of this range. Consequently, lyophilicity and lyophobicity can be regarded as properties of the system.

Most of the examples discussed are related to reversible (equilibrium) conditions. The behavior of the adsorption layer formed on hydrophobic methylated surfaces in aqueous surfactant solutions is shown in Figure 4.26a. Compression of the particles resulted in the retraction of the adsorption layer, which returned once the particles were separated. The situation is principally different in the case of adsorption from a nonpolar phase on polar particles, as is schematically shown in Figure 4.27b. The fixation of the adsorption layer by chemisorption, as in the case of the adsorption of amines on silicate glass, resulted in an adsorption layer that had its own strength, and a certain

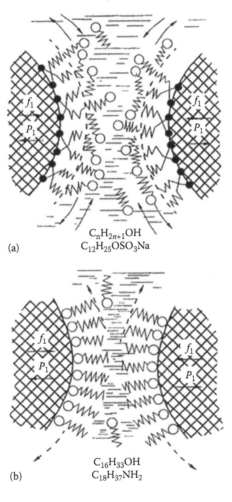

(a) $C_nH_{2n+1}OH$
           $C_{12}H_{25}OSO_3Na$

(b) $C_{16}H_{33}OH$
           $C_{18}H_{37}NH_2$

FIGURE 4.26 Schematic illustration of the reversible behavior of the adsorption layer under conditions of reversible adsorption from an aqueous medium on methylated surface (a), and under the condition of specific adsorption (chemisorption) from a hydrocarbon medium on the polar surface (b). In the latter case, there is a critical compressive force that one needs to apply in order to rupture the adsorption layer.

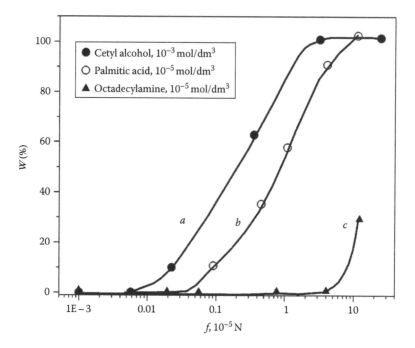

FIGURE 4.27 Probability of the adsorption layer rupture (W, %) between polar surfaces (glass beads) as a function of the compressive force (f, dyn) in heptane solutions of 10^{-3} mol/dm³ of cetyl alcohol (a), of 10^{-5} mol/dm³ of palmitic acid (b), and of 10^{-5} mol/dm³ of octadecylamine (c).

critical compressive force, f_c, was required to rupture the layer and displace it from the gap with a certain probability (Figures 4.26b and 4.27).

The experimental technique described utilizing solid surfaces in contact with various surfactant solutions in various electrolytes provides a means for addressing other important questions [52,53] (see Section 4.4 for more details).

Let us now assume that the system has reached thermodynamic equilibrium and the conditions of reversibility. The question of interest is whether the known thermodynamic relationships always adequately characterize such systems, especially when the size reaches molecular dimensions. This question can be addressed using molecular dynamic (MD) simulations of the contact between the particles. A series of pictures from the MD experiments originally conducted by Shchukin [53] and Yushchenko et al. [54] is shown in Figure 4.28. The figure shows the experiment with a convex particle in lyophilic system. The conditions selected are those corresponding to complete wetting: the work of adhesion equals the work of cohesion in a liquid phase. Consequently, the liquid phase must spontaneously penetrate into the gap, and the particle must separate from the substrate. Indeed, this is what was observed in approximately 60% of cases after the MD experiment had run a reasonably long time (Figure 4.28b through d). In the rest of cases, instead of separating off, the particle turned and strongly adhered to the substrate via the monolayer (Figure 4.28e through h). If the particle was flat, there was no spontaneous separation at all, despite the completely lyophilic conditions, which seems to be in contradiction with what was stated earlier. Here, it is not so important that the particle was small—it is important that the contact was flat. This situation is illustrated in Figure 4.29.

Let us now change the conditions of the experiment by applying a constant force to the particle. The results for a flat particle using Zhurkov coordinates, the log of the time until failure versus the applied force, were presented in the work of Shchukin et al. [50–54]. This dependence is linear, which is in line with a thermally activated process. Nevertheless, in a wetting liquid, the time it took to rupture the contact was longer than in a vacuum, which again was not consistent with the expectations.

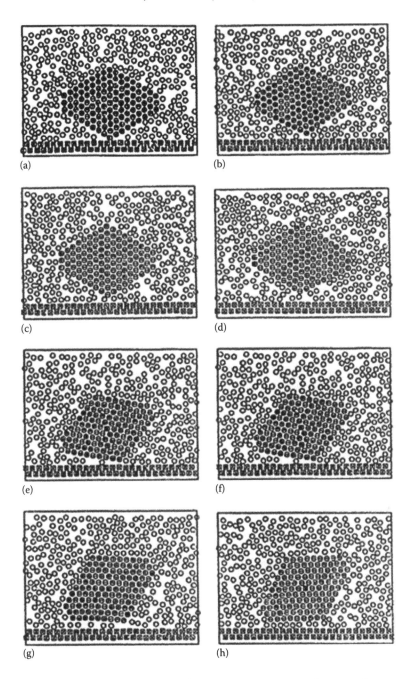

FIGURE 4.28 Molecular dynamics simulation of the spontaneous rupture of a contact between a *large* (convex) particle and a substrate in a lyophilic system: initial condition (a), the rupture of a contact (25, 30, 35 units of time, respectively) (b–d), the turning of the particle (10, 20, 30, and 40 units of time, respectively) (e–h). (Redrawn from Yushchenko, V.S. et al., *Colloids Surf.,* 110, 63, 1996.)

The rupture of a contact in the case when the particle moved with a constant velocity under the applied external force is illustrated in Figure 4.29. Here, we encounter gaps between the particle and the substrate that are on the order of molecular dimensions. In order to realize the work of wetting, the molecules of the medium must penetrate into the gap. This was possible only after the gap became comparable in size to the molecules. However, by this point in time, the attractive forces had

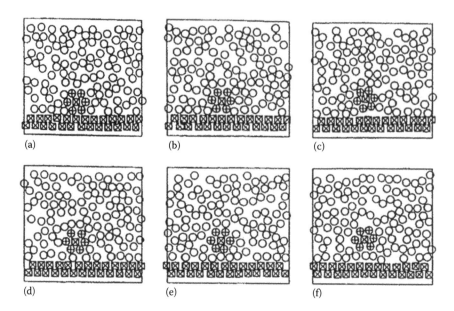

FIGURE 4.29 The separation of a *small* (flat) particle from a substrate in lyophilic system in the course of particle motion with a constant velocity of 0.005 units. The diagrams (a–f) correspond to the sequence of times 0, 0.15, 0.5, 0.8, 0.95, 1,125 units, respectively. These times correspond to the following situations: initial state (a), maximum of an external force (b), maximum of spent work (c), the first minimum of external force (d), the first minimum of work (e), second maximum of external force (f). (Redrawn from Yushchenko, V.S. et al., *Colloids Surf.,* 110, 63, 1996.)

already been overcome. The work in resisting molecular attraction is conducted by the external force up to the point when the contribution from the wetting starts to play a role. After the molecules of the media penetrate into the gap and form a monolayer, the situation repeats itself, that is, the second monolayer must overcome its own barrier. This leads to oscillations in the repulsive and attractive forces. For a *large* convex particle, such oscillations are shown in Figure 4.30. It is noteworthy that the period of such oscillations was about half the diameter of a molecule. This is a consequence of the gap geometry: the media enters like a 30° "wedge." It turns out that the repulsion (the positive component of the disjoining pressure) did not originate in the zone of direct contact, still free from the molecules of the medium, but rather in the area where the molecules of the medium were present.

It has been demonstrated that the factors determining the mechanisms of resistance of the interfacial layers to rupture and displacement from the gap, and the factors determining colloid stability (i.e., surface lyophilicity) could be considered both comprehensively and separately from each other. At the same time, one has to be cautious once molecular dimensions are being approached.

In the section that follows, we will address in some detail the stability of emulsions against coalescence and discuss the experimental results within the framework of the structural–mechanical barrier. We will devote special attention to systems with fluorinated interfaces. The most noteworthy result was that the most effective stabilizers for fluorinated hydrocarbons in aqueous solutions were nonfluorinated surfactants, while fluorinated surfactants were very effective for stabilizing emulsions of nonfluorinated hydrocarbons.

4.2.1 STABILITY OF FLUORINATED SYSTEMS: STRONG STABILIZATION BY THE STRUCTURAL–MECHANICAL BARRIER

The use of fluorinated hydrocarbons in various applications, including biomedical ones (e.g., blood substitutes), has inspired the need to study in detail interfacial phenomena at the interfaces between fluorinated organic liquid phases and aqueous surfactant solutions. It has been established that

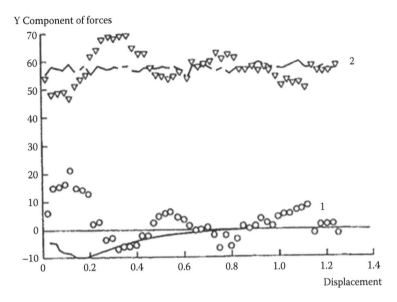

FIGURE 4.30 The forces acting on a *large* (convex) particle in the course of its separation from the substrate with a constant velocity of 0.005 units: the external force acted from the side of the surface (solid line) and was equal to the external force in a vacuum (1); the force acting from the side of a liquid on the *lower* surface of the particle (2). The resulting force acting on the particle from the side of a liquid (i.e., the disjoining pressure) can be seen from a comparison of these two curves. (Redrawn from Yushchenko, V.S. et al., *Colloids Surf.,* 110, 63, 1996.)

EO and PO block copolymers, nonionic surfactants with molecular weight of ~8000 amu, that is, Pluronics and Proxanols, are good emulsifiers for fluorocarbons. However, their effectiveness varies for different types of fluorinated organic liquids [55–58].

One would have expected that fluorinated surfactants, akin to the fluorinated organic liquids, would be better emulsifiers than conventional nonionic surfactants. However, even the preliminary initial studies with fluorosurfactants showed the opposite: these substances were a lot less effective stabilizers for fluorocarbon emulsions than conventional EO/PO surfactants. These findings have prompted an interest in investigating the various properties of the adsorption layers formed with conventional surfactants and with fluorinated surfactants at various interfaces between their aqueous solutions and nonpolar phases. The concept of Rehbinder's structural–mechanical barrier formed at interfaces has been essential in the interpretation of the experimental data. The current section contains a brief review of critical studies relating to fluorinated interfaces conducted at the Chemistry Department of Moscow State University and the Institute of Physical Chemistry of the Russian Academy of Science by Shchukin et al. [34,42–45]. In the described studies, block copolymers of ethylene and PO were used as nonionic surfactants, specifically Pluronic F-68 with a molecular weight of 8,700 amu, Proxanol 168 (MW = 8,000 amu), and Proxanol 268 (MW 13,000 amu). The perfluorodiisononylene-polyethylene glycol, C_9F_{17}–O–$(EO)_{20}$–C_9F_{17}, or φ-PEG, was used as a water-soluble fluorinated surfactant. Different fluorinated organic compounds with cyclic and linear backbones were used as liquid phases. In the present review, we provide mainly the data obtained for moderately water-soluble PFD ($C_{10}F_{18}$), less soluble PFMCHP (CF_3–(C_6F_{10})–(NC_5F_{10})), and practically insoluble PFTBA (($C_4F_9)_3N$).

The properties of these systems were studied with the experimental methods that we have already mentioned earlier in this chapter: (1) the rheological properties of the interfacial layers using the torsion pendulum device (Figure 4.11); (2) the investigation of the compression of two nonpolar droplets immersed in an aqueous surfactant solution and the measurement of the force, f_{coal}, necessary for their coalescence; and (3) the estimation of the free energy of interaction between the nonpolar groups of the interfacial adsorption layers and various nonpolar fluids by measuring the contact rupture force between two methylated (or fluorinated) smooth solid particles immersed into a fluid medium.

4.2.1.1 Rheological Properties of Interfacial Adsorption Layers in Fluorinated Systems

The rheological behavior of interfacial adsorption layers formed between the nonpolar phase (fluorinated or nonfluorinated) and the aqueous fluorinated or *regular* nonionic surfactant solution can be studied using the torsion pendulum instrument shown in Figure 4.11. The results of such studies for the adsorption layer of Pluronic F-68 formed at the interface between its aqueous solution and three different nonpolar phases are illustrated in Figure 4.31, which shows the shear stress, τ, as a function of the time of deformation, t. The shear stress was applied to the entire thickness of the adsorption layer, and the time, t, was proportional to the shear deformation at a constant angular velocity, $\Omega = 0.084$ rad/s.

The rheological behavior observed at the interface between an aqueous solution of Pluronic and heptane (i.e., the hydrocarbon surfactant/hydrocarbon liquid, or HS/HL system) is significantly different from the behavior at the interface of the same surfactant solution with PFD and PFTBA, that is, from the behavior of the hydrocarbon surfactant/fluorocarbon liquid (HS/FL) systems. In the former case (HS/HL), the development of the deformation is analogous to the viscous flow of the fluid, while in the latter case (HS/FL), nonlinear solid-like behavior is observed. In particular, after a period of elastic (reversible) deformation, the rheological curve showed a maximum, which is indicative of the yield stress, which is the evidence of the *mechanical strength* of the interfacial adsorption layer. Along with the principal differences between the HS/HL and HS/FL systems, there were also noticeable differences between the pairs of HS/FL systems. For PFTBA, the strength was higher, the free energy of interaction (F) was higher, but the work of adhesion (W) was lower than for PFD. Consequently, the nature of the nonpolar liquid phase plays an important role in determining the rheological properties of interfacial adsorption layers.

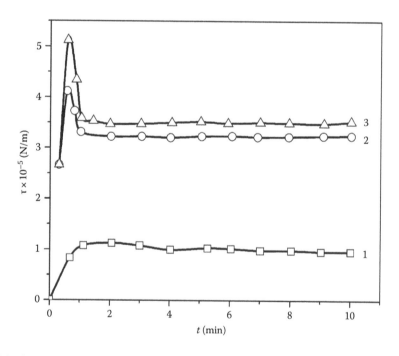

FIGURE 4.31 Shear stress, τ (10^{-5} N/m), as a function of the time of deformation, t (min), for the adsorption layer formed at the interface between Pluronic F-68 10^{-3} M aqueous solution and three nonpolar liquids: heptane (1), perfluorodecalin (2), and perfluorotributylamine (3), obtained for constant angular velocity, $\Omega = 0.084$ rad/s. (Redrawn from Shchukin, E.D. et al., *Colloids Surf.*, 176, 35, 2001.)

Stress/deformation curves similar to those shown in Figure 4.31 also were observed for the interfacial adsorption layers formed at the interfaces between PFD and various Proxanol solutions. All of these systems demonstrated a nonlinear rheological behavior with typical maxima in the $\tau = \tau(t)$ curves corresponding to the yield stress (i.e., the strength of the interfacial adsorption layer). The heights of these maxima were different for different surfactants, with the highest strength corresponding to the most developed hydrophilic portion of the molecules of the nonionic surfactants.

4.2.1.2 Study of the Rupture and Coalescence of Individual Droplets

These studies were carried out using the above described method of measuring the force necessary to cause the coalescence of two droplets of a liquid and subsequently the rupture of the resulting liquid column, as schematically shown in Figure 4.14. Two droplets of a nonpolar HL or FL fluid were placed on two base surfaces of cylindrical rods so that they formed hemispherical segments. The rods had a very smooth flat surface. For the measurements to be reproducible, the flat surfaces needed to be exactly perpendicular to the rod axis. The rods were of identical diameter, $2r = 0.5$–0.7 mm. Rods with mounted droplets were then immersed into an aqueous solution of a surfactant. The holders were made such that their movement resulted in the movement of the mounted specimen along the axis normal to the mounting rod surfaces. The fixation of the droplets was achieved by making the mounting rod surface hydrophobic. In the course of the measurement, two droplets were brought into contact and compressed against each other until they coalesced into a single segment. Then, a stretching force was applied in the opposite direction. This force resulted in a gradual elongation of the liquid column. The force, p_{cyl}, necessary to maintain the cylindrical shape of the liquid column was measured with high precision. This measurable force is directly related to the interfacial tension, $\sigma_{12} = p_{cyl}/\pi r$. The time during which the cylindrical shape of the fluid column could be maintained was varied from several seconds to several hours, which allowed one to study the kinetics of the adsorption. The stretching force was then increased, and the column was eventually separated into two separate droplets. The elementary act of dispersion (the rupture of the asymmetric monolayer) took place when the stretching force reached the value of p_{br}. It was observed that the liquid column always split into two identical spherical segments.

The droplets formed as a result of the rupture of the liquid column were then again brought into direct contact, became flattened, and eventually coalesced. In the absence of any surface active matter (i.e., in pure water), the droplets of both HL and FL coalesced spontaneously. In this case, the adsorption bilayer needs to be ruptured. In a surfactant solution, a particular compression force f_{coal} must be applied for the droplets to coalesce. Hydrostatic effects were avoided in these experiments because very small volumes of nonpolar phases were used—the ratio of the liquid column length to the base radius, l/r, was between 1.0 and 1.2.

A significant feature of the results obtained is the dependence of both the values of p_{br} and f_{coal} on the nature of the organic fluid used. Figure 4.32 shows the values of f_{coal} as a function of the concentration, C, of the Pluronic F-68 solutions for PFTBA, PFD, and heptane. In pure water, there was no initial resistance to coalescence, that is, $f_{coal} = 0$ in all these cases. The value of f_{coal} reached a maximum at $C \sim 10^{-7}$–10^{-6} mol/dm^3 and decreased at higher concentrations. These trends are in good agreement with general observations of emulsion stability. However, there is a significant difference in the magnitude of the maxima. The data become even more informative if the values of f_{coal} corresponding to the same low concentration of Pluronic F-68 are compared, as illustrated in Table 4.5. At a Pluronic F-68 concentration of 5×10^{-9} mol/dm^3, the resistance of the droplets to coalescence in the HS/FL was higher than in HS/HL system by up to two orders of magnitude. Among the HS/FL systems, the resistance to coalescence in PFTBA was higher than in PFD.

The observed behavior is related to the behavior of the Pluronic molecules in the adsorption layer and between the same molecules in the nonpolar phase. The perfect cohesion and mutual solubility of the hydrophobic tails of the surfactant molecules in organic liquid causes deep immersion of

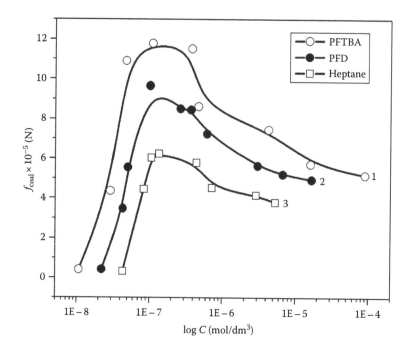

FIGURE 4.32 The compressive force needed to cause coalescence, f_{coal} (10^{-5} N) as a function of the concentration, C, of Pluronic F-68 for droplets of perfluorotributylamine (1), perfluorodecalin (2), and heptane (3). (Redrawn from Shchukin, E.D. et al., *Colloids Surf.*, 176, 35, 2001.)

these groups into the organic liquid without the formation of a dense adsorption layer. Conversely, the lower physical-chemical affinity between the surfactant and the organic liquid results in the larger displacement of the surfactants toward the aqueous phase in which the attraction between the hydrophobic portions of the Pluronic molecules is higher than in the organic phase. This in turn results in an increase in the density, viscosity, and strength of the interfacial adsorption layers.

The opposite ("mirrorlike") behavior can be expected in the case of fluorinated surfactants. Indeed, the results obtained for ϕ-PEG solutions indicate that the resistance to coalescence was higher for heptane (fluorocarbon surfactant/hydrocarbon liquid, FS/HL system) than for PFD (fluorocarbon surfactant/fluorocarbon liquid, FS/FL): f_{coal} is 1.2 dyn for heptane and only 0.1 dyn for PFD (Table 4.5).

Table 4.5 also shows the lowering of the interfacial tension at the interfaces between the aqueous and organic liquids, that is, the values of $\Delta\sigma = \sigma_0 - \sigma(C)$. The difference between these values for the different systems was rather small; however, a general trend is visible: the higher the resistance to coalescence, the less the interfacial tension lowering. This agrees with the concept that resistance

TABLE 4.5

Force f_{coal} Required to Cause Coalescence between Droplets of Various Nonpolar Liquids in a Solution of Pluronic F-68 (5×10^{-9} mol/dm³) and an Aqueous Solution of Perfluorodiisononylene-Polyethylene Glycol, ϕ-PEG (5×10^{-6} wt %)

	In a Pluronic F-68 Solution				In a ϕ-PEG Solution	
Nonpolar Phase	Heptane	PFD	PFMCHP	PFTBA	Heptane	PFD
f_{coal}, 10^{-5} N	<0.01	2.5	9	>10	1.2	0.1
$\Delta\sigma$, mJ/m²	15	17	12	10	4	14

Note: The interfacial tension lowering at the corresponding interfaces is also shown.

to coalescence is related to a deficiency in the adhesion between the nonpolar phase and the hydrophobic tails of the surfactants, or, according to Rehbinder, to the "imperfect compensation" of the intermolecular interactions across the interface.

4.2.1.3 Interactions between Hydrophobized Solid Surfaces in Nonpolar Liquids

The cohesion between the hydrophobic part of the interfacial adsorption layer and the adjacent nonpolar phase can be modeled using the cohesion between model hydrophobic surfaces in the same liquid. In such a simulation, the hydrophobic solid surfaces represent the hydrophobic tails of the surfactant molecules. This approach allows one to overcome the difficulties associated with the mutual solubility of the components (see Chapter 1). For the solid/liquid/solid interface, the main parameter characterizing the interactions is the free energy of interaction, F (or $\Delta\sigma_f$), which can be established experimentally using Derjaguin's theorem, that is, $p = \pi RF$, where p is the cohesive force in a direct contact between two spherical particles immersed in a liquid medium. Suitable model systems include spherical molecularly smooth glass beads with a radius $R \sim 1$–1.5 mm and hydrophobized surfaces of different natures, namely, HS and HL, immersed into the hydrocarbon and fluorocarbon liquids, HL and FL. Only dispersion forces are present in such systems, which makes the quantitative description of their interaction well defined and not complicated by the presence of various polar components.

The data on the free energy of interaction, the work of adhesion, and the interfacial free energy obtained for the model HS/FL and HS/HL systems are summarized in Table 4.6. The free energy of interaction is given in terms of the absolute value of $F/2$. The values of F for these systems differ by orders of magnitude. In the HS/HL system, where the two phases are very much alike, the observed low values of F characterize complete lyophilicity. For the fluorinated HS/FL systems, the values of $F/2$ were around 5 mJ/m^2, which is an indication of significant differences in the affinity of these phases for each other. In these systems, the values of F decreased from the highest (in the case of PFTBA) to lower ones for the more soluble PFMCHP and PFD. The latter two systems showed the lowest resistance to coalescence. Similar behavior was also observed in the *mirrorlike* systems of the FS/HL type, as well as for the FS/FL and FS/HL systems, where the values of $F/2$ reached several mJ/m^2. The stabilization against coalescence is related to some excess in the interaction free energy, that is, to the lack of similarity between the nonpolar portions of the surfactant molecules and the adjacent nonpolar phase.

It is worthwhile to relate the measured values of the free energy of interactions to the theoretical and experimental estimates for the work of adhesion and the interfacial free energy for the neighboring nonpolar phases. Following Gibbs, let us compare the values of the free energy of a liquid (L) film, σ_{film}, filling the gap between two identical solid (S) surfaces, the free interfacial energy,

---

TABLE 4.6

Free Energy of Interaction F/2 (i.e., Δσ$_f$/2), Work of Adhesion, W, and Interfacial Free Energy, σ$_{SL}$, of Solid Hydrocarbon Surfaces Immersed in Various Nonpolar Liquids

Liquid Phase	$F/2$, mJ/m^2	W, mJ/m^2	W^a, mJ/m^2	$W_{(A)}$, mJ/m^2	$W_{(F)}$, mJ/m^2	σ_{SL}^a, mJ/m^2	$\sigma_{SL(A)}$, mJ/m^2	$\sigma_{SL(F)}$, mJ/m^2
PFTBA	7.0	33	33	34	39	6.5	5	0.3
PFMCHP	6.8	35	38	39	41.5	3.5	2/5	<0.1
PFD	4.7	37	38	39	42	4	2	<0.1
Heptane	<0.01	42	41	41	42	<0.01	<1.0	0

Note: The symbols without a subscript are actual experimental data [34,43]. The data marked with (A) and (F) represent interfacial free energies calculated using Antonow's and Fowkes' approximations [59]. The data obtained from contact angle measurements are marked with ^{a}.

σ_{SL}, and the free energy of interaction, $F_{SLS} = F$; $\sigma_{film} = 2\sigma_{SL} - F$. From the Dupré rule for the work of adhesion at the interface, $W_{SL} = W$; $\sigma_{SL} + W = \sigma_S + \sigma_L$ (provided that no adsorption takes place). In the case of HS/FL and HS/HL model systems, σ_S is the surface energy of the methylated surface in air and σ_L is the surface free energies (surface tensions) of various studied liquids (it is worth recalling here that, in the HS and FS abbreviations, the "S" refers to the surfactant and not the solid phase).

Since for methylated surfaces, the value $\sigma_{film} \ll \sigma$, the Derjaguin theorem yields $\sigma_{SL} \approx F/2 \approx p/2R \approx 22$ mJ/m², which is a reasonable estimate of the surface free energy of a solid hydrocarbon. The Dupré rule then provides a means to get an estimate for $W = \sigma_S + \sigma_L - F/2$, where all three terms are measured independently. According to Fowkes [59], the dispersion component of the work of adhesion, $W_{(F)} = 2(\sigma_S^d \sigma_L^d)^{1/2}$, and $\sigma_{SL(F)} = [(\sigma_S^d)^{1/2} - (\sigma_L^d)^{1/2}]^2 = F_{(F)}/2$. The geometric mean approximation is utilized, as in the theory of Hilderbrand solubility parameters and in the estimation of the complex Hamaker constant.

A principally different estimate is obtained if one extends the empirical Antonow's rule for two liquids to the solid/liquid interface, that is, $\sigma_{SL} \approx \sigma_S - \sigma_L$, $W_{(A)} = 2\sigma_L$, and $F_{(A)}/2 \approx \sigma_S - \sigma_L$. This means that that the "plane of fracture" between the phases is not located at the interface but inside the weakest phase.

In all of these approximations, the value of W describes the "cohesive strength" between the molecules of the two phases, while σ_{SL} and $F/2$ both describe the "unsaturated interactions."

These approaches were applied to the methylated solid surface in contact with four characteristic nonpolar liquids. The values of the work of adhesion, W, were estimated as $W = \sigma_S + \sigma_L - F/2$ and were also obtained from independent contact angle measurements, $W^* = \sigma_L(1 + \cos \theta)$. The estimates of the interfacial free energy, σ_{SL}^*, were obtained as, $\sigma_{SL}^* = \sigma_S - \sigma_L \cos \theta$, and $W_{(A)}$, $W_{(F)}$, $\sigma_{SL(A)}$, $\sigma_{SL(F)}$ are the values estimated on the basis of Antonow and Fowkes relations, as indicated by the corresponding indices. The estimates summarized in Table 4.6 are rather crude: it was assumed that the value of σ_{film} was small and could be neglected, and that a typical spread in σ_S values was on the order of 3%. Nevertheless, the following trends are seen in these data: while for the heptane (HS/HL system), all estimated work of adhesion values are in good agreement with each other; the value $W_{(F)}$ for all HS/FL systems has been overestimated. Consequently, the $\sigma_{SL(F)}$ value for the same surface–liquid pairs was underestimated. This indicates that the geometric mean rule used in the Fowkes equation is not applicable to hydrocarbon/fluorocarbon systems, despite the fact that only dispersion interactions are involved in both the hydrocarbon/hydrocarbon and the hydrocarbon/fluorocarbon systems. The adhesion in the molecular HS/FL systems is weaker than in the HS/HL systems (imperfect compensation of molecular forces), and the traditional geometric mean approach yielding very low values of σ_{SL} and $F/2$ is not applicable.

The interactions between the hydrophobic parts of the surfactant molecules and the nonpolar liquid phases play an important role in controlling the stability of emulsions in these systems [13–20,56]. While each particular system requires an individual approach, it can be concluded that the high stability of fluorocarbon emulsions against coalescence is related to a deficit in the adhesion in the HS/FL system resulting in the "squeezing out" of hydrophobic chains from the nonpolar liquid phase.

These results can also be analyzed and generalized from the standpoint of Rehbinder's structural–mechanical barrier concept. The mechanical strength of the interfacial adsorption layers in the FS/HL and HS/FL systems is responsible for strong stabilization in the fluorocarbon–hydrocarbon emulsions and their resistance to coalescence. This mechanical strength originates from a specific dense structure of the interfacial adsorption layers. In the case of water-soluble surfactants, the hydrophilic tails and EO loops cannot be viewed as the source of such strength. Being strongly hydrated, these moieties extensively penetrate into an aqueous phase and propagate through it like tentacles. These "tentacles" do provide some stability against coagulation, but they play a very limited role in manifesting the resistance of an interfacial film to thinning. The "tentacles" not only provide a long-range effect but also contribute to a "pulling out" of the hydrophobic moieties of the surfactant molecules from the nonpolar phase toward the geometric interface. The more developed the hydrophilic parts (as in Pluronics) and the less their similarity with a nonpolar phase (as in the case of PFTBA), the more significant this effect is. In this aspect, the most important factor is the interaction between the hydrophobic parts of the surfactant molecules and the nonpolar liquid phase. Where there is high similarity between them, as in the HS/HL and FS/FL systems (good adhesion, very low values of interfacial free energy of interaction, $F/2$), the hydrophobic tails and loops may also show tentacle-like behavior and are not capable of forming a mechanically strong structure. Conversely, if the adhesion between the hydrophobic tails of the surfactants and the non-polar phase is relatively weak (i.e., they are more dissimilar, characterized by high values of F), the hydrophobic loops and tails turn out to be displaced from the bulk of the nonpolar liquid to the interface with water. At the organic liquid–water interface the attraction between the surfactant tails and loops is enhanced, which allows them to aggregate into a compact structure, which is a real carrier of mechanical strength.

It is worth emphasizing here that while the structural–rheological properties (i.e., mechanical strength) of the interfacial adsorption layer play a determining role in the stability of the system toward coalescence, they alone may not be sufficient for complete stabilization. The prevention of coagulation also requires that the structural–mechanical barrier formed is lyophilic (hydrophilic) with respect to the surrounding polar liquid. The latter can be achieved by the introduction of common surfactants, for example, sodium dodecyl sulfate (SDS).

4.3   CONDITIONS OF SPONTANEOUS DISPERSION AND THE FORMATION OF STABLE COLLOID SYSTEM

Let us now address an interesting problem in colloid science and physical-chemical mechanics related to the contact interactions in disperse systems, namely, the possibility of spontaneous dispersion and the formation of thermodynamically stable colloid system. Originally, this problem was formulated by Max Volmer in 1927 [60,61] and later addressed by Rehbinder and Shchukin. Shchukin has made two principal contributions to the analysis of this problem [33,62–69]. The first is the detailed analysis of the conditions that make the process of spontaneous dispersion (at constant volume of the disperse phase, constant particle size, or constant number of particles) possible. Second, he proposed incorporating the entropy of mixing into the description of the conditions of spontaneous dispersion. The latter allows one to quantitatively estimate the concentration of the disperse phase in the disperse system formed. The analysis of the thermodynamics of spontaneous dispersion has important implications in the analysis of colloidal stability and in the control of various technological processes.

The dispersion of the macroscopic phase into n spherical particles with radius r and interfacial energy σ is accompanied by a change in the free energy of the system, ΔF. If ΔF is negative, the dispersion process is thermodynamically favorable, that is, $\Delta F = n4\pi r^2 - T\Delta S < 0$, where T is the temperature and ΔS is the change (increase) in entropy due to the participation of the particles formed in Brownian motion. If there is a factor present that prevents the dispersion down to molecular dimensions, b, then there is the possibility that $\Delta F < 0$ at $r^{(min)} \gg b$, that is, there is a minimum in free energy, and the formation of a thermodynamically stable colloidal system becomes possible (the so-called Rehbinder–Shchukin criterion) [69]. In the following analysis, we will examine the behavior of the $\Delta F = \Delta F(r, \sigma, n)$ function under three conditions: (1) at constant volume of the disperse phase, v (or concentration, C); (2) at constant particle radius, r; and (3) at a constant number of particles, n. In all cases, monodisperse systems with a constant volume of the dispersion medium, V, and various values of the interfacial energy, σ, are assumed. For clarity, it is worth emphasizing that, in the present analysis and in the context of Section 4.4, ΔF has the meaning of the *change in the free energy of the system*, so it should not be confused with the previously used energy of interaction in a thin film, $F = \Delta \sigma_f$, or the force, F, in Chapters 5 and 7.

In all three of the aforementioned situations, the $\Delta F = 0$ condition is the *necessary* condition for spontaneous dispersion to take place and for a stable lyophilic colloidal system to form. At room temperature and low concentrations of the disperse phase, this condition requires that both the particle size and the interfacial tension are sufficiently small, that is, $r \sim 10^{-6}$ cm and $\sigma \sim 10^{-2}$–10^{-1} mJ/m². These conditions become "more relaxed" for the dispersion of aggregates (e.g., σ is on the order of units of mJ/m²) and "more stringent" for highly concentrated systems (i.e., ultralow σ, $\sim 10^{-3}$ mJ/m² or less).

Before we immerse ourselves in a detailed discussion, let us first briefly review the historic developments of approaches to this problem.

In 1927, M. Volmer proposed that the thermodynamic stability of a microheterogeneous system near the critical temperature of mixing can be explained by the thermal motion of colloidal-size droplets with low surface energy [60,61]. Further development of this approach was hampered due to the lack of experimental data on ultralow interfacial tension. However, despite of this limitation, the idea that critical emulsions in fact represent microheterogeneous systems was recognized and accepted [70,71]. On the basis of Volmer's original idea, Rehbinder examined the possibility of a spontaneous (i.e., induced by thermal fluctuations) separation of microblocks of size δ from the surface of a solid body under conditions where there was a substantial lowering of the interfacial free energy, σ_s—all the way down to a critical value σ_{cr}—determined by the work $\delta^2\sigma \ll kT$. The lowering of the interfacial tension can be achieved by the immersion of the material into a surface active medium [1].

The conditions needed for a spontaneous dispersion of the condensed phase and the formation of a lyophilic colloidal system were analyzed by Rehbinder and Shchukin back in 1958, and a quantitative description of this problem was proposed. That original analysis was based on the estimation of the changes in the free energy, ΔF, upon dispersing a condensed phase in a given dispersion medium [62]. Let's assume that, as a result of dispersion, n particles have separated from the condensed phase. Due to the participation of these separated particles in Brownian motion, the work of dispersion, $n\alpha\delta^2\sigma$, is balanced by the gain in entropy, ΔS

$$\Delta F = n\alpha\delta^2\sigma - T\Delta S$$

where the numerical coefficient α depends on the particle shape. Within the limits of the theory of regular solutions, the entropy increase resulting from the mixing of n particles or $N_1 = n/N_A$ moles of particles with N_2 moles of dispersion medium is given by

$$\Delta S = R\left\{N_1 \ln\left[\frac{(N_1 + N_2)}{N_1}\right] + N_2 \ln\left[\frac{(N_1 + N_2)}{N_2}\right]\right\} \tag{4.1}$$

In dilute colloid dispersions, $N_1 \ll N_2$, and

$$\Delta S = R\left\{N_1 \ln\left[\frac{N_2}{N_1}\right] + N_1\right\} = nk\left\{\ln\left[\frac{N_2}{N_1}\right] + 1\right\} \tag{4.2}$$

Consequently,

$$\Delta F = n\alpha\delta^2\sigma - nkT\{\ln[N_2/N_1] + 1\} = n(\alpha\delta^2\sigma - \beta kT) \tag{4.3}$$

In real dilute systems, the parameter $\beta = \ln[N_2/N_1] + 1$ can be on the order of 10–15 or higher.

The value of ΔF is positive for coarse dispersions (with respect to the initial compact phase) and drops sharply as the size of the separated particles decreases. The transition from positive to negative values of ΔF constitutes a *necessary* condition for the thermodynamically favorable formation of a dispersed system consisting of fine particles. Consequently, the $\Delta F = 0$ represents a critical condition for spontaneous dispersion, known as the Rehbinder–Shchukin criterion:

$$\sigma_c \leq \frac{\beta kT}{\alpha\delta^2} \tag{4.4}$$

At some even smaller particle sizes and under certain other conditions (e.g., an increase in σ on approaching molecular dimensions), a negative minimum in the free energy can be observed. This minimum corresponds to the formation of a thermodynamically stable, lyophilic disperse system. These concepts were further developed in several directions in theoretical and experimental studies by Shchukin, Pertsov, and Kochanova [33,66–69], and in works by Rusanov et al. [63–65].

A certain amount of controversy has been caused by the question as to what specific characteristics of the ΔF function represent the universal conditions for spontaneous dispersion? To address this subject, we will analyze changes in the free energy, ΔF, associated with the dispersion, as a function of the size and number of particles, their concentration, and the value of the free interfacial energy at the interface between the disperse phase and the dispersion medium. This analysis is performed for three characteristic conditions: (1) varying the particle size at a constant volume of the disperse phase, (2) varying the number of particles at a constant particle size, and (3) varying the particle size at a constant number of particles. The analysis will be restricted to systems that are monodisperse at every stage of the dispersion process and consist of spherical particles with radius r. The volume of the dispersion medium is assumed to be constant, for example, $V = 1000$ cm^3. At 300 K, the kT value is 4.14×10^{-21} J.

The work needed to isolate a spherical particle from a stable equilibrium macroscopic phase equals $w = 4\pi r^2\sigma$, and Equations 4.1 and 4.3 then take the following form:

$$\Delta F = n4\pi r^2\sigma - nkT\left\{\ln\left[\frac{1}{C}\right]+1\right\} \tag{4.5}$$

Note that for an *equilibrium* macroscopic phase, the expression of work does not contain the 1/3 multiplier [26].

In the aforementioned expression, the ratio N_2/N_1 is replaced by a generalizing parameter—the dilution $1/C$—where C is the disperse phase concentration. Generally speaking, C can be defined in several different ways: as a ratio of the number of disperse phase particles, n, to the number of particles (molecules) of the dispersion medium, N, $(C = n/N)$; as a ratio of the corresponding numbers of moles $(C = m/M)$; or as a ratio of the total volume of the disperse phase particles, v, to the volume, V, of the dispersion medium, $C = v/V$. Because of ambiguity in the definition of C, there is also an ambiguity in the calculation of the entropy of mixing. The latter reflects the known controversy in the estimation of the chemical potential encountered in the thermodynamics of small systems in the case of significantly different sizes of disperse phase particles and dispersion medium molecules [26,47,72–76]. The latter corresponds to the definition of what is a *mole* of the disperse phase, that is, whether it is a mole of particles or a mole of molecules. Both definitions correspond to the number of *units* equal to Avogadro's number, N_A. A detailed discussion of this subject is beyond the scope of this book. Here, we will consider two *limiting* cases: the representation of C as a number ratio, that is, $C = n/N$ (in an aqueous medium, $N = 3.37.10^{25}$ molecules/10^3 cm^3), and as a volume ratio, $C = v/V$ assuming $V = 10^3$ cm^3.

For the $C = v/V$ approach, expressions in Equations 4.3 and 4.5 can be written as

$$\Delta F = n4\pi r^2\sigma - nkT\left\{\ln\left[\frac{V}{v}\right]+1\right\} \tag{4.6}$$

and

$$\frac{\Delta F}{kT} = \left(\frac{v}{(4/3)\pi r^3}\right)\left\{\frac{4\pi r^2\sigma}{kT} - \ln\left[\frac{V\exp}{v}\right]\right\} \tag{4.7}$$

or

$$\frac{\Delta F}{kT} = n\left\{\frac{4\pi r^2\sigma}{kT} - \ln\left[\frac{V\exp}{(4/3)\pi r^3 n}\right]\right\} \tag{4.8}$$

Within the framework of the $C = n/N$ approach, the same expressions in Equations 4.3 and 4.5 are as follows:

$$\Delta F = n4\pi r^2\sigma - nkT\left\{\ln\left[\frac{N}{n}\right]+1\right\} \tag{4.9}$$

and

$$\frac{\Delta F}{kT} = \left(\frac{v}{(4/3)\pi r^3}\right)\left\{\frac{4\pi r^2\sigma}{kT} - \ln\left[\frac{(V/v_m)\exp}{v/(4/3)\pi r^3}\right]\right\} \tag{4.10}$$

or

$$\frac{\Delta F}{kT} = n\left\{\frac{4\pi r^2\sigma}{kT} - \ln\left[\frac{N\exp}{n}\right]\right\} \tag{4.11}$$

where v_m is the volume of a molecule of the dispersion medium.

If the starting point is not a compact phase, but an aggregate of insoluble or partially insoluble particles, the term $4\pi r^2\sigma$ is replaced by the product of the energy of cohesion in a contact between particles and the average coordination number.

These expressions are further used in the analysis of the behavior of ΔF as a function of the various parameters. The challenge here is to overcome the mental barrier associated with the traditional formal approach to this problem. Particular examples of real physical processes and states reflecting various types of behavior of ΔF include the peptization–coagulation transition in colloid sols and particle bridging in the course of the hydration hardening of mineral binders.

4.3.1 Behavior of $\Delta F(r)$ at v = const

Let us examine the behavior of $\Delta F(r)$ when v = const. This situation reflects the process of a stepwise (virtual) grinding of the initial volume v into identical spherical particles whose size decreases with every step of the grinding. Such an approach was used in both the original and the subsequent studies [61,66–68]. In this case, the number of particles, n, is inversely proportional to the third power of the particle size, that is, $n = v/(4/3\pi r^3)$, and Equations 4.6 and 4.7 acquire a simplified form, namely,

$$\frac{\Delta F}{kT} = c_1\left(\frac{1}{r}\right) - c_2\left(\frac{1}{r^3}\right) \tag{4.12}$$

Equations 4.9 and 4.10 are then written as

$$\frac{\Delta F}{kT} = c_1\left(\frac{1}{r}\right) - \left(c_2 + c_3 + c_4\ln r\right)\left(\frac{1}{r^3}\right) \tag{4.13}$$

where c_1, c_2, c_3, and c_4 are constants.

Both expressions show that as the grinding yields smaller and smaller particles, the value of ΔF grows slowly at first as $+1/r$ from zero up to some maximum value, ΔF_{max}, corresponding to $r^{(max)}$, and then rapidly decays as $-1/r^3$. The function then passes through zero at $r = r^{(0)}$ and reaches negative values. For the critical value of $r = r^{(0)}$, one can write

$$4\pi r^2 \sigma = kT\left\{\ln\left[\frac{1}{C}\right]+1\right\} \tag{4.14}$$

or

$$\sigma = \sigma_c = \frac{\beta kT}{4\pi r^2}$$

Both $C = v/V$ and $C = n/N$ approaches show that a positive maximum of the ΔF function is observed when $d\Delta F/dr = 0$, which corresponds to the particle radius $r^{(max)} = \sqrt{3}r^{(0)}$. Such a maximum is of a *virtual* nature [26]: a *continuous* approach to this maximum from the side of a macroscopic phase implies that a separation of a single molecule from each particle and further combination of these molecules into a new particle of the same size take place. Such process is not kinetically possible. Consequently, it is important to realize that in examining changes in ΔF at $v = $ const, we are not exploring a real physical process in which the system disperses down into one with smaller and smaller particles. We are only restricting ourselves to a *virtual* comparison of such states so that the estimates for $r^{(0)}$ and σ_c can be obtained, as shown by Pertsov et al. [66–68].

Viewing a maximum in $\Delta F(r)$ as a real potential barrier, as in [72], means that in a real dispersion process, such as grinding, the process must go on spontaneously, once the size $r^{(max)}$ has been reached. However, such a process of *discrete* transition from one monodisperse system to another can't be materialized. In the case of a spherical particle, such a process would imply dividing a single particle into two equal particles with equal volumes, which would result in a decrease in the radius and an increase in the surface area by a factor of 1.26. For example, using the critical values from [72], $\sigma = 2 \times 10^{-2}$ mJ/m^2 for $r = 5 \times 10^{-6}$ cm, or $\sigma = 5 \times 10^{-3}$ for $r = 10^{-5}$ cm, one gets a surface energy for a spherical particle of 152 kT. Dividing such a particle into two particles of equal size is associated with a surface area increase of 26%, which requires work equal to 39.5 kT, that is, for the entropic factor one can write that $39.5 = \ln(1/C) + 1$, which produces a value of $C = 7 \times 10^{-18}$ (in the $C = v/V$ scheme). This means that for 1 cm^3 of a disperse phase, the corresponding volume of the dispersion medium reaches 1.4×10^{17} cm^3, which corresponds to the volume of a lake that is 14 m deep and has an area of 100×100 km^2. It is also worth pointing out that it is not possible to achieve sizes on the order of 1/100 of a micron using a mill. Such are the critical sizes corresponding to realistic values of σ.

It was discussed in [62,68] that a transition to particle sizes that are less than $r^{(max)}$ or $r^{(0)}$ does not by itself constitute a sufficient condition for spontaneous dispersion to take place and for a stable colloid-disperse system to form. The combination of the factors considered so far, $c_1/r - c_2/r^3$, does not place any restriction on the dispersal of the system down to molecular sizes. Consequently, to form a stable disperse system, one needs to introduce another factor, that is, there is a need for a "terminating condition" that would prevent dispersion down to the level of individual molecules. Such a factor is the dependence of surface energy on size, that is, the increase in σ at small r. This phenomenon may be related to an asymmetric structure in the disperse phase molecules, particularly that of surfactants in the micelles. Other reasons may include a domain-like structure of the solid phase, a globular structure with open porosity with coagulation or phase contacts between particles, and a polymer system in which the particles are macromolecules. Other physical-chemical factors and mechanisms influencing the disperse system's stability, including long-range interactions between particles, are beyond the scope of this section.

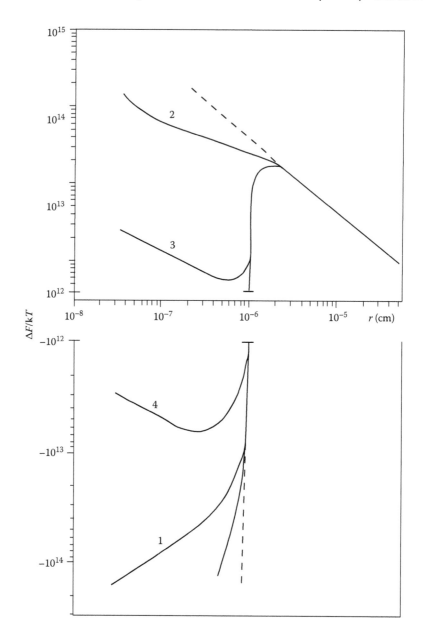

FIGURE 4.33 Schematic illustration of the behavior of $\Delta F/kT$ as a function of the particle radius, r, on a system approaching molecular dimensions, b: a monotonous decrease when there is no factor that prevents dispersion down to molecular dimensions (1); the trend in the presence of a factor resisting dispersion (2); the appearance of the minimum in the ΔF function: positive minimum (3) and negative minimum (4), corresponding to the formation of a stable lyophilic colloidal system. (From Shchukin, E.D., *J. Dispers. Sci. Technol.*, 25, 875, 2004.)

The influence of this third factor on the trends in ΔF may differ significantly depending on its nature, as shown in Figure 4.33. The two limiting cases here correspond, respectively, to the complete absence and the "very strong presence" of this factor, namely, the dispersion down to individual molecules or ions (i.e., dissolution) on the one hand and a monotonous increase in the ΔF on the other hand. The latter represents a case of high σ values, which are typical in lyophobic systems, for example, a solid body in the absence of a surface-active medium. In between these two extreme situations, a minimum in ΔF may be observed for the particles of colloid dimensions. If this minimum is positive, the system is metastable and tends to return to the original compact state. In the case when $\Delta F = 0$ is achieved at some critical size, $r = r^{(0)}$, for a "moderately strong third factor," a negative minimum in ΔF function can exist at a particular radius, $r^{(eq)} < r^{(0)}$. This corresponds to a stable equilibrium colloidal system. It is essential that the critical size, $r^{(0)}$, significantly exceeds the molecular dimensions, b. Since $r^{(0)} = (\beta kT/4\pi\sigma)^{1/2}$, this condition can be formulated in terms of a Rehbinder–Shchukin criterion

$$RS = \frac{r^{(0)}}{b} = \left(\frac{\beta kT}{\alpha\delta^2\sigma}\right) \gg 1 \qquad (4.15)$$

where $\alpha = \pi$ for spherical particles with diameter $\delta = 2r$.

Consequently, for the case of $v = $ const, reaching $\Delta F = 0$ has a significance for establishing conditions leading to a spontaneous dispersion and the formation of a stable colloidal system. At the same time, one has to keep in mind that these critical values of $r^{(0)}$, σ_c, $w_c = 4\pi r^{(0)2}\sigma$ (and the work w_c in general) do not, by themselves, represent the absolute and self-sufficient characteristics of a spontaneous dispersion process, even for monodisperse systems. In the case of an equilibrium system, the concentration also represents the critical parameter, which for a dilute system is given by the factor $\beta = \ln[1/C] + 1$. In the case of polydisperse systems, the critical parameters can be represented by distribution histograms.

For actual calculations of the ΔF function within the $C = v/V$ scheme, it is convenient to present Equations 4.6 and 4.7 in the form

$$\frac{\Delta F}{kT} = \left(\frac{v}{(4/3)\pi r^{*3}}\right)\left(\frac{r*}{r}\right)^3\left\{\left[\frac{4\pi r^{*2}\sigma}{kT}\right]\left(\frac{r}{r*}\right)^2 - \ln\left[\frac{V\exp}{v}\right]\right\}$$

In the $C = n/N$ scheme, Equations 4.9 and 4.10 can be written as

$$\frac{\Delta F}{kT} = \left(\frac{v}{(4/3)\pi r^{*3}}\right)\left(\frac{r*}{r}\right)^3\left\{\left[\frac{4\pi r^{*2}\sigma}{kT}\right]\left(\frac{r}{r*}\right)^2 - \ln\left[\frac{V\exp}{v}\right] - \ln\left[\frac{(4/3)\pi r^{*3}}{v_m}\right] - 3\ln\left[\frac{r}{r*}\right]\right\}$$

Here, $r*$ is a typical small dimension. It is introduced for illustrative purposes as a *reference point* that corresponds to a sharp ΔF dependence on r in the vicinity of $r^{(0)}$. The dimensions that are of practical interest are rarely outside of the range from ~ several 10^{-7} cm to ~several 10^{-6} cm, and thus it is convenient to assume that $r* = 10^{-6}$ cm.

Some typical quantitative data for $\Delta F(r)$ calculation with $v = $ const are summarized in Table 4.7 and are also shown in Figure 4.34. Table 4.7 shows the critical value of the radius, $r^{(0)}$, as a function of the concentration, C, and the interfacial energy σ, and the C values and $r^{(0)}$ values are all within approximately one order of magnitude.

Figure 4.34 shows the $\Delta F(r)/kT$ function for both calculation schemes, $C = v/V$ and $C = n/N$, for $v = $ const $= 10^{-5}$ cm^3. For ease of comparison, the σ values are selected in such a way that the particle critical radius is the same: $r^{(0)} = 10^{-6}$ cm. Both schemes produce qualitatively similar results. The $\Delta F(r)/kT$ curves also have similar shapes when $v = 10$ cm^3, with a quantitative difference in their values of six orders of magnitude. Same figure also shows $\Delta F(r)/nkT$ per single particle. This is a monotonous function, showing rapid growth in the range of positive values.

The sharp dependence, $1/r^3 < 0$, predetermines an abrupt transition from positive values of ΔF to negative ones as r decreases. Consequently, the critical positions of $r^{(0)}$ and $r^{(max)} = \sqrt{3}r^{(0)}$ are very well defined. The transition to a spontaneous dispersion (i.e., to *lyophilicity*) corresponds to

TABLE 4.7

Critical Radius of a Disperse Phase Particle, $r^{(0)}$, ($\times 10^{-6}$ cm) Corresponding to $\Delta F = 0$ as a Function of the Interfacial Free Energy, σ (mJ/m²), and the Disperse Phase Concentration, C, Defined as $C = v/V$

$C = v/V$ ⟍ σ	0.001	0.01	0.1	1.0
10^{-2}	4.3	1.36	0.43	0.14
10^{-4}	5.8	1.84	0.58	0.18
10^{-6}	7.0	2.2	0.70	0.22
10^{-8}	8.0	2.5	0.80	0.25

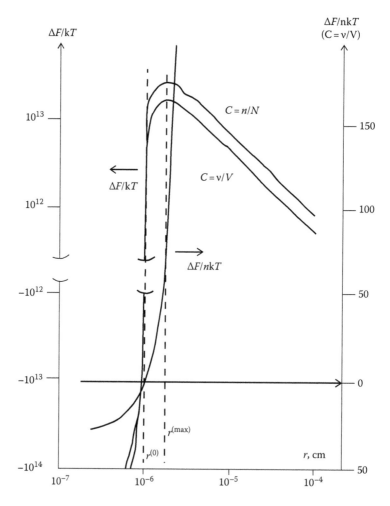

FIGURE 4.34 Change in the free energy $\Delta F/kT$ as a function of the particle radius, r (cm), upon the dispersion of a constant volume $v = 10^{-5}$ cm³ of a compact solid phase in $V = 1000$ cm³ for the $C = v/V$ approach with $\sigma = 0.0064$ mJ/m² and the $C = n/N$ approach with $N = 3.37 \times 10^{25}$ molecules with $\sigma = 0.103$ mJ/m². The values of $r^{(0)}$ and $r^{(max)}$ are 10^{-6} and 1.73×10^{-6} cm, respectively, for both approaches. The value of ΔF for the reference monodisperse systems abruptly changes by several orders of magnitude upon a transition from positive to negative values at $r = r^{(0)}$. (Redrawn from Shchukin, E.D., *J. Dispers. Sci. Technol.*, 25, 875, 2004.)

the range of possible values of the interfacial energy σ between 10^{-3} and 10^{-1} erg/cm^2 (mJ/m^2). The dilutions, that is, $1/C$, can fall into a very broad range, for example, 10^1–10^8 or even broader (one, however, needs to remember the logarithmic dependence of ΔF on this parameter). Thus, the critical value of r is inversely proportional to $\sqrt{\sigma}$, and, as a result, at room temperature, $r^{(0)}$ values are on the order of magnitude of 10^{-6} cm. The latter dimension corresponds to a *nanodisperse system*. Depending on the physical-chemical conditions, that is, on how low σ can be, both high and very low disperse phase concentrations are possible.

By examining the behavior of $\Delta F\,(r)$ at $v = \text{const}$, we have obtained correct estimates for critical parameters, that is, $r^{(0)}$ and σ, which are needed for the possibility of spontaneous dispersion to take place. However, the constraint of $v = \text{const}$ in the $C = v/V$ approach seemingly dictates the value for the concentration of disperse phase and hence does not correspond to a realistic spontaneous dispersion process producing an *equilibrium* colloidal system.

Using either the $C = v/V$ or the $C = n/N$ approach for the evaluation of the entropic factor, one obtains qualitatively similar results in both cases; namely, in both cases, the abrupt transition from positive to negative $\Delta F\,(r)$ values is observed, with a clearly expressed correlation between $r^{(0)}$ and σ (or the work $w_c \sim [r^{(0)}]^2\sigma$). Although both approaches yield qualitatively similar data, the quantitative difference may be very significant. For example, within the $C = n/N$ approach, for the same values of σ, the $r^{(0)}$ is positioned more to the right (the critical value of $w(r^{(0)}) = w_c$ becomes larger) than in the $C=V/v$ approach. Indeed, $N/n \gg V/v$, and this difference increases with the increase in the particle size at a given v. The comparison of the *concentrations* corresponding to a given value of the $\ln(1/C) + 1$ term in the entropy factor shows that the difference is very large. For $r^{(0)} = 10^{-6}$ cm and $\sigma = 0.064$ mJ/m^2, $w_c/kT = 19.4$. Consequently, $19.4 = \ln(1/C) + 1$, which yields a dilution $1/C = 10^8$. In the $C = v/V$ approach, when $V = 10^3$ cm^3, this yields $v = 10^{-5}$ cm^3 and the number of particles with a given size, $n^{(0)}$, of 2.39×10^{12}. At the same time, with employing the $C = n/N$ approach for $N = 3.37 \times 10^{25}$ water molecules per L, the same dilution, $1/C = 10^8$, yields a very different number of particles, that is, $n^{(0)} = N/10^8 = 3.3 \times 10^{17}$, which is 1.4×10^5 times greater. Despite these enormous differences, there is no need to dramatize them: with proper experimentation, it is possible to figure out which scheme produces realistic estimates for the entropic ΔS factor and thus corresponds to the proper way to estimate the chemical potential in colloidal systems.

4.3.2 Behavior of $\Delta F(n)$ at $r = $ const

Let us now turn to the analysis of the $\Delta F(n)$ dependence for the case involving a constant degree of dispersion, that is, for $r = \text{const}$ [33,69]. This situation does not permit unlimited dispersion down to molecular dimensions and is realistic for a number of real systems. The examples of the latter include coagulates consisting of insoluble particles, disperse system with phase contacts, solids with a microdomain structure, or multiphase solids (alloys or composites) with a fine structure of grains. Other examples include swelling and dispersion of clays, and surfactant micelles and polymer macromolecules in particular.

Expressions 4.3, 4.5, 4.6, and 4.9 in the $C = v/V$ scheme are represented by a single equation, namely, Equation 4.8, and in the $C = n/N$ scheme by Equation 4.11. In the latter scheme, the competition between the work of dispersion needed to isolate spherical particles from a macroscopic phase and the entropy gain due to the participation of the released particles in Brownian motion can be represented by the following relation:

$$\frac{\Delta F}{kT} = n(c' - c'' \ln n) \tag{4.16}$$

where c' and c'' are constants depending on fixed values of r and σ. The aforementioned function passes through zero at a given $n^{(0)}$ as n decreases and, at even smaller value when $n = n^{(eq)}$, has a minimum corresponding to a thermodynamically stable, equilibrium *lyophilic* colloidal system. In both $C = n/N$ and $C = v/V$ schemes, the values of $n^{(eq)}$ and $n^{(0)}$ differ from each other by the same constant factor, that is, $n^{(eq)} = n^{(0)}/\exp$.

In the $C = n/N$ scheme, the point $\Delta F(n) = 0$ corresponds to a number of particles, $n^{(0)} = N\exp-(4\pi r^2\sigma/kT)$. The equilibrium number of particles, that is, the *colloid solubility* of particles at the minimum of $\Delta F(n)$

$$n^{(eq)} = N\exp\left(-\left(\frac{4\pi r^2\sigma}{kT}\right)\right) \tag{4.17}$$

This minimum is quite shallow: $\Delta F(r^{(eq)})/kT = n^{(eq)}$, that is, it constitutes only one kT per particle. Such a system may exist only in dynamic equilibrium between a colloidal solution and the corresponding macroscopic phase.

In order to obtain quantitative estimates, it is convenient to present relations show in Equations 4.8 and 4.11 in the $C = n/N$ scheme as

$$\frac{\Delta F}{kT} = n\left\{\left[\frac{4\pi r^{*2}\sigma}{kT}\right]\left(\frac{r}{r^*}\right)^2 - \ln\left[N\exp\right] + \ln n\right\}$$

and in the $C = v/V$ scheme as

$$\frac{\Delta F}{kT} = n\left\{\left[\frac{4\pi r^{*2}\sigma}{kT}\right]\left(\frac{r}{r^*}\right)^2 - \ln\left[\frac{V\exp(r^*/r)^3}{(4/3)\pi r^{*3}}\right] + \ln n\right\}$$

Some examples of such quantitative estimates are shown in Figures 4.35 through 4.37. For realistic concentrations of the disperse phase, the correlation between the particle size, r, and the interfacial energy, σ, (or the work, w) is very critical, that is, the transition from a macroscopic state to a colloid solution of monodisperse particles is abrupt.

Figure 4.35 shows the $\Delta F(n)/kT$ function for the $C = n/N$ scheme, with $r = \text{const} = 10^{-6}$ cm and $\sigma = 0.064$ mJ/m^2. The function has a characteristic negative minimum at $n^{(eq)} = n^{(0)}/\exp$. In this case, $\Delta F(n)$ changes abruptly by 30 orders of magnitude with the transition from positive to negative values of n at $n = n^{(0)}$. The $\Delta F/nkT$ function is monotonous; for an equilibrium (saturated) system at $n^{(eq)}$, this function equals -1 (Figure 4.35). In the $C = v/V$ scheme, the appearance of the $\Delta F(n)/kT$ graph is similar but with substantially lower absolute values (see Figure 4.39 for comparison).

Typical examples of real physical systems and processes conforming to this approach ($r = \text{const}$) include colloid solutions of micelle-forming surfactants and the dispersion (peptization) of a globular structure with a given strength of contacts between the particles, p_1, and the work of the contact rupture, u_1.

The structure and properties of micellar systems are thoroughly covered in the literature [76–79]. It would be worthwhile to make a semiquantitative comparison. Let us assume the value of $r = 0.25 \cdot 10^{-6}$ cm as a typical radius of a micelle. According to Equations 4.11 and 4.8, realistic concentrations of a surfactant in a micelle correspond to relatively high σ values at the interface between a micelle and a medium: these σ values are on the order of fractions of a mJ/m^2 or even higher, as it is seen in Figure 4.36. These estimates are in agreement with published values [14,47].

The case of dispersion in a globular structure was examined by Pertsov [80], and independently by Martynov and Muller [81,82]. In this case, the work required to isolate a particle from the initial phase can be represented in a different way, that is, as $w = \frac{1}{2}zu_1$; where u_1 is the free energy of cohesion in an individual contact (absolute value) and z is the average coordination number.

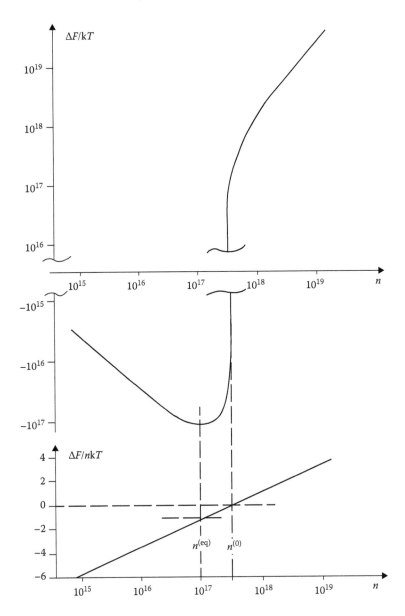

FIGURE 4.35 Change in the free energy $\Delta F/kT$ as a function of the number of particles, n, in a monodisperse system of spherical particles with radius $r = 10^{-6}$ cm and $\sigma = 0.064$ mJ/m^2 in the $C = n/N$ scheme with $N = 3.37 \times 10^{25}$ molecules per 10^3 cm^3 and the corresponding values of $\Delta F/nkT$ per particle.

Let us restrict the discussion to the case of coagulation contacts that are mechanically reversible. The cohesion in such contacts is governed by the surface dispersion interactions. The cohesion force in an individual contact, $p_1 = \pi r \Delta \sigma_f$, where $\Delta \sigma_f$ is the free energy of interaction of these surfaces in a given medium. By assuming $\Delta \sigma_f \approx 2\sigma_{12}$, we can write that $w \approx (z/2)p_1 h_0 = (z/2)\pi r 2\sigma/h_0$ with h_0 being the gap in the equilibrium contact (on the order of molecular dimensions). For a relatively loose coagulation structure with $z \approx 4$, one obtains the estimate

$$w \approx 4\pi r \sigma h_0 = 4\pi r^2 \sigma \frac{h_0}{r} \tag{4.18}$$

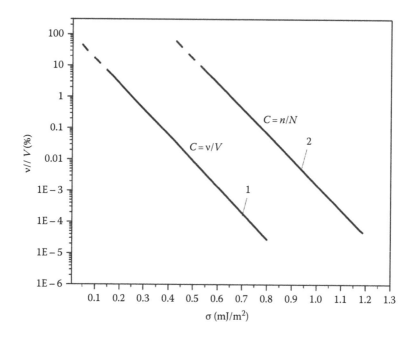

FIGURE 4.36 The concentration (volume fraction) of dispersed spherical particles of a constant radius $r = 2.5 \times 10^{-9}$ cm as a function of the interfacial free energy corresponding to a minimum in the $\Delta F/kT$ function under the condition of equilibrium with a macroscopic phase: in the $C = v/V$ scheme ($V = 10^3$ cm^3) (1) and in the $C = n/N$ scheme with $N = 3.37 \times 10^{25}$ water molecules (2). (Redrawn from Shchukin, E.D., *J. Dispers. Sci. Technol.*, 25, 875, 2004.)

This rather crude estimate reveals substantial differences in the energy of dispersion of a compact phase and that of agglomerate consisting of aggregated particles. In the latter case, the work of dispersion is r/h_0 times lower than in the former case, and the difference increases with an increase in particle size. Consequently, for a particle aggregate, the conditions under which spontaneous dispersion takes place are quantitatively different from those for a compact phase (Figure 4.37).

For $r \sim 10^{-6}$ cm, the disperse phase concentration can reach 1% or more at relatively high values of interfacial free energy, such as several mJ/m^2. These are the conditions under which the peptization–coagulation transition takes place in hydrophobized silica dispersions in mixtures of ethanol and propanol with water, as well as in dispersions of unmodified silica in aqueous solutions of CPB surfactant [50–53]. In these experiments alongside with the studies on coagulation (monitored as an increase in turbidity with a change in the aqueous media polarity), the free energy of interaction between similar macroscopic surfaces, $\Delta\sigma_f$, in the same media was independently measured. Indeed, the critical value of the $\frac{1}{2}\Delta\sigma_f \approx \sigma_c$ reached several mJ/m^2.

Similar to the $\Delta F(r)$ behavior at $v = $ const, the $\Delta F(n)$ analysis at $r = $ const may reveal significant differences between the $C = v/V$ and $C = n/N$ schemes. In this case, these differences are also quantitative: the positions of $r^{(0)}$ and, correspondingly, of $n^{(0)}$ (i.e., ln $n^{(0)}$) are shifted. In both of these schemes, the dependence of ΔF on the relation between the work of dispersion in an individual act of particle isolation, w, and the entropy factor is similar. It is worth emphasizing here that this potential barrier, w, is a true and universal physical characteristic of the described systems, differing from a *virtual* maximum in the $\Delta F(r)$ function at $v = $const.

4.3.3 Behavior of $\Delta F(r)$ when $n = $ const

Let us now briefly address another case of $\Delta F(r)$ behavior, that is, when $n = $ const (Figure 4.38) [33,69]. It is apparent that in this case, at the specified dilution, which is constant in the $C = n/N$ scheme, the value of w changes rapidly as r^2 in the vicinity of a critical value $4\pi r^2\sigma = kT \ln[1/C + 1]$.

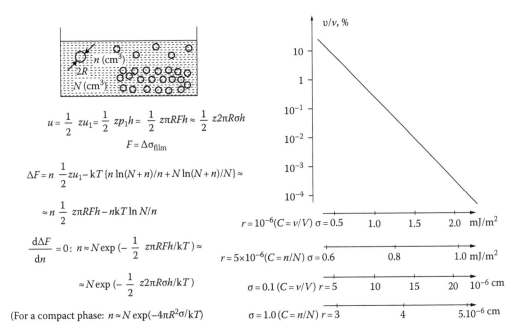

FIGURE 4.37 The concentration (% by volume) of the disperse phase (spherical particles of radius r [or R]) that is insoluble or sparingly soluble in a given dispersion medium in equilibrium with a concentrated structure at a negative minimum of the $\Delta F/kT$ function. The dependence of particle radius, r (cm), at $\sigma = $ const (1); and of the interfacial energy, σ (mJ/m^2), at $r = $ const (2). The calculations are provided for both the $C = v/V$ scheme ($V = 10^3$ cm^3) and the $C = n/N$ scheme with $N = 3.37 \times 10^{25}$.

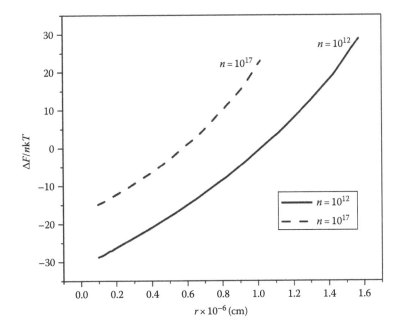

FIGURE 4.38 Free energy per particle, $\Delta F/nkT$, as a function of the particle radius, r (cm), in monodisperse systems at $\sigma = 0.064$ mJ/m^2 for the cases of $n = 10^{12}$ and $n = 10^{17}$ in the $C = v/V$ scheme ($V = 10^3$ cm^3). The $\Delta F/kT$ functions are monotonous and change abruptly by several orders of magnitude upon crossing the $r^{(0)}$ value. (Redrawn from Shchukin, E.D., *J. Dispers. Sci. Technol.*, 25, 875, 2004.)

The description of the entropy factor in this case is similar to that in the case of v = const in the $C = v/V$ scheme. However, the $\Delta F(r)$ dependence for n = const in the $C = v/V$ scheme also takes into account changes in the disperse phase volume, $v = n\,^4/_3\pi r^3$, which is an additional factor contributing to the facilitated dispersion process with decreasing r.

The $\Delta F(r)$ curve decays *monotonously* with a decrease in r and passes through zero at the same ratio of critical parameters as in the aforementioned described cases. In the present case, the condition of $\Delta F = 0$ also constitutes the necessary condition for spontaneous dispersion leading to the formation of a stable colloidal system.

It may initially appear that such an analysis of ΔF in systems that are monodisperse at each given stage is of a somewhat *virtual* nature, similar to the case described for v = const. However, in the present case, there is a direct link to various physical processes. These include the gradual dissolution (or evaporation) of the particles due to changes in the medium content (i.e., transition from coagulation to peptization), as well as the opposite processes, in which the growth of the disperse phase nuclei particles is retarded. In the latter case, a transition from a free-disperse system to a connected-disperse system takes place. Depending on the circumstances, this can be either desirable or undesirable. This is of relevance in many industrial processes based on sol–gel transitions or the hydration hardening of mineral binders [31,53,83].

4.3.4 Influence of the Interfacial Tension, σ, and Temperature, T, on ΔF

Let us now focus on a detailed discussion of the free interfacial energy, σ, which can also be treated as a parameter in the ΔF expression. While there is no need to supply additional quantitative data for analyzing the role of σ, it is worth emphasizing the conditions under which changes in σ can take place.

The interfacial tension can undergo significant changes if the polarity of the medium is altered, such as in the stability/coagulation transition caused by the addition of water to hydrophobic silica dispersions in propanol or ethanol [44,52,53]. Also, the addition of small additives of various surface-active substances can have a dramatic effect on the structure and properties of disperse systems and the conditions of transitions [14,16,17,26]. The formation and structure of stable micellar systems and various surfactant association colloids, such as microemulsion systems and liquid crystalline phases formed in various multicomponent water/hydrocarbon/surfactant/alcohol systems with varying compositions and temperatures, have been described in numerous publications [14–22,78,79,84–88]. These studies provide a detailed analysis of the phase equilibria under various conditions and cover all kinds of systems with all levels of disperse phase concentration. Special attention is devoted to the role of low and *ultralow* values of the surface energy at the interfaces. The author's first observations of areas of stable microheterogeneity in two-, three-, and four-component systems were documented in [66–68].

It is worth pointing out here that the problem of establishing the quantitative criteria for spontaneous dispersion should not be viewed as an isolated area of research. It remains an integral part of colloid and surface science with the preservation of its unique characteristic features, such as addressing the solid phases, including both the compact ones, and others consisting of particle structures [31,44,48,51–53,83]. The discussion of the influence of surfactants on the strength of interparticle contacts is addressed in various parts of this book.

In Rehbinder's concept of the stability of emulsions and other disperse systems, the focus is on the lyophilic *structural–mechanical barrier* as a factor responsible for the strong stabilization of disperse systems. This barrier is manifested with the interfacial surfactant adsorption layer formed predominantly with high molecular weight substances (the so-called protective colloids). This barrier on the one hand promotes the formation of a system with substantial mechanical strength that is capable of resisting coalescence and the rupture of the droplets and, on the other hand, is lyophilic with reference to the dispersion medium. The lyophilic nature of the barrier is characterized by a low value of the interfacial energy, σ, on the side facing the dispersion medium. One can thus say that the layer is characterized by small values of the complex Hamaker constant and the free energy of interaction, which are the factors preventing particle coagulation. Under these conditions, a decrease in σ down to 1/100 and 1/1000 of mJ/m² can provide stabilization even in dispersions of large micron-sized particles, such as milk, natural latex, and crude oil–water emulsions.

There is also an extreme case in which a substantial decrease in the interfacial energy, σ, occurs upon the contact of a solid body with a liquid phase that is strongly surface active in relation to that solid. A particular example of such a situation is the contact between a metal and a metallic melt with similar physical-chemical properties. This is the effect of liquid metal embrittlement comprising a dramatic decrease in the strength and plasticity of metals and alloys, leading to a tendency toward spontaneous dispersion [89–91]. Similar phenomena are also possible in the contact of rocks and minerals with some low melting point salts [91]. A typical example of a system that approaches the conditions of a spontaneous dispersion is highly hydrophilic montmorillonite clay in a state of swelling [92,93].

In the context of physical-chemical mechanics, spontaneous dispersion can be viewed as "the other end" of a continuous spectrum of phenomena that involve overcoming the cohesive forces in the condensed phases. In particular, all such phenomena can be viewed as filling a continuous range from a purely mechanical, medium-independent fracture to a spontaneous dispersion induced by the active medium at low values of σ. A significant role played by the active medium in facilitating and significantly accelerating the processes of *mechanical* dispersion, degradation, and wear is discussed in Chapter 7.

Finally, let us briefly address the role of temperature. It is directly included in the equations for the conditions of spontaneous dispersion as a factor to which the impact of σ and of elementary work, w, are compared. In this aspect, it would be interesting to consider dispersion phenomena at very high temperatures, for example, the dispersion of graphite (or diamond) in liquid iron (melted lava).

At the same time, significant changes in the state of a system can result from fairly moderate deviations in T, for example, the changes in the mutual solubility of both the disperse phase and the dispersion medium components, leading to a radical decrease in σ. Typical examples include studies on systems approaching the critical point, T_c (and yet still below the T_c), such as those carried out with binary mixtures of paraffins with moderately polar organic substances, such as oxyquinoline [26,67,68]. In these works, the formation of direct, inverse, and bicontinuous microemulsions had been described and analyzed in comparison with the independently determined values of σ down to 10^{-1}–10^{-2} mN/m.

4.3.5 Concluding Remarks

Within a restriction to the case of dilute monodisperse systems, the analysis of the ratio between the elementary work of dispersion, w, conceived as the work of particle isolation from a compact phase or disperse structure, and the entropy factor, $kT\{\ln [1/C]+1\}$, represents a general, universal approach to the evaluation of the *possibility* of the spontaneous dispersion of a macroscopic phase into colloid size particles (Figure 4.39). In the analysis of the behavior of the ΔF function, while setting the various variables, r, σ, C, T, v, and n constant, the function is determined by the "work of the dispersion factor," which is equal to the "entropy factor." The transition in the dispersion process from positive values for the system free energy to negative ones determines the possibility of the formation of a thermodynamically stable system.

The analysis of the $\Delta F(r)$ function for the case of a constant volume of the disperse phase, v = const, reveals that there is a maximum in the ΔF function. The analysis of the $\Delta F(n)$ function at constant dispersion, r = const, shows that the formation of a stable system corresponds to a negative minimum. For n = const, the $\Delta F(r)$ function is monotonous. The condition $\Delta F = 0$ is realized at a given critical ratio of the system parameters using either of the analytical approaches. If the values of these parameters are physically realistic (r > b), then the $\Delta F = 0$ condition is the

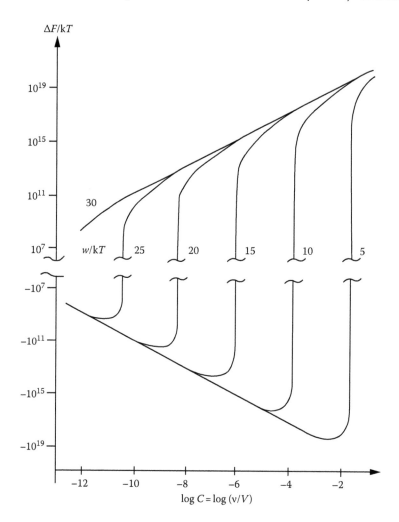

FIGURE 4.39 The relationship between a change in the system free energy, $\Delta F/kT$, for a monodisperse system and the particle concentration for various values of the work of dispersion, w (in kT units), needed to isolate an individual particle from a macroscopic phase, which is either a compact phase ($w = \alpha\delta^2\sigma$) or a globular structure composed of identical particles ($w = 1/2zu_1$). The characteristic steep transitions from positive values of ΔF to negative ones and negative minima at larger dilutions are clearly visible. (Redrawn from Shchukin, E.D., *J. Dispers. Sci. Technol.*, 25, 875, 2004.)

general condition for spontaneous dispersion leading to the formation of a thermodynamically stable (*lyophilic*) colloidal system. This criterion essentially establishes the *necessary* condition for spontaneous dispersion.

There are no principal complications in shifting to an approximate description of spontaneous dispersion in concentrated systems. In this case, both terms in the expression for the entropy of mixing, ΔS, need to be taken into account. The minimal gain in entropy per particle occurs at $N_1 = N_2$. Such an approach can be used for a description of the spontaneous dispersion of polymer phases (compare with the concept of a *good solvent* [94]). Within a set of approximations, a microheterogeneous system containing equal amount of two phases with particle size $r = 0.25 \times 10^{-6}$ cm would be stable at $\sigma \sim 0.04$ mJ/m^2.

Among other factors, the internal stresses inside a solid phase can be subjected to quantitative examination. These stresses constitute a factor of initial metastability facilitating dispersion. This is directly related to the subject of physical-chemical mechanics. Of special interest are biological systems, especially because they allow one to explore the role that the thermodynamic stability of disperse systems plays in living matter, in particularly with regard to fluctuations in temperature.

A thorough quantitative description of polydisperse systems is of principal importance. Such an analysis leads to substantial complications in the numerical treatment but will make the existing approaches more general. It is essential to establish the theoretical grounds for deciding which approach to the calculation of concentrations is more reasonable to use in estimating the contribution of the entropic factor. One can also work backward, that is, obtain an estimate for an entropic factor from experimental studies and then back calculate the concentration of the disperse phase in a lyophilic system. It would be interesting to extend such an approach to disperse systems that are lyophobic under normal conditions but stable at very high dilutions, such as aerosols.

Finally, at the end of this section, we include an essay that illustrates the universal nature of our approach to the determining role played by low interfacial tensions in obtaining finely dispersed systems. This particular example addresses somewhat unusual systems, that is, emulsions formed in a three-component metallic melt at temperatures close to that of complete mutual mixing. The rapid hardening of such an emulsion leads to a substantial decrease in grain size. The pioneering work by Zanozina et al. [95] has pointed to the possibility of effecting fast hardening on a cold metallic surface with high thermal conductivity. When the original work was published, there were no technical capabilities allowing one to move forward in this direction. Later, this approach was realized in technologies that produce "amorphous-like" metallic foils by hardening melts on a copper drum cooled by nitrogen.

4.3.6 Possibility of Obtaining Fine Disperse Structures in Melts by Hardening Melt Emulsions

The preparation of alloys with a given set of mechanical and physical-chemical properties is closely linked to establishing a scientific basis for understanding and controlling the formation of structures in such melts. An increase in the strength of metals and melts is typically achieved by creating a fine disperse microstructure with uniformly distributed ultramicroscopic defects (grain boundaries, block boundaries, phase inclusions, etc.). The formation of such a structure always takes a place in the region of the phase diagram where the solid phase is already present, that is, below the liquidus point and in most cases below the solidus points. At the same time, it should be possible to fine-tune the metallic system in the stage where the system is still liquid. It is not possible to obtain an emulsion solely by treating melts with ultrasound: one needs to operate under conditions of very low interfacial tension between the liquid phases. If the interfacial tension is lowered to ~0.1 erg/cm^2 (1 erg/cm^2 = 0.001 J/m^2) or less [62], then spontaneous formation of fine disperse systems with particle size on the order of 10^{-6} cm is possible. If the interfacial tension is not that low (i.e., ~1 erg/cm^2), the emulsion does not form spontaneously. However, effective emulsification by mechanical means (e.g., by ultrasound) may already be quite effective in producing a stable finely disperse emulsion.

It is known that if substances A and B are partially immiscible in a liquid state, then near the critical point M of this binary system, the interfacial tension between the two phases of close composition can be very low. An example of such a binary system is shown in Figure 4.40. A critical emulsion can be obtained by cooling a homogeneous melt of composition C_m to the vicinity of temperature T_m. Emulsion formation can also be induced by vibration in a temperature range near T_m where the interfacial tension is not too high. The structure of this emulsion may then be preserved by rapid hardening. One may expect that the degree of dispersion in the formed solid will be as fine as that in the starting emulsion or even finer, due to the separation of the emulsion droplets into smaller ones in the crystallizing system. However, this strategy is difficult to implement experimentally, because hardening always take place at a finite rate and the temperature interval between the point M (zero interfacial tension) and the liquidus line is rather broad. The coalescence of the emulsion droplets does therefore take place and can't be prevented by vibration.

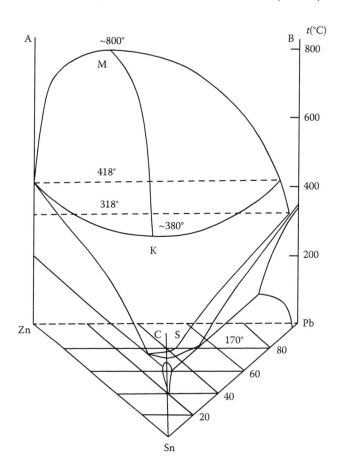

FIGURE 4.40 A schematic phase diagram for a zinc–tin–lead system. (Redrawn from Shchukin, E.D. et al., *Doklady AN SSSR*, 160, 1355, 1965.)

The temperature at which the formation of a critical emulsion takes place can be significantly lowered in a three-component system, such as one in which the components A and B undergo partial phase separation, while being completely miscible with the third, more fusible component C. A zinc–tin–lead system is a particular example of this. In this case, the "dome" (Figure 4.40) limiting from the top the region where the two liquid phases coexist has a critical "ridge," connecting point M with the point K. The latter point already belongs to the liquidus surface and corresponds to a composition of 38% zinc, 28% lead, and 34% tin at the temperature of 380°C. The temperature at point M is not lower than 800°C.

The experiments were conducted in the following way. Samples of 99.99% pure Zn, Pb, and Sn with a total mass of 40 g were placed into cylindrical crucibles (inner diameter 20 mm, height 70 mm, and a wall thickness of 1.5 mm). The crucibles were then mounted on a bar connected to either a low-frequency mechanical vibrator producing vibrations with frequency of ~50 Hz or the conical concentrator of an ultrasonic vibrator operating at a frequency of about 20 kHz. A cylindrical furnace equipped with a type K thermocouple (Chromel/Alumel) allowed one to maintain the desired temperature with an accuracy of 3°C. In certain control experiments, another thermocouple was inserted directly into the melt. The furnace was mounted on vertical guides, which allowed one to move it away quickly at the onset of the hardening process. The temperature in the crucible was first raised to a point 50°C–100°C above the temperature corresponding to the phase separation between the liquid phases. At that point, intensive vibrational mixing of the homogeneous liquid melts was conducted. The temperature was then lowered to a selected point T_1 typically lying within the phase separation region at various distances from the separation "dome." The system was allowed to equilibrate thermally for 30 min. After 28 min had elapsed, the vibration was turned on in most of the experiments. The furnace was then removed, and the matrix was cooled down using a circular shower or tub at a rate of 15°C/s. The solidified matrix was then cut with a saw along the ruled surface and studied by means of metallography. After polishing and electrolytic etching in a chloric/acetic acid mixture for 1–2 min at a current density of 9–15 A/dm² with a tin-plated cathode, the specimens were examined under a microscope. The resulting microphotographs corresponding to various conditions are shown in Figures 4.41 through 4.43.

FIGURE 4.41   Structure of the emulsions with a constant ratio of Zn and Pb concentrations, $C_{Zn}:C_{Pb} = 2.5:1$ and various concentrations of Sn, C_{Sn}: 0% (a), 5% (b), 15% (c), and 20% (d). The hardening temperature was 450°C.

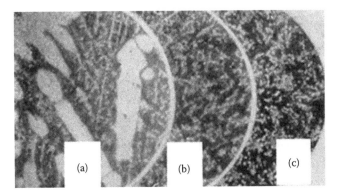

FIGURE 4.42   Structure of the emulsions at a constant composition of the melt (38% Zn, 29% Pb, 33% Sn) and various temperatures of hardening: $T_1 = 375°C$ (a), $T_1 = 400°C$ (b), and $T_1 = 600°C$ (c).

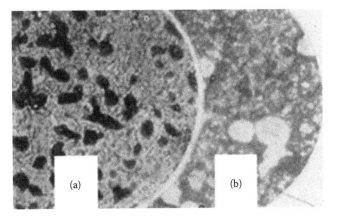

FIGURE 4.43   Structure of the emulsions at a constant tin content, C_{Sn} of 20% and varying Zn to Sn concentration ratios, $C_{Zn}:C_{Sn}$ 6:1 (a), 1:2.3 (b), the hardening temperature was 550°C.

Sets of experiments were conducted at a constant temperature, T_1; a constant ratio of concentrations, $C_{Zn}:C_{Pb}$; and varying Sn concentrations at a fixed concentration of all three components and varying temperature T_1 and at T_1 = const, C_{Sn} = const and varying concentration ratios $C_{Zn}:C_{Pb}$. Various structures, formed upon the hardening of the emulsions over a broad range of concentrations and temperatures both inside and at the surface of the phase separation "dome," as well as the hardening of the systems at point T_1 outside of the dome, were observed.

The structures formed upon the hardening of the emulsions at T_1 = 450°C at low-frequency vibration for a set of compositions with a constant Zn to Pb ratio, $C_{Zn}:C_{Pb}$ = 2.5:1. As the concentration of Sn increased, the starting point moved closer to point K, the interfacial tension at the interface between the two liquid phases decreased, and the structure of the emulsion became finer (Figure 4.41).

The structures formed at a constant composition (38% Zn, 29% Pb, 33% Sn) but at varying temperatures, T_1, are shown in Figure 4.42. In the case shown in Figure 4.42a, the temperature T_1 = 375°C is 5° below the liquidus line, and large dendrites of the zinc-rich phase are seen. The case in Figure 4.42b corresponds to a hardening with low-frequency vibration, which took place from the two-phase liquid system (T_1 = 400°C, which is ~10°C–15°C lower than the phase separation boundary). Finally, Figure 4.42b corresponds to a system with T_1 = 600°C located above the phase separation "dome." The rate of hardening used did not allow the formation of a colloid emulsion nor the preservation of the initial homogeneous structure as the system moved through the two-phase liquid region existing in the temperature interval between 415°C and 380°C. Along with the microphotographs shown in Figure 4.41, this result indicates that in order to obtain and preserve fine emulsion structures, one needs to start the hardening process at a temperature T_1 that is very close to point K in the phase separation region and perform the hardening process rapidly. Rapid hardening is not feasible in the case of a massive large-scale specimen but can be used to make thin films, coatings, and threads.

At constant temperature T_1 = 550°C and constant Sn concentration, C_{Sn} = 20%, rather coarse emulsions were obtained with varying Zn to Sn ratios, as shown in Figure 4.43. These conditions correspond to a horizontal movement along the line located rather far away from the "ridge." These microphotographs illustrate phase inversion phenomena that occurred as the concentration of one of the components was increased. Light droplets correspond to the zinc-enriched phase, while dark droplets represent the phase enriched with lead.

The described results clearly illustrate the transfer of the general concepts of colloid science and spontaneous dispersion to the area of metal science. Using the model fusible system, we have illustrated the possibility of controlling the disperse structure of a solid body that is forming under conditions yielding low interfacial tensions, while it is still in the emulsion state. Conducting similar studies with refractory systems, such as cobalt–copper, molybdenum–copper, or chromium–copper with nickel or iron as the third component, would be of interest.

4.4 CONTACT INTERACTIONS AND THE STABILITY OF SOLS FORMED WITH PHASES OF DIFFERENT NATURE

In the preceding section, we described the general conditions of spontaneous dispersion leading to the formation of a stable colloidal system. In this section, which concludes the current chapter, we will address some specific examples of physical-chemical transitions and equilibria in the coagulation and peptization processes.

The interfacial energy, σ_{12}, at the interface between the disperse phase and the dispersion medium, or in a more general sense—the free energy of interaction, $\frac{1}{2}\Delta\sigma_f(h_0)$, in a direct contact between the particles immersed in a given medium—is the central parameter of the thermodynamic condition $\Delta F(r, \sigma...) < 0$. As in the previous section, ΔF has the meaning of the change in the free energy of the system. At the same time, the estimate of the entropy of dispersion, that is, the entropy of mixing of disperse phase particles and dispersion medium molecules, is not rigorously defined.

For this reason, it is of interest to compare the experimentally determined coagulation conditions with the independently obtained values of the free energy of dispersion, that is, ΔF. To address this, we will illustrate the results of two series of experiments conducted with particles of similar nature immersed in similar media [50–53,69].

The experiments in the first series were conducted with methylated surfaces immersed in aqueous solutions of alkylbenzenesulfonates. The free energy of interaction between methylated glass and quartz spherical surfaces having radii of about 1 mm was studied as described in Section 1.2. Parallel to these studies, the colloid stability of 10 nm particles of methylated Aerosil (i.e., hydrophobized nanoparticles of quartz) suspensions was monitored via turbidity measurements. A characteristic sharp increase in turbidity was observed at the coagulation threshold. This behavior was reversible: an increase in the surfactant concentration resulted in a decrease in turbidity, while dilution of the solutions caused the turbidity to increase.

Both curves, that is, $|\Delta\sigma_f(h_0)| = \Delta\sigma_f$ and $\tau = \tau(C)$, are shown in Figure 4.44. The surfactant concentration corresponding to the *critical coagulation concentration* is obtained from the $\tau = \tau(C)$ curve. This concentration also corresponds to the critical value of the free energy of interaction, $\frac{1}{2}\Delta\sigma_{f(c)}$. The latter has a rather high value—on the order of mJ/m^2 (compare with Section 4.3). Assuming that for methylated surfaces a direct contact with no dispersion medium left in the gap was established, the value of $\frac{1}{2}\Delta\sigma_{f(c)}$ can be used as an estimate of the critical value of the interfacial energy at the solid/liquid interface, σ_c.

At sufficiently high surfactant concentrations, the system becomes hydrophilic, and the methylated Aerosil suspension is stable, that is, the following condition is valid:

$$\Delta F = n\frac{1}{2}z\pi r^2\sigma \, h_0/r - nkT \ln(N/n) < 0$$

$$n \sim N \exp\left[-\frac{(\frac{1}{2}z\pi r^2\sigma \, h_0/r)}{kT}\right]$$

where n and N are the number of particles per unit volume in the colloidal solution and the coagulate, respectively. If one assumes that in the coagulate the aqueous medium drains out of the gap between the particles, then $\sigma \approx \frac{1}{2}\Delta\sigma_f$.

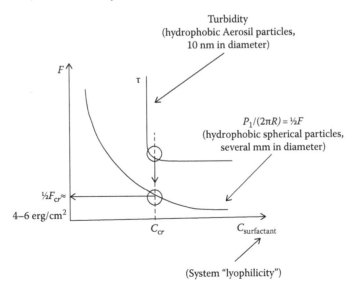

FIGURE 4.44 The measured interaction free energy, F, that is, $\Delta\sigma_f(C)$, and the coagulation of the colloid system monitored by changes in the turbidity, $\tau(C)$, in macroscopic and nanoscale systems of similar nature lyophilized by the addition of surfactant.

In the other words, n is the concentration of the particles in a colloid solution corresponding to an equilibrium between the two disperse phases: a colloid solution of low concentration, n, and an aggregate (i.e., coagulate or precipitate) with a much higher concentration, N. In this latter case, the ratio of $n/N \sim 10^{-4}$ has an exponent of around 10. Assuming that for a loose packing of the particles in the aggregate $\frac{1}{2}z\pi \sim 5$ and $\sigma \approx \frac{1}{2}\Delta\sigma_f \sim 4$ mJ/m^2, an estimate for the last undefined parameter can be obtained, that is, $h_0 \sim 0.2$ nm. The latter value agrees with the probable distance in a contact between methyl groups and thus confirms the validity of the model used. The possible aggregation of the Aerosil introduces a certain degree of uncertainty in the estimate of r.

In another series of experiments, hydrophobic surfaces were used: the free energy of interaction between the hydrophobic macroscopic surfaces, $\frac{1}{2}\Delta\sigma_f$, with a radius ~ 1 mm was compared to the results obtained in the coagulation measurements of a hydrophobic Aerosil sol with particles ~ 10 nm in size. However, this time the experiments were conducted in a hydrocarbon medium, so that the system changed from lyophilic to lyophobic. This transition was caused by the addition of ethanol or propanol, which increased the polarity of the medium. The $\frac{1}{2}\Delta\sigma_f$ value is low in pure hydrocarbon (see Section 1.3) and increases with increases in the alcohol concentration. At a particular critical value, $\frac{1}{2}\Delta\sigma_{f(c)}$, which is on the order of several mJ/m^2, the turbidity in the system increases due to coagulation, in complete agreement with the previous series of experiments conducted in surfactant solutions.

It was of interest to evaluate the extent to which the results in these experiments were in agreement with DLVO theory. Figure 4.45 shows the results of experiments in which rather large amounts of KCl salt were added to aqueous solutions of propanol.

As seen in Figure 4.45, the addition of KCl shifted the onset of coagulation in the $\tau(C)$ curve toward lower water concentrations, that is, the suppression of the electrostatic barrier took place at a higher starting lyophilicity of the system. At the same time, the addition of electrolyte did not cause any changes in the $\frac{1}{2}\Delta\sigma_f$ (C) curve. In order to explain this behavior, let us recall the basics of DLVO theory and the application of Derjaguin's theorem, $\Delta\sigma_f = \pi R p_1$, to the analysis of particle–particle interactions, that is, let us revisit the integrals of the disjoining pressure. As has been discussed already in Section 4.1, the ability of a Brownian particle to overcome a potential barrier is at the center of DLVO theory. This implies the possibility of a transition from either a free state in a colloidal solution or a shallow secondary potential energy minimum to a primary potential energy minimum. The latter is an irreversible process. The framework of DLVO theory is restricted to the use of one positive and one negative component of both of the parameters, that is, $\Pi(h)$ and $\Delta\sigma_f(h)$, and by itself does not describe the potential energy minimum. At the same time, detailed data on the depth of the potential energy minimum are very important for interpreting the results obtained in the measurement of the cohesive forces in a contact and the observed reversibility of the coagulation–peptization transitions. Equally important is the interpretation of the observed absence of the influence of electrolyte on the values of the interaction free energy.

Indeed, measurement of the cohesive force and the free energy of interaction in the equilibrium contact provides one with information on the depth of the primary minimum. The minimum can be either deep, as in lyophobic systems, or shallow as in lyophilic ones. However, this interpretation is not rigorous, as one must interpret what is the exact meaning of "deep" and "shallow." Obviously, the data on the free energy of interaction in a flat–parallel gap do not provide any means for answering this question. A general thermodynamic answer was provided in the preceding discussion of spontaneous dispersion. Namely, it is the relationship between the free energy (i.e., work, w) of the particle separation from a macroscopic phase or from an aggregate and the change in the free energy when a separated particle enters into Brownian motion (i.e., the entropic factor, $-kT \ln(1/C) < 0$). Realistic values for this logarithmic term, when the dilution in the colloidal solution is accounted for, are typically on the order of 10–20 (or even 10–15) and do not fall outside of the range from 5 to 25. Indeed, the lower limit of 5 would correspond to essentially unachievable high concentrations in extremely lyophilic colloidal sols, while the upper boundary of 25 would correspond to negligibly low concentrations in lyophobic systems. It is worth referring to the experimental data in order to obtain a quantitative assessment of the work w. From the preceding text, one can get some

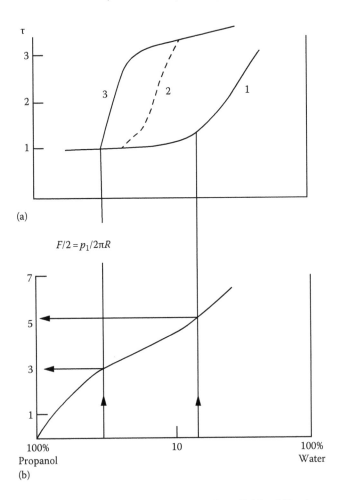

FIGURE 4.45 The free energy of interaction F (i.e., $\Delta\sigma_f$) and the colloid stability (represented by the change in turbidity) in comparable macroscopic and *nanoscopic* systems lyophilized by a gradual increase of the water content. The addition of electrolyte promotes coagulation (transition from curve 1 to curve 3) but does not have any significant impact on the value of F. (From Shchukin, E.D. and Yaminskiy, V.V., *Colloids Surf.*, 32, 19, 1985.)

idea of the real range of w values for the reversible transition between coagulation and peptization. The latter corresponds to an equilibrium between the "macroscopic phase" (coagulation structure) and a colloidal solution of a realistic concentration. The values of w thus cover a relatively narrow spectral band, that is, $10kT < w < 15–20kT$, or $5–25kT$ in the most extreme cases.

One may ask whether the specific data needed to estimate w directly are available. In case of the separation of spherical particles from the macrophase $w = 4\pi R^2\sigma$, the principal difficulty was to obtain an estimate for the interfacial free energy $\sigma = \sigma_{12}$. In contrast to this, the present case corresponds to a colloidal dispersion of insoluble particles and a macroscopic phase composed of the aggregate of such particles. In this case, the work of particle isolation is in fact determined by the work of cohesion of the particles in an individual contact, w_1, and the average number of neighboring particles in the aggregate, that is, the coordination number z: $w = \frac{1}{2}zw_1$. The multiplier $\frac{1}{2}$ indicates that each bond is shared by two neighboring particles. Examples of typical structures with various values of z are discussed in the next chapter. In the case of a moderately lyophobic system with strong particle cohesion, fairly loose structures with $z \sim 3–4$ form. Here, we must again focus on what thermodynamics provides us with in efforts to estimate the value of w_1, that is, we are interested in particular in the *free energy of cohesion between two particles* and not the specific free energy of interaction per unit area. In principle, in contrast to the quadruple integration of the interaction potential between molecules utilized in the determination of $\Delta\sigma_f(h)$, this requires a sixfold integration over the particle volume. Formal calculations yield expressions conforming to the models chosen; however, for nanosize particles, one encounters difficulties associated with the effective area of contact (Section 4.1).

Let us examine the result that is obtained by applying Derjaguin's theorem to the contact interactions, $p_1(h) = \pi R \Delta\sigma_f(h)$. This is the case for coagulation contacts, that is, only the *short-range* forces essential in physical-chemical mechanics are considered.

The starting point in the description of the thermodynamics of a contact between the surfaces is the disjoining pressure per unit area in a flat–parallel gap, that is, $\Pi(h)$, where $h = h_0$ is some characteristic distance between the molecules in the zone of contact. The value of h_0 is on the order of several tenths of a nanometer. For dispersion interactions, the relationships between this starting characteristic and its integrals are as follows:

$$\Pi(h) = -\frac{A*}{6\pi h^3}$$

and

$$\left|\Pi(h_0)\right| = \frac{A*}{6\pi h_0^3}\,(\mathrm{mJ/m^3 = mN/m^2})$$

$$-\int_\infty^h \Pi(h)dh = \Delta\sigma_f(h) = -\frac{A*}{12\pi h^2}$$

$$\left|\Delta\sigma_f(h_0)\right| = \Delta\sigma_f = \frac{A*}{12\pi h_0^2}\,(\mathrm{mJ/m^2})$$

The free energy of interaction (free energy of cohesion for dispersion forces), $\Delta\sigma_f$, is the main invariant with the respect to the surface geometry characteristic. Integrating $\Pi(h)$ twice over the cross section of the spherical particles in the zone of contact yields the cohesive force between two particles, that is, a parameter that can be directly measured in experiments with macroscopic bodies with spherical or cylindrical surfaces, that is,

$$p_1(h) = \pi R \Delta\sigma_f(h) = -\frac{A*R}{12h^2}$$

and

$$\left|p_1(h_0)\right| = -\frac{A*R}{12h_0^2} = p_1$$

The main parameter correlating the coagulation–peptization transition is the third integral of $\Pi(h)$

$$w_1(h) = -\int_\infty^h p_1(h)dh = -\frac{A*R}{12h}$$

which defines the energy. For a direct contact, this energy is the energy of cohesion:

$$\left|w_1(h_0)\right| = \frac{A*R}{12h_0} = w_1$$

This relationship requires an independent estimate for h_0. Alternatively, it can be used to estimate the value of h_0 from approximate estimates of $w_1(h_0)$ determined experimentally for the coagulation–peptization transition. In the general case of nondispersion interactions, we only have

$$-\int_{\infty}^{h} p_1(h)\mathrm{d}h = p_1(h_0)h_{\mathrm{eff}} \qquad (4.19)$$

where the values of h_{eff} may cover a broad range depending on the nature of the forces. It can range from less than 10 Å to hundreds or thousands of Angströms. In contrast to the second integral, the quantitative estimation of the third integral becomes impossible if additional data are not available. This is the so-called problem of the third integral.

Here, we can again turn to Derjaguin's theorem. The following is a summary of what one can achieve by using it:

1. First of all, it allows one to obtain an invariant characteristic, $\Delta\sigma_f(h)$, for the interaction forces of all natures for the gaps $h \ll R$ from a direct force measurement.
2. The $\Delta\sigma_f$ value where $\sigma_f = 0$, that is, for the ideal contact with no residual film present, yields an estimate for the interfacial free energy, $\sigma = \sigma_{12} \approx \frac{1}{2}\,\Delta\sigma_f$.
3. A very important feature of the $p_1(h) = \pi R \Delta\sigma_f(h)$ dependence is the linearity with respect to the particle size, R. This feature allows one to extrapolate the forces measured between macroscopic particles, $p_1(R_m)$, and to obtain the forces between nanoparticles that cannot be directly measured, that is, $p_1(R_n) = p_1(R_m)\,R_n/R_m$.
4. By employing Equation 4.19 and using an additional estimate for h_0 or h_{eff}, one can determine the value of w_1, which, in comparison with kT, can be used as a criterion for strong or weak cohesion, that is, a measure of the depth of the primary *potential energy minimum.*

In order to establish reasons for the limited sensitivity of $\Delta\sigma_f$ (i.e., of the measured force p_1) to the presence of electrolytes in experiments with *hydrophobic* particles in hydrocarbon/alcohol mixtures, one can compare these results with the results of measurements in the aqueous medium. The latter represent the main subject of DLVO theory. The final discussion in this chapter is devoted to addressing the relationship between the contact interactions (i.e., cohesion forces at the primary potential energy minimum) and the results of DLVO theory (i.e., mainly long-range forces). Some of these experiments were conducted by Yaminskiy [30,50–52]. In addition to the contact forces, the $p_1(h)$ and $\Delta\sigma_f(h)$ isotherms shown in Figure 4.46 were also determined.

The experiments were conducted with hydrophilic surfaces (glass or quartz) in aqueous solutions of a cationic surfactant, CPB. The cohesive force $p_1(h_0) = p_1$ and the free energy of interaction $|\Delta\sigma_f(h_0)| = \Delta\sigma_f$ are negligible in pure water (see Sections 1.2 and 2.2) and increase as the surfaces become more hydrophobic with the addition of surfactant. The addition of a *neutral* electrolyte does not significantly influence these parameters; however, the hydrophilic Aerosil sols reveal coagulation at a particular concentration of the surfactant, in agreement with DLVO theory. This transition took place at values of $\Delta\sigma_f$ on the order of several mJ/m^2, which is similar to the $\Delta\sigma_f$ value reported for experiments with hydrophobic surfaces.

The discovered *insensitivity* of $p_1(h)$ and $\Delta\sigma_f(h)$ to electrolytes can be explained using the $\Delta\sigma_f(h)$ isotherms of DLVO theory. By examining the isotherm on a proper scale, one can see sharp quantitative differences in the geometry of the primary minimum and the electrostatic barrier. The $\Delta\sigma_f(h)$ isotherms are analogous to the isotherms of force for particles of radius R in view of Derjaguin's theorem. The depth of the primary minimum, $|\Delta\sigma_f(h_0)| = \Delta\sigma_f = \Delta\sigma_f^{(\mathrm{prim})}$, is on the order of several mJ/m^2, and its effective length, $h_{\mathrm{eff}} = \Delta h^{(\mathrm{prim})}$, does not exceed fractions of a nanometer. The length of electrostatic barrier, $\Delta h^{(\mathrm{el})}$, may stretch to tens of nm, while its height, $\Delta\sigma^{(\mathrm{el})}$, is on the order of only 1/100 of a nm. The ratio $\Delta\sigma_f^{(\mathrm{prim})}/\Delta\sigma_f^{(\mathrm{el})}$ reflects both the strength of the molecular attraction in

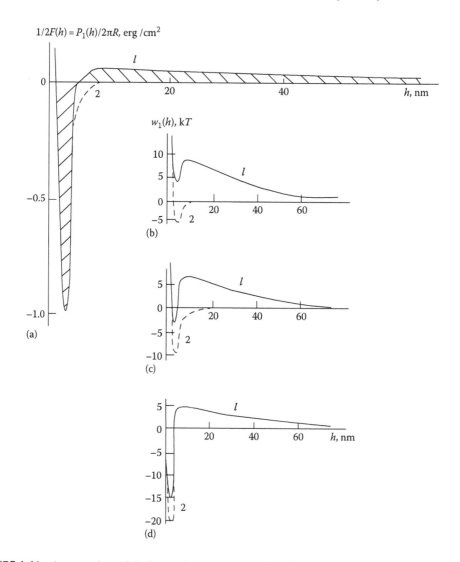

FIGURE 4.46   A comparison of the force–distance curve, $p_1(h)$, and the free energy of interaction, $F(h) = \Delta\sigma_f$ (h) (a), with the energy of interaction between particles, $w_1(h)$ (b–d). These results explain why $F(h)$ and $p_1(h)$ are insensitive to the addition of electrolyte, while the latter makes a significant impact on the $w_1(h)$ curve and the primary minimum depth. This impact is visible when the minimum is ~ $10kT$. (Data from Shchukin, E.D. and Yaminskiy, V.V., *Colloids Surf.*, 32, 19, 1985.)

the contact and the strength of the electrostatic repulsion. The latter is well below the precision of the measurement and hence is not *visible*. This is the case with the first and second integrals of the disjoining pressure. A completely different behavior of the electrostatic component of the disjoining pressure is observed in the third integral, $w_1(h) = -\int_{\infty}^{h} p_1(h)dh$. In this case, instead of the depths and heights of minima and maxima, one deals with *areas* that are on the order of $p^{(\text{prim})}\Delta h^{(\text{prim})}$ and $p^{(\text{el})}\Delta h^{(\text{el})}$. These areas can be both comparable to each other and different (larger or smaller) from each other.

It is therefore possible to provide a quantitative solution to the *third integral problem*, that is, to estimate the energy of cohesion, w, between nanoparticles and to assess the influence of the electrolyte concentration on it. The latter corresponds to the depth of the primary minimum in the $w(h)$ plot. Experiments indicate that the DLVO electrostatic barrier can't have a significant influence on the value of $\frac{1}{2}\Delta\sigma_f$, even if the latter is lowered to the order of several mJ/m^2—it is still much larger than the height of the electrostatic barrier, $\Delta\sigma_f^{(el)} = U^{(el)}$, which does not exceed several hundredths of mJ/m^2 (0.1 mJ/m^2 in the most extreme case). For this reason, measurements of the force, $p_1(h)$, yield nearly identical results regardless of the presence or absence of salt. At the same time, the integral of the force, $p_1^{(el)}(h) = U^{(el)}(h)\pi R_a$, at a large distance $\Delta h^{(el)}$ may significantly contribute to $w^{(el)}$. For instance, if we assume that $U^{(el)} = 0.1$ mJ/m^2, the radius of the Aerosil particles, $R_a = 10$ nm and $\Delta h^{(el)} = 15$ nm, we find for $w_1^{(el)} \sim U^{(el)}(h)\pi R_a \Delta h^{(el)} \sim 4\times10^{-13}$ erg $= 4\times10^{-20}$ J $\sim 10kT$. The estimate of the energy of cohesion in the first primary minimum $w_1 \sim \frac{1}{2}\Delta\sigma_f \approx 2\pi R_a\, h_{eff}$ (in the absence of a potential energy barrier), at $R_a = 10$ nm, $h_{eff} = \Delta h^{(prim)}$ is on the order of several Å, and $\frac{1}{2}\Delta\sigma_f = \frac{1}{2}\Delta\sigma_f^{(prim)} \sim$ mJ/m^2. This yields $w_1 \sim (10\text{–}20)\ kT$.

The aforementioned result means that, as long as the free energy of interaction does not exceed a few mJ/m^2, the system is stable, even in the presence of an electrolyte. This condition is still quite far from that of *complete lyophilicity*. In the absence of electrolyte, that is, in the presence of a potential barrier, the stability is maintained at higher degree of lyophilization (controlled by the surfactant), because the value of $w_1^{(el)}$ is subtracted from w_1.

Here, it is worth highlighting a principally new view on the problem: we are considering a *thermodynamic factor*. We are not focusing on the kinetic role of the DLVO potential barrier, but on its possible influence on the depth of the primary potential energy minimum of the $w(h)$ function. That is, we are examining the role of the potential barrier on the energy of particle cohesion.

The results of experiments in which the ionic strength was altered by the addition of NaCl are presented in Figure 4.46. This figure shows three characteristic types of effects caused by simultaneous changes in the depth of the primary minimum owing to surface hydrophobization and by changes in the potential barrier height owing to the addition of electrolyte. It is worth noting here that the results presented in Figure 4.46 are not representative of a flat–parallel barrier, that is, $\Delta\sigma_f$, but of the cohesive energy between individual nanoparticles (in the units of kT). This figure shows three curves corresponding to three different CPB concentrations, both in the absence of any electrolyte and in the presence of 10^{-1} mol/dm^3 NaCl. The lowest CPB concentration, $C = 10^{-6}$ mol/dm^3, corresponds to the lowest degree of hydrophobization, which produces a shallow potential energy minimum $\sim 3\text{–}5kT$. In the absence of any electrolyte, that is, in the case of a well-developed electrostatic barrier, the curve in the vicinity of h_0 is lifted by $\sim 10kT$, and the minimum is positive. When the potential barrier is suppressed, the minimum is negative, but its depth is only $3\text{–}5kT$. This depth is lower than the entropic factor $kT \ln (N/n) \sim 10kT$ and is not sufficient to cause an irreversible coagulation. Consequently, under these conditions, the system remains lyophilic and stable toward coagulation, irrespective of the presence of or the amount of electrolyte.

The lower curve corresponds to a high CPB concentration, $C \sim 2 \times 10^{-5}$ mol/dm^3. Under these conditions, the potential energy minimum is rather deep, that is, it is on the order of *tens* of kT. Under these conditions, the curve is *lifted* around h_0 by a positive amount, $w_1^{(el)} \sim 10kT$. This *lifting* does not significantly influence the curve position, as the depth of the minimum still exceeds $10kT$. Consequently, the thermal motion is not capable of moving the particle out of the deep primary minimum, and the system is thermodynamically unstable, that is, does coagulate in the presence of electrolyte.

An interesting case is that of a moderate degree of lyophilization, when the depth of potential energy minimum, w_1, is comparable to the height of potential energy barrier, $w_1^{(el)}$. This case corresponds to a CPB concentration, $C = 5 \times 10^{-6}$ mol/dm^3. In the presence of an electrolyte, the minimum is substantially deep and can ensure irreversible coagulation, but in the absence of an electrolyte, the system is lyophilized and is stable.

The described approach introduces a principally new way of looking at the problem of the transition between lyophilicity and lyophobicity in fine disperse colloidal systems, both free disperse and connected disperse.

Finally, let us again turn to the central problem of the physical-chemical mechanics of concentrated disperse systems, namely, the role of the contact interactions in the structural–rheological properties of coagulates and the possibility of controlling these properties in various disperse systems. In Chapter 3, we discussed the role of the contact interactions as a factor responsible for the resistance of connected-disperse systems to mechanical destruction. This resistance can manifest itself in many different ways, namely, as viscosity, plasticity, or mechanical strength. In general, the strength of the interparticle contacts is responsible for the stability of concentrated connected-disperse colloidal systems against deformation and destruction. This situation is in principle the opposite of the case of dilute free-disperse systems, in which *stability* means the tendency of preserving the free-dispersed state.

Let us now briefly address the phenomenon of *sedimentation,* which is closely related to the interaction between the particles in suspensions, sols, and gels formed from sols. In the course of sedimentation, a sediment or a gel is formed as a result of either gravitational or centrifugal forces. It was noted long ago that the sedimentation process is strongly dependent on the properties of the dispersion medium and on the disperse phase particles. When the cohesion is strong, the particles in the prepared suspension quickly form a loose 3D network with low coordination numbers between adjacent particles. Typical coordination numbers may be three to four or less, but they are always greater than two. Such a loose structure will settle as a whole and with time become dense. A distinct boundary between the clear liquid and the turbid sediment is developed. This boundary gradually moves downward until an equilibrium *sediment volume* has been reached. Such a process may be referred to as *structural sedimentation* and is a characteristic of the lyophobic nature of the system. One can quantitatively compare the sediment volume to the particle–particle interactions studied in similar media.

Sedimentation occurs differently when the cohesive forces between particles are weak (in comparison to kT). Under these conditions, coagulation does not take place, and the particles settle individually. Large particles settle first and are followed by the smaller particles. The sediment forming at the bottom gradually grows upward, and the dispersion medium in the upper portion of the vessel gradually clears up. The sediment in this case is two to three times denser than that formed in *structural sedimentation.* The force of gravity (i.e., particle weight) competes with the cohesive force in the contacts, and the coordination number is much higher, that is, six (primitive cubic packing) or even eight (volume-centered cubic packing). This situation is typical of lyophilic systems. This type of settling process may be referred to as *nonstructured sedimentation,* because no 3D network is formed in the bulk. However, that definition is not precise, because in both cases, one ends up with a concentrated connected-disperse system—more loose in the first case and more dense in the second case. These two types of sediment may have very different properties. Strong contacts between the particles in lyophobic sediment result in the stronger resistance of the sediment to the action of external forces—it has a higher effective viscosity and plasticity. Lyophilic sediment maintains high mobility and low viscosity, even at high coordination numbers due to the weak contacts between the particles.

The methods and quantitative characteristics used in the analysis of the contact interactions between the particles can be applied to the processes involving the formation of sediments. This can be verified experimentally in sedimentation experiments with hydrophobic spherical particles (e.g., methylated glass or fluoropolymers) in a medium of a different polarity. The sedimentation of hydrophobic particles in low-polarity media, such as propanol or butanol, is of *lyophilic type.* The sedimentation of such particles in polar media, such as water or ethylene glycol, follows the *lyophobic scheme.* Similar to the case of experimental contact interaction studies, one can also observe a gradual transition from one type of sedimentation to the other. One particular example is the sedimentation of hydrophobic particles in aqueous solutions of alcohols. As the system becomes more lyophilic with the shift to alcohols with lower polarity or with an increase in the alcohol concentration, a transition from lyophobic sedimentation to lyophilic sedimentation takes place. A gradual decrease in the volume of the sediment is observed. Small additives of surfactants that make the surface of the particles hydrophilic have a similar effect. The concentration of an alcohol or a surfactant at which the transition between hydrophilic and hydrophobic sedimentation takes place can be used as a quantitative physical-chemical metric characterizing the transition.

This brief description of the processes of the formation of particle sediments is important in numerous applications. Many of these abundant processes are governed by contact interactions and their critical role in controlling the properties of the dense or porous structure that is being formed.

REFERENCES

1. Rehbinder, P. A. 1978. *Selected Works. Surface Phenomena in Disperse Systems: Colloid Chemistry.* Moscow, Russia: Nauka.
2. Adamson, A. 1979. *Physical Chemistry of Surfaces.* Moscow, Russia: Mir.
3. Gibbs, J. W. 1982. *Thermodynamics: Statistical Mechanics.* Moscow, Russia: Nauka.
4. Derjaguin, B. V. 1935. Studies on the external friction and adhesion. V. Theory of adhesion. *Zh. Fizicheskoy Himii.* 6: 1306–1319.
5. Churaev, N. V., Derjaguin, B. V., and V. M. Muller. 1987. *Surface Forces.* New York: Plenum Publishing Corporation.
6. Rusanov, A. I. 1967. *Phase Equilibria and Surface Phenomena.* Leningrad, Russia: Himiya.
7. Sonntag, H. and K. Strenge. 1972. *Coagulation and Stability of Disperse Systems.* New York: Wiley.
8. Vold, R. D. and M. J. Vold. 1983. *Colloid and Interface Chemistry.* London, U.K.: Addison-Wesley.
9. Fridrihsberg, D. A. 1984. *Colloid Chemistry*, 2nd edn. Leningrad, Russia: Himiya.
10. Israelashvili, J. N. 1985. *Intermolecular and Surface Forces.* Orlando, FL: Academic Press.
11. Lyklema, J. 1991–2000. *Fundamentals of Interface and Colloid Science,* Vols. 1–4. Orlando, FL: Academic Press.
12. Hunter, R. J. 1994. *Introduction to Modern Colloid Science.* Oxford, U.K.: Oxford University Press.
13. Van Oss, C. J. 1994. *Interfacial Forces in Aqueous Media.* New York: Marcel Dekker.
14. Joensson, B., Lindman, B., Kronberg, B., and K. Holmberg. 1998. *Surfactants and Polymers in Aqueous Solution.* New York: Wiley.
15. Evans, D. F. and H. Wennerstrom. 1999. *The Colloidal Domain: Where Physics, Chemistry, Biology, and Technology Meet,* 2nd edn. New York: Wiley-VCH.
16. Mittal, K. L. and P. Kumar. (Eds.). 2000. *Emulsions, Foams, and Thin Films.* CRC Press/Taylor & Francis Group.
17. Holmberg, K. (Ed.). 2002. *Handbook of Applied Surface and Colloid Chemistry,* Vol. 2. New York: Wiley.
18. Norde, W. 2003. *Colloids and Interfaces in Life Sciences.* CRC Press/Taylor & Francis Group.
19. Cosgrove, T. (Ed.). 2005. *Colloid Science: Principles Methods and Applications.* New York: Wiley.
20. Zana, R. (Ed.). *Dynamics of Surfactant Self-Assemblies: Micelles, Microemulsions, Vesicles and Lyotropic Phases.* CRC Press/Taylor & Francis Group.
21. Stechemesser, H. J. and B. Dobias. (Eds.). 2005. *Coagulation and Flocculation,* 2nd edn. CRC Press/Taylor & Francis Group.
22. Somasundaran, P. (Ed.). 2006. *Encyclopedia of Surface and Colloid Science.* CRC Press/Taylor & Francis Group.
23. Gregory, J. 2006. *Particles in Water.* Boca Raton, FL: CRC Press.
24. Zana, R. and E. W. Kaler. (Eds.). 2007. *Giant Micelles: Properties and Applications.* Boca Raton, FL: CRC Press.
25. Soloshonok, V. A., K. Mikami, T. Yamazaki, J. T. Welch, and J. Honek. (Eds.). 2007. *Current Fluoroorganic Chemistry.* Oxford, U.K.: Oxford University Press.
26. Shchukin, E. D., Pertsov, A. V., Amelina, E. A., and A. S. Zelenev. 2001. *Colloid and Surface Chemistry.* Amsterdam, the Netherlands: Elsevier.
27. Miller, C. A. and P. Neog. (Eds.). 2007. *Interfacial Phenomena: Equilibrium and Dynamic Effects,* 2nd edn. CRC Press/Taylor & Francis Group.
28. Fanun, M. (Ed.). 2009. *Microemulsions: Properties and Applications.* CRC Press/Taylor & Francis Group.
29. Birdi, K. S. 2009. *Surface and Colloid Chemistry: Principles and Applications.* CRC Press/Taylor & Francis Group.
30. Steblin, V. N., Shchukin, E. D., Yaminskiy, V. V., and I. V. Yaminskiy. 1991. Hydrodynamic interaction of surfaces in electrolyte solutions. A new method of surface forces studies using capacitor ultradynamometer. *Kolloidnyi Zh.* 53: 684–687.

31. Shchukin, E. D. 2002. Surfactant effects on the cohesive strength of particle contacts: Measurements by the cohesive force apparatus. *J. Colloid Interface Sci.* 256: 159–167.

32. Shchukin, E. D., Amelina, E. A., and V. N. Izmaylova. 2003. The lyophilic structure-mechanical barrier as a factor of dispersion strong stabilization. In *Role of Interfaces in Environmental Protection*, S. Barany (Ed.), pp. 81–90. Amsterdam, the Netherlands: Kluwer Academic Publishers.

33. Shchukin, E. D. and A. V. Pertsov. 2007. Thermodynamic criterion of spontaneous dispersion. In *Colloid and Interface Science Series*, T. Tadros (Ed.), Vol. 1, pp. 23–47. New York: Wiley-VCH.

34. Shchukin, E. D., Amelina, E. A., and A. M. Parfenova. 2008. Stability of fluorinated systems: Structure-mechanical barrier as a factor of strong stabilization. In *Highlights in Colloid Science*, D. Platikanov and D. Exerova (Eds.), pp. 41–53. New York: Wiley-VCH.

35. Izmaylova, V. N., Alexeeva, I. G., Shchukin, E. D., and P. A. Rehbinder. 1972. Rheological properties of interfacial adsorption layers of proteins and surfactants. *Doklady AN SSSR.* 206: 1150–1153.

36. Izmaylova, V. N., Alexeeva, I. G., Shchukin, E. D., and P. A. Rehbinder. 1973. Rheological properties of interfacial adsorption layers of gelatin on the boundary with benzene. *Kolloidnyi Zh.* 35: 860–866.

37. Izmaylova, V. N., Yampolskaya, G. P., and B. D. Summ. 1988. *Surface Phenomena in Protein Systems.* Moscow, Russia: Himiya.

38. Shchukin, E. D., Amelina, E. A., Parfenova, A. M. et al. 1988. The influence of the nature of the non-polar phase on the stabilizing effect of adsorbed layers of nonionic surfactants. *Kolloidnyi Zh.* 50: 790–794.

39. Parfenova, A. M., Amelina, E. A., Vitvitskiy, V. M. et al. 1990. The effect of various Pluronics on the resistance to coalescence of perfluorodecaline droplets. *Kolloidnyi Zh.* 52: 800–803.

40. Amelina, E. A., Kumacheva, E. E., Pertsov, A. V., and E. D. Shchukin. 1990. Stability of perfluorocarbon emulsions towards Ostwald ripening and flocculation. *Kolloidnyi Zh.* 52: 216–220.

41. Shchukin, E. D., Amelina, E. A., and A. M. Parfenova. 2001. Influence of the nature of non-polar phase on the mechanical stability of adsorption layers of hydrocarbon and fluorocarbon surfactants at the interface between their aqueous solutions and non-polar media. *Colloids Surf.* 176: 35–51.

42. Amelina, E. A., Kumacheva, E. E., Chalyh, A. E., and E. D. Shchukin. 1996. The structure of interfacial adsorption layers in systems of an aqueous solution of a block copolymer of ethylene oxide and propylene—Fluorocarbon. *Kolloidnyi Zh.* 58: 437–441.

43. Kabalnov, A. S. and E. D. Shchukin. 2002. Ostwald ripening theory: Applications to fluorocarbon emulsions stability. *Adv. Colloid Interface Sci.* 38: 69–97.

44. Shchukin, E. D. and E. A. Amelina. 2003. Surface modification and contact interaction of particles. *J. Dispers. Sci. Technol.* 24: 377–395.

45. Shchukin, E. D., Krasnov, M. M., Izmaylova, V. N. et al. 1994. On the irreversible stretching of eye biopolymers. *Kolloidnyi Zh.* 56: 463–464.

46. Shchukin, E. D., Izmaylova, V. N., Krasnov, M. M. et al. 1997. The effect of the active medium on the eye sclera creep. *Vestnik Ophtalmol.* 3: 3–4.

47. Shah, D. O. (Ed.). 1998. *Micelles, Microemulsions, and Monolayers.* New York: Marcel Dekker.

48. Claesson, P. M., Ederth, T., Bergeron, V., and M. W. Rutland. 1996. Techniques for measuring surface forces. *Adv. Colloid Interface Sci.* 67: 119–183.

49. Christensen, H. K. and P. M. Claesson. 2001. Direct measurements of the force between hydrophobic surfaces in water. *Adv. Colloid Interface Sci.* 91: 391–436.

50. Shchukin, E. D., Amelina, E. A., and V. V. Yaminskiy. 1982. Thermodynamic equilibrium between coagulation and peptization. *J. Colloid Interface Sci.* 90: 137–142.

51. Yaminskiy, V. V., Pchelin, V. A., Amelina, E. A., and E. D. Shchukin. 1982. *Coagulation Contacts in Disperse Systems.* Moscow, Russia: Himiya.

52. Shchukin, E. D. and V. V. Yaminskiy. 1985. Thermodynamic factors in the sol-gel transition. *Colloids Surf.* 32: 19–55.

53. Shchukin, E. D. 1996. Some colloid-chemical aspects of the small particles contact interactions. In *Fine Particles Science and Technology*, E. Pelizzetti (Ed.), pp. 239–253. Amsterdam, the Netherlands: Kluwer Academic Publishers.

54. Yushchenko, V. S., Edholm, O., and E. D. Shchukin. 1996. Molecular dynamics of the particle/substrate contact rupture in a liquid medium. *Colloids Surf.* 110: 63–73.

55. Banks, R. E. 1982. Preparations, properties and industrial applications of organofluorine compounds. In Ellis Horwood Series in Chemical Science. Ellis Horwood Ltd, New York, Brisbane, Chichester, Toronto: John Wiley and Sons Ltd.

56. Davis, S. S., Perwall, T. S., Buscal, R. et al. 1987. *Proceedings of the International Conference on Colloid and Interface Science*, Vol. 2, p. 265. New York: Academic Press.

57. Kiss, E. (Ed.). 2001. Fluorinated surfactants and repellants. In Surfactants Science Series, CRC Press/ Taylor & Francis Group.
58. Kirsch, P. 2004. *Modern Fluoroorganic Chemistry.* New York: Wiley-VCH.
59. Fowkes, F. M. 1972. Donor-acceptor interactions at interfaces. *J. Adhes.* 4: 155–159.
60. Volmer, M. 1927. Zur Theorie der lyophilen Kolloiden. *Z. Phys. Chem.* 125: 151–157.
61. Volmer, M. 1957. Die kolloidale Natur von Fluessigkeitsgemischen in der Umgebung des kritischen Zustandes. *Z. Phys. Chem.* 207: 307–320.
62. Shchukin, E. D. and P. A. Rehbinder. 1958. The formation of new surfaces during deformation and fracture of a solid in a surface active medium. *Kolloidnyi Zh.* 20: 645–654.
63. Rusanov, A. I., Shchukin, E. D., and P. A. Rehbinder. 1968. The theory of dispersion. 1. Thermodynamics of monodisperse systems. *Kolloidnyi Zh.* 30: 573–580.
64. Rusanov, A. I., Kuni, F. M., Shchukin, E. D., and P. A. Rehbinder. 1968. The theory of dispersion. 2. Dispersion in vacuum. *Kolloidnyi Zh.* 30: 735–743.
65. Rusanov, A. I., Kuni, F. M., Shchukin, E. D., and P. A. Rehbinder. 1968. The theory of dispersion. 3. Dispersion in liquid phase. *Kolloidnyi Zh.* 30: 744–753.
66. Fedoseeva, N. P., Kuchumova, V. M., Kochanova, L. A., and E. D. Shchukin. 1978. Sodium dodecyl-sulfate influence on the behavior of the ternary system of water-heptane-tertiary butyl alcohol near the critical point. *Kolloidnyi Zh.* 40: 578–580.
67. Shchukin, E. D., Kochanova, L. A., and A. V. Pertsov. 1979. The nature of the lyophilic colloidal stability of emulsions. In *Physical-Chemical Mechanics and Lyophilic Properties of Disperse Systems*, Ovcharenko F. (Ed.), pp. 15–31. Kiev, Ukraine: Naukova Dumka.
68. Shchukin, E. D. and L. A. Kochanova. 1983. Physical-chemical fundamentals of microemulsions preparation. *Kolloidnyi Zh.* 45: 726–736.
69. Shchukin, E. D. 2004. Conditions of spontaneous dispersion and formation of thermodynamically stable colloid systems. *J. Dispers. Sci. Technol.* 25: 875–893.
70. Frenkel, Y. I. 1959. *Kinetic Theory of Liquids.* Leningrad, Russia: Izd. AN SSSR.
71. Reiss, H. 1975. Entropy-induced dispersion of bulk liquids. *J. Colloid Interface Sci.* 53: 61–70.
72. Rusanov, A. I. 1999. Thermodynamics of dispergation: Development of Rehbinder's ideas. *Colloids Surf.* 160: 79–87.
73. Hill, T. 1963. *Thermodynamics of Small Systems.* New York: Benjamin Inc.
74. Rusanov, A. I. 1992. *Micelle Formation in Solutions of Surfactants.* Saint-Petersburg, Russia: Himiya.
75. Davis, H. T. 1996. *Statistical Mechanics of Phases, Interfaces and Thin Films.* New York: Wiley.
76. Friberg, S. E. 1997. Emulsion stability. In *Food Emulsions*, S. E. Friberg and K. Larsson (Eds.). New York: Marcel Dekker.
77. Shinoda, K., Nakagava, T., Tamamushi, B., and T. Isemura. 1966. *Colloidal Surfactants.* Moscow, Russia: Mir.
78. Markina, Z. M., Bovkun, O. P., and P. A. Rehbinder. 1973. The thermodynamics of the formation of micelles of surfactants in an aqueous medium. *Kolloidnyi Zh.* 35: 833.
79. Mittal, K. L. (Ed.). 1997. *Micellization, Solubilization and Microemulsions.* New York: Plenum Press.
80. Pertsov, A. V. 1967. The studies of dispersion processes under conditions of the strong decrease in the free interfacial energy. Cand. Sci. thesis. Moscow, Russia: Izd. MGU.
81. Martynov, G. A. and V. M. Muller. 1972. On the role of disintegration in the mechanism of aggregation stability of colloid systems. *Doklady AN SSSR.* 207: 370–373.
82. Muller, V. M. 1983. Theory of aggregation transformations and stability of hydrophobic colloids. Science thesis. Leningrad, Russia: Leningrad State University.
83. Shchukin, E. D. and E. A. Amelina. 1979. Contact interactions in disperse systems. *Adv. Colloid Interface Sci.* 11: 235–287.
84. Hubbard, A. T. (Ed.). 1996. *Encyclopedia of Surface and Colloid Science.* New York: Marcel Decker.
85. Dobias, B., X. Qiu, and W. von Rybinski. (Eds.). 1999. *Solid-Liquid Dispersions.* New York: Marcel Dekker.
86. Wilson, P. M. and C. F. Brandner. 1977. Aqueous surfactant solutions which exhibit ultra-low tension at the oil/water interface. *J. Colloid Interface Sci.* 60: 473–479.
87. P. Kumar and K. L. Mittal. (Eds.). 1999. *Handbook of Microemulsion Science and Technology.* New York: Marcel Dekker.
88. Sjoblom, J. (Ed.). 2001. *Encyclopedic Handbook of Emulsion Technology.* New York: Marcel Dekker.
89. Likhtman, V. I., Shchukin, E. D., and P. A. Rehbinder. 1962. *Physicochemical Mechanics of Metals.* Moscow, Russia: Izd. AN SSSR.

90. Pertsov, A. V., Mirkin, L. I., Pertsov, N. V., and E. D. Shchukin. 1964. On the spontaneous dispersion under conditions of the strong decrease in the free interfacial energy. *Doklady AN SSSR.* 158: 1166–1168.

91. Rehbinder, P. A. and E. D. Shchukin. 1972. Surface phenomena in solids during deformation and fracture processes. In *Progress in Surface Science*, S. G. Davison (Ed.), pp. 97–188. Oxford, U.K.: Pergamon Press.

92. Giese, R. F. and C. J. van Oss. 2001. *Colloid and Surface Properties of Clays and Related Materials.* New York: Marcel Dekker.

93. Rehbinder, P. A. 1979. *Selected Works: Surface Phenomena in Disperse Systems. Physicochemical Mechanics.* Moscow, Russia: Nauka.

94. Tager, A. A. 1978. *Physical Chemistry of Polymers.* Moscow, Russia: Himiya.

95. Shchukin, E. D., Zanozina, Z. M., Kochanova, L. A., Lichtman, V. I., and P. A. Rehbinder. 1965. On the possibility of formation of fine structures in alloys by hardening their emulsions. *Doklady AN SSSR.* 160: 1355–1357.

Section II

Surface Phenomena in the
Structures with Phase Contacts
and in Continuous Solid Bodies

5 Deformation and Degradation of Solid Bodies and Materials: Description and Measurements

This chapter is devoted to the basic description of strains and stresses existing in solids and addresses principal experimental methods and techniques used to measure them. This may be regarded as the introduction into the resistance of materials, which is one of the most difficult engineering disciplines to understand and master. There are many classic textbooks on this subject, some of which we are citing and referring to [1–24], alongside with numerous priority publications relevant to physical-chemical mechanics [25–41].

The material is presented in such a way that it is oriented toward the audience that does not have extensive specialized knowledge in the area of mechanics of a rigid body. Yet, knowledge and understanding of the principles discussed in this chapter is essential in making a transition from the discussion of the physical-chemical mechanics of coagulation structures to the discussion of physical-chemical mechanics of solid-like systems.

5.1 FORCES AND DEFORMATIONS IN A UNIFORM STRESSED STATE

5.1.1 FORCES AND STRESSES

We will start the discussion by introducing the concepts of *force* and *stress*, which will lead to a discussion on the comparison of a *vector* to a *tensor*. In physics, the term "force" automatically implies that there are means to measure it.

Let a force F be applied to a body. Let us draw an origin through the point at which the force is applied and then replace the force with two forces, F_x and F_y, *measured* using two calibrated dynamometers along the two chosen axes. This scheme allows us to introduce the concept of the *components* of a force.

Now, let us turn the coordinate system by an arbitrary angle, θ. The projections of the force components on to the new axes, x′ and y′, can be calculated from the original force components and direction cosines as

$$F_{x'} = F_x \cos(x'x) + F_y \cos(x'y)$$

$$F_{y'} = F_x \cos(y'x) + F_y \cos(y'y)$$

The transform matrix contains the following components:

$$\cos(x'x) = \cos\theta = a_{11} \quad \cos(x'y) = \sin\theta = a_{12}$$
$$\cos(y'x) = -\sin\theta = a_{21} \quad \cos(y'y) = \cos\theta = a_{22}$$

where indices 1 and 2 correspond to x and y, respectively. The first index is related to the new axes and the second one to the old axes. Using the components of the transformation matrix, the transformations of the force components can be written as

$$F_{x'} = F_1' = a_{11}F_1 + a_{12}F_2 = \sum_j^{1-2} a_{1j}F_j$$

$$F_{y'} = F_2' = a_{21}F_1 + a_{22}F_2 = \sum_j^{1-2} a_{2j}F_j$$

The above expressions can be generalized as $F_i' = \sum_j a_{ij}F_j = a_{ij}F_j$, where the repeating index automatically implies summation.

The essence of this transformation is that the components of force transform linearly with the turn of the axes, that is, they can be represented by the sums of the initial components, with the coefficients given by the *direction cosines to the first power*. This establishes the characteristic of the force as a physical variable. We have also defined a vector as a *first-order tensor*. The first order of the tensor is defined in this case by the first power of the cosines in the transformation equation. A vector can be multidimensional while still remaining a first-order tensor. A scalar can be defined as a tensor of *zeroth* order.

Force is a *field vector*, characterizing the presence of an external force field. Field vectors do not represent specific characteristics of materials, but they describe interactions in the materials, the motion of materials, etc. There are also *material vectors*, one example of which is the pyroelectric vector showing polarization in the body upon heating.

Let us turn again to the decomposition of a force into components (Figure 5.1). Does one need to have exactly two components (in 3D space, three components), or can this representation be simplified? In general, one *does* indeed need two components. However, one of them may disappear when the axes are turned in a certain way. Indeed, one of the components becomes equal to zero when the origin is turned by an angle θ such that

$$F_{y'} = F_x(-\sin\theta) + F_y\cos\theta = 0$$

$$\tan\theta = \frac{F_y}{F_x} \quad \text{or} \quad \theta = \arctan\left(\frac{F_y}{F_x}\right)$$

In such (and only in such) a coordinate system, one will have $F_{x'} = (F_x^2 + F_y^2)^{1/2} = |F|$. The coordinate system in which this is true is said to be formed with *principal axes*. In these coordinates, the length of a vector, that is, its modulus $|F| = \left(\sum F_i^2\right)^{1/2}$, is the *invariant (Inv)* of a vector in all axes.

In order to fully understand the meaning of a vector, it is essential that we examine all these transformations in detail in various coordinate systems. A force can be easily identified using a picture of a load hanging on a rope. The rope shows the force direction, while the load indicates the force magnitude.

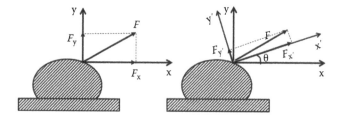

FIGURE 5.1   Decomposition of force acting on a body into the components.

The situation is more difficult if one is measuring the strength of a magnetic field after removing a piece of wire from a coil: in this case, one can only observe the direction in which the galvanometer arrow moves rather than the direction of the field. The situation is even more difficult in the case of multidimensional vectors that one can't represent with a drawing in a 3D space. Nevertheless, the concepts of a vector variable and a nonvector variable are quite rigorous: if the components reflecting the action imposed on a body or the properties of a body follow the described transformation scheme, then the property is a vector quantity and one can make a transition to the principal coordinate system in which the "arrow" will symbolize the magnitude and direction of a given physical quantity.

Let us go back to the case where a body undergoes deformation or strain as a result of an applied force. All interatomic bonds get deformed, and the atoms are displaced from their equilibrium positions. If originally the body had a rectangular shape, after the deformation, its shape has changed, that is, the body was stretched and sheared. As a result of the strain, a field of mechanical stresses has been generated. This field has produced localized stresses that penetrate the entire body. These stresses are compensated overall by the applied force, F. How can one describe these internal forces? To do this, let us employ the same approach as was used in describing the components of a force vector. Within a given coordinate system associated with the body, one can select elementary flat *platforms* that have a unit area. One now needs to find a way to estimate the forces that act on these areas. Let us use the symbols x and y to denote platforms perpendicular to the x and y axes, respectively. Let us make incisions along these platforms (Figure 5.2). In order to maintain the material on both sides of the incisions in place, one needs to apply an additional force to compensate for the broken ("cut") bonds. The forces that need to be applied can be measured with spring dynamometers. Let us restrict ourselves to a 2D model. Let us mount a dynamometer in a cut x in the direction x. This dynamometer will measure the stress component σ_{xx}. In this notation, the first index refers to the platform and the second one indicates the force direction. It is rather remarkable that in colloid chemistry, σ is commonly used for surface tension, while in mechanics, it is used to denote stress, and in physical-chemical mechanics, it is used for both! Similarly, the dynamometer in a scission y applied in the direction y will measure the stress σ_{yy}.

However, it is not sufficient to use just the two springs that we have so far utilized. One needs to also compensate for the tangential (shear) forces. To do this, two more spring dynamometers parallel to the plane of scission are needed. These dynamometers will measure the stress components σ_{xy} and σ_{yx}. These components are measured in the plane x in the direction y and in plane y in the direction x, respectively. We have thus obtained a combination of four variables that completely describe the *stressed state* in the selected 2D model.

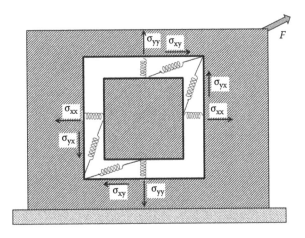

FIGURE 5.2 The description of the stress tensor components.

Our model will be more informative if we completely separate out a unit cube and apply to the unit –x and –y planes (in the "left" and "lower" scissions) an equilibrating system of four forces that are pairwise symmetric with respect to components in x and y planes. This will allow us to exclude both cube translation and rotational motion in the plane of the figure (around the z-axis, which we have not yet shown). The latter means that the shear components are identical, that is, that $\sigma_{xy} = \sigma_{yx}$ and their net momentum around the z-axis is zero. The matrix containing the components of the stressed state is thus symmetric: only three components in it are independent:

$$\begin{pmatrix} \sigma_{xx} & \sigma_{xy} \\ \sigma_{yx} & \sigma_{yy} \end{pmatrix}$$

At the same time, we have concluded that we need four spring dynamometers. The dynamometers measuring the tangential forces are equally stretched (or compressed), but these forces belong to different platforms (as required by Newton's third law). This does not look like a vector at all.

When the components of the stressed state are the same, that is, completely coincide with each other, one can talk about a *uniform* stressed state. If any one of these components is different from the others, one has a nonuniform stressed state. In this section, we will address only the uniform stressed state.

Now, let us make a transformation to a different coordinate system by turning the axes by an angle θ, similar to how we did for a vector. The peculiarities of this transformation reveal the specifics of the stressed state as the *second-order tensor*.

Let us use the following two-step example (Figure 5.3). Let us cut our cube parallel to the new axis y′, that is, the cut is in the y′Oz plane. Such a cut separates a prism out of the cube. In the plane of the figure, this prism is represented by a right-angled triangle. Let us choose the section in such a way that the length of the hypotenuse and consequently the area of a section in the perpendicular plane are equal to one. Let us mount two springs in this cut—one normally and the other one tangentially—that would compensate the components σ'_{xx} and σ'_{xy}. In Figure 5.3, the two components that are depicted by arrows are not applied to the hypotenuse (left and bottom) but to the opposite side of the section (right and top). This is done in order to provide a more clear illustration of the summation procedure that we are about to consider.

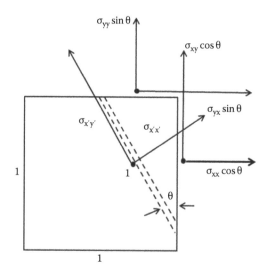

FIGURE 5.3 Description of a stressed state in a different coordinate system.

We know the four forces (stress components) acting on the catheti: by projecting them on the hypotenuse, one finds both the σ'_{xx} and the σ'_{xy} components in the new coordinates. First, one needs to recognize that not the entire force is applied to the right cathetus but only part of that force, which is proportional to the cathetus length and the cosine of the angle θ between the x and x' axes, that is, $\sigma_{xx} \cos \theta$. Consequently, the normal force applied to the upper cathetus equals $\sigma_{yy} \sin \theta$. The same is true for the tangential components. The transformation matrix for the projection of all four "partial" components on the new x' axis can be written as

$$
\begin{pmatrix} \cos(x'x) & \cos(x'y) \\ \cos(x'x) & \cos(x'y) \end{pmatrix} = \begin{pmatrix} \cos\theta & \sin\theta \\ \cos\theta & \sin\theta \end{pmatrix} + \begin{pmatrix} a_{11} & a_{12} \\ a_{11} & a_{12} \end{pmatrix}
$$

In this transformation, the first index corresponds to the new axis and the second one to the old axis. From this, one gets

$$
\frac{\sigma_{x'x'}}{\sigma'_{xx}} = \sigma_{xx} \cos(x'x)\cos(x'x) + \sigma_{xy} \cos(x'x)\cos(x'y) + \sigma_{yx} \cos(x'y)\cos(x'x) + \sigma_{yy} \cos(x'y)\cos(x'y)
$$

which is equivalent to

$$
\sigma'_{11} = a_{11}a_{11}\sigma_{11} + a_{11}a_{12}\sigma_{12} + a_{12}a_{11}\sigma_{21} + a_{12}a_{12}\sigma_{22} = \sum_{k,l}^{1-2} a_{1k}a_{1l}\sigma_{kl}
$$

Generalizing this approach for all four components in the new coordinate system and omitting the summation sign for repeating indices, one gets

$$
\sigma'_{ij} = \sum_{k,l}^{1-2} \frac{a_{ik}a_{jl}\sigma_{kl}}{a_{ik}a_{jl}\sigma_{kl}} \tag{5.1}
$$

So, what does the "stress state" concept really mean? We have applied a single force ("arrow") to the body. This force is a vector that undergoes linear transformation when the coordinate system is rotated. Linear transformation implies that the old components are summed with the transformation coefficients given by the direction cosines to the first power. Within a volume of a solid body, we have encountered a stressed state that can't be described using a single vector. We had to isolate a "cube with scissions" and introduce a system of arrows to depict the acting forces. Each arrow represents a component that is a vector with a particular direction and absolute value. At the same time, a system of arrows is a completely different physical representation. The peculiarities of this representation are associated with how the "system of arrows" undergoes linear transformation. We have rigorously carried out these transformations and have shown that a new physical variable that we have referred to as the stressed state has several components that undergo linear transformation when the coordinate system is rotated. In the new coordinates, these components are represented by the linear combination of the old components with the corresponding transformation coefficients. These coefficients are the cosines between the new and old axes raised to the *second power*. This power constitutes the principle difference between stresses and vectors. Equation 5.1 represents a definition of a *second-order tensor*. Within the given planar example, this tensor has two dimensions, and in 3D space, it has three dimensions. In the 2D form, this tensor consists of four components with only three components being independent. In the 3D form, the tensor has nine components. These tensors are *symmetric* with respect to their diagonals.

Is it possible to further simplify the expression of the stressed state, similar to how it was done for a vector? Is there a coordinate system in which it can be shown with only one arrow? The former is possible, while the latter is not! Simplification in this case implies a transformation to the coordinate system in which the shear components (tangential to the coordinate planes) will be zero. This means that all components with mixed indices are zero, that is, $\sigma'_{ij} = \sigma'_{ij} = 0$.

In agreement with the summation rules, one gets for σ'_{12}

$$\sigma'_{12} = \sigma'_{21} = 0 = a_{11}a_{21}\sigma_{11} + a_{11}a_{22}\sigma_{12} + a_{12}a_{21}\sigma_{21} + a_{12}a_{22}\sigma_{22} = \cos\theta(-\sin\theta)\sigma_{11} + \cos\theta\cos\theta\sigma_{12}$$
$$+ \sin\theta(-\sin\theta)\sigma_{21} + \sin\theta\cos\theta\sigma_{22}$$

This indicates that the condition $\sigma'_{12} = \sigma'_{21} = 0$ is required in order to preserve only the normal components in the equation of the stressed state. This condition can be met only if the coordinate system is turned by an angle θ:

$$\theta = \frac{1}{2}\arctan\left[\frac{\sigma_{12}}{\frac{1}{2}(\sigma_{11} - \sigma_{22})}\right] = \frac{1}{2}\arctan\left[\frac{\sigma_{xy}}{\frac{1}{2}(\sigma_{xx} - \sigma_{yy})}\right]$$

As one can see, out of four components only two remain present:

$$\begin{pmatrix} \sigma'_{11} & 0 \\ 0 & \sigma'_{22} \end{pmatrix} = \begin{pmatrix} \sigma_1 & 0 \\ 0 & \sigma_2 \end{pmatrix}$$

The right-hand notation is the conventional shortened form of the *principal components* of a tensor in its principal coordinate system.

It is worth recalling here that each tensor has an *order* (I, II, III, IV, etc.). Tensor order reflects the physical properties of a tensor and is determined by the power of the direction cosines product, that is, the power of the product of linear transformation coefficients. The tensor order physically reflects the possibility of visualizing the various properties of a field or a body from different viewpoints. Tensor order is also an indicator of the different ways in which spatial anisotropy is revealed. Scalar quantities, that is, temperature, mass, and amount of heat, are zeroth-order tensors; the vectors of velocity or force are the first-order tensors; mechanical stresses and strains are second-order tensors, while the elasticity modulus is a fourth-order tensor, as will be shown in the following text.

At the same time, when a physical property is represented by a tensor of a given order, it can be characterized by a *particular number of components in a given space*. For example, the three spatial components and time form a 4D field vector, which is a first-order tensor. Another noteworthy example is that of the fourth-order elasticity tensor, which in an isotropic medium is degenerated into two scalar quantities: the Young's modulus and the Poisson ratio.

Any physical quantity along with its spatial and geometric dimensions can also be characterized by the *physical dimensionality of the components*, which is the same for all components. The components of a second-order stress tensor have the dimensions of $N/m^2 = Pa = 10$ dyn/cm$^2 = 10^{-5}$ kg-force/cm^2. At the same time, a second-order strain tensor's components are dimensionless (of zeroth-order physical dimensionality). This is true for both 2D and 3D models.

One can thus state a critical difference between a tensor and a vector. The aforementioned linear transformation, when applied to a vector in the principal coordinates leaves only one component. In the case of a tensor, two components are left when the transformation is applied. These components can be expressed numerically or shown graphically with arrows, but they can't be combined, because they are applied to *different* platforms (planes). Without considering the 3D case here, let us give several particular examples of stressed states in a 2D model (Figure 5.4).

    a. While a *uniaxial extension* represents the simplest case of a stressed state, it requires the most careful and detailed consideration. The term applies to the extension of a rod, a plate, or a cylinder but *does not* apply to the extension of a hook or a chain. In these latter cases, the extensions are nonuniform.

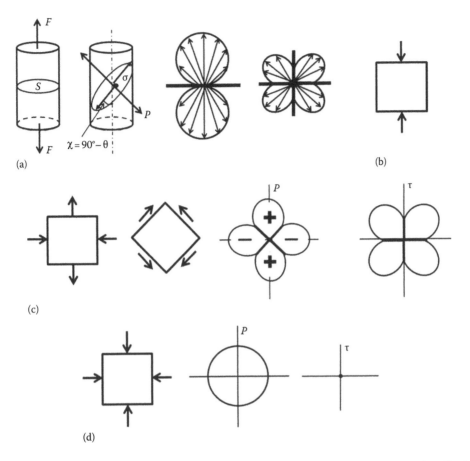

FIGURE 5.4 Examples of common stressed states: uniaxial extension (a); uniaxial compression (b); simple shear (c); hydrostatic compression (d); equilateral extension (d).

Let the object for uniaxial extension be a cylindrical rod with a cross-sectional area S to which an extension force F is applied. The ratio $F/S = P$ is the *extensional stress*. Alternatively, this is a component of a stress tensor in the selected coordinate system, $\sigma_{yy} = \sigma_{22} = \sigma_2 = P$. The action of the force resulted in a stressed state that is uniform in all directions (we are neglecting nonuniformities and local concentrations of stresses in certain special areas, such as the mounting points of the load or of the rod itself). The selected coordinate system in this case is the principal one. From symmetry considerations at $P = \sigma_{yy}$, this tensor is as follows:

$$\begin{pmatrix} 0 & 0 \\ 0 & P \end{pmatrix}$$

It appears that we have ended up with only a single component ("arrow"). But in fact, this is not the case. To illustrate this, let us rotate the axes by an angle θ (e.g., counterclockwise; a turn can also be characterized by either the "+" or the "−"sign). The summation yielding the new components takes place over the single present component, $\sigma_{yy} = P$, which yields a new form of the same tensor:

$$\begin{pmatrix} P\sin^2\theta & P\sin\theta\cos\theta \\ P\sin\theta\cos\theta & P\cos^2\theta \end{pmatrix}$$

For a vector F, there are only two components, $F \cos \theta$ and $F \sin \theta$.

In Chapter 7, where we will discuss the Rehbinder effect in detail, we will be looking only at the influence of the active media on the mechanical properties of single metallic crystals grown out of metal wires. In this discussion, we will use notations that are common for such systems. Within a given crystallographic plane, a single elliptical platform oriented by an angle χ relative to the longitudinal axis is selected. Relative to the selected directions of axes in the preceding text, $\chi = 90° - \theta$. This chosen crystallographic plane is typically a sliding or cleavage plane specific to a given crystal. With respect to this plane, there are two stress components: the extension (typically positive, "+") component, p, perpendicular to the plane and the maximal tangential shear component, τ, applied in the direction of the larger axis of the ellipse, that is,

$$p = \sigma'_{22} = P \sin^2 \chi,$$
$$\tau = \sigma'_{21} = P \sin \chi \cos \chi.$$

For crystals with various orientations, with χ between 0 and 90°, as determined from x-ray diffraction, p and τ both depend on χ as $\sin^2 \chi$ and $\sin \chi \cos \chi = 1/2 \sin 2\chi$, respectively. The shear components are present in any stressed state with the exception of the hydrostatic one. In the case of the uniaxial extension, the magnitude of these components may reach one-half of the extension components, acting at a 45° angle.

When the diagrams of $\sin^2 \chi$ and $1/2 \sin^2 \chi$ are plotted in polar coordinates, they represent a good example of the orientation dependence of tensor components. In this case, particular components are plotted in every direction along the polar axis, which results is a family of analytical *lemniscate* curves with two "petals" for p and four "petals" for τ. In the case of a uniaxial extension, the shear components are twice shorter than the extension components.

b. *Uniaxial compression* (e.g., of a pillar). In this case, one also has $F/S = P$, but the value of P is negative. The p and τ stress rosettes in this case have the same appearance, with the only exception that the p "petals" have an inverse sign:

$$\begin{pmatrix} 0 & 0 \\ 0 & -P \end{pmatrix}$$

c. *Pure (or simple) shear* represents an interesting case in which the tensor components in the principal coordinate system have *the same absolute value but differ in their sign*:

$$\begin{pmatrix} -P & 0 \\ 0 & P \end{pmatrix}$$

If this coordinate system is turned by a 45° angle, it will transform into the following tensor:

$$\begin{pmatrix} 0 & \tau \\ \tau & 0 \end{pmatrix}$$

where $\tau = |P|$. In different coordinate systems, one can see the same stressed state from different sides. This is especially well seen in the present example. In the principal coordinate system, one only needs the springs that are perpendicular to the sides of a scission, while in the second case, only the tangentially oriented springs are needed. The stress rosettes now have a different appearance: they have four "petals" for both p and τ. These "petals" are oriented at a 45° angle with respect to each other and are all of the same length. In the stress rosette for p, the "+" petals alternate with the "−" petals.

This summarizes the case of pure shear. We have shown one more time that the simple summation of the components is not possible. Nevertheless, there is a need to carry out the summation of the components with pairwise indices located on the main diagonal. The algebraic sum of these components is the same in all coordinate systems. This is the *invariant* of a tensor given by its *trace*, Sp. In our case, Sp = 0, which corresponds to the absence of hydrostatic (equilateral) extension or compression. In other words, the equilibrium of a force pair in octahedral coordinates corresponds to the components in the principal coordinate system with the same absolute values and opposite signs.

d. *Hydrostatic compression*: $\sigma_1 = \sigma_2 = P < 0$; $\Sigma\sigma_i = $ Sp < 0 represents an exceptional case of the complete absence of shear stresses. In this case, the τ rosette degenerates into a single point, while the p rosette is represented with a circle (or with a sphere in the 3D case). The tensor representing this situation is referred to as the *spherical tensor*.

e. *Equilateral (equiaxial) extension*: $\sigma_1 = \sigma_2 = P > 0$; Sp $= \Sigma\sigma_1 > 0$.

Let us once again turn to an arbitrary stressed state characterized by the tensor in the principal coordinate system,

$$\begin{pmatrix} \sigma_1 & 0 \\ 0 & \sigma_2 \end{pmatrix}$$

where the components can have any sign. Let us now turn the coordinate system by a $45°$ angle with respect to the principal coordinate system. The new coordinate system is referred to as *octahedral* (edges of an octahedron are {111}), and the tensor in the new coordinate system can be written as

$$\begin{pmatrix} \frac{1}{2}(\sigma_1 + \sigma_2) & \frac{1}{2}(\sigma_2 - \sigma_1) \\ \frac{1}{2}(\sigma_2 - \sigma_1) & \frac{1}{2}(\sigma_1 + \sigma_2) \end{pmatrix} = \begin{pmatrix} p & 0 \\ 0 & p \end{pmatrix} + \begin{pmatrix} 0 & \tau_{max} \\ \tau_{max} & 0 \end{pmatrix}$$

In the new coordinate system, one essential feature that was not visible in the principal coordinate system became revealed. Namely, an arbitrary stressed state can be viewed as the sum of two tensors: the *spherical* tensor characterizing the equilateral extension or compression and the shear stress *deviator tensor*. Either of these tensors (or both of them) can be zero.

It is worth turning here to the rheological Maxwell model showing stress relaxation. When we discussed rheology, we represented the Maxwell model by the elastic and viscous elements connected in series, representing an isotropic shear in the 2D model. Here, we have encountered a different representation of the *same* model. To illustrate this, let us consider the following geological example. Deformation and relaxation processes are continuously going on inside our planet's crust. These deformation processes can sometimes be quite devastating and unpredictable, while the intensity of the relaxations is a function of the temperature and the nature of the mechanical stresses and consequently can significantly depend on depth. Under these conditions, the thermodynamic equilibrium corresponds to the relaxation of the shear components, that is, to a gradual dissipation of the deviator tensor in the course of microscopic and macroscopic creep. At the same time, the spherical tensor is preserved and tends to become the sole carrier of stress. This corresponds to Pascal's law: at equilibrium at a given hydrostatic pressure, the pressure is the same at every point of an arbitrary oriented place, that is, $\sigma_1 = \sigma_2 = \sigma_3$.

5.1.2 Strain, Displacements, and Turns

Up to this point, we have solely focused on the stresses and have not considered any change in the shape of a body. Let us now do the opposite: take a close look at the deformation without putting too much emphasis on stresses, while staying away from "the chicken-or-the egg" dilemma.

Let us consider an object that can change both shape and volume. Such an object can be a piece of clay or a piece of rubber. Let us look at the 2D model shown in Figure 5.5. Let us separate a small

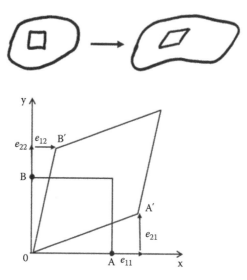

FIGURE 5.5   Schematic illustration of strain (deformation).

cube with a side of unit length. While undergoing deformation, the cube transforms in such a way that the sides that were parallel remain parallel and the square transforms into a parallelogram. How many numbers does one need for the description of such a change involving the transformation of two starting points, A and B, into the new ones, A′ and B′. The latter can be described with four numbers in the x–y coordinate system: the displacement of point A located on the x axis by a distance $e_{xx} = e_{11}$ along the x axis and by a distance $e_{yx} = e_{21}$ along the y axis; and for point B located on the y axis by a distance $e_{yy} = e_{22}$ along the y axis and by a distance $e_{xy} = e_{12}$ along the x axis. In these notations, the first index indicates from where the point is to be moved, while the second one shows where it is to be moved to. These four numbers form a square matrix

$$\begin{pmatrix} e_{11} & e_{12} \\ e_{21} & \sigma_{22} \end{pmatrix}$$

In general, the aforementioned matrix describing relative displacements is not symmetric, that is, $e_{12} \neq e_{21}$ in contrast to the stress tensor, σ_{ij}.

To understand the nature of asymmetry, one can write the tensor e_{ij} as the sum of symmetric and *antisymmetric* parts:

$$\begin{pmatrix} e_{11} & e_{12} \\ e_{21} & e_{22} \end{pmatrix} = \begin{pmatrix} e_{11} & \tfrac{1}{2}(e_{12} + e_{21}) \\ \tfrac{1}{2}(e_{21} + e_{12}) & e_{22} \end{pmatrix} + \begin{pmatrix} 0 & \tfrac{1}{2}(e_{12} - e_{21}) \\ \tfrac{1}{2}(e_{21} - e_{12}) & 0 \end{pmatrix} = \begin{pmatrix} \varepsilon_{11} & \varepsilon_{12} \\ \varepsilon_{21} & \varepsilon_{22} \end{pmatrix} + \begin{pmatrix} 0 & \omega \\ -\omega & 0 \end{pmatrix}$$

The left-hand side of the aforementioned expression is symmetric, $\varepsilon_{12} = \varepsilon_{21}$. This is the *strain tensor*, ε_{ij}, which is a second-order tensor. Indeed, since the sides OA and OB of the square both equal 1, each value ε_{ij} represents a *relative deformation* of extension (compression) and shear in a given direction. It is assumed here and further on that deformations are small in magnitude.

The right-hand side of the expression corresponds to the turn (i.e., the slope) of the selected elementary square. This part of the e_{ij} tensor shows no change in shape of the square. To illustrate this, let us reproduce here the two displacements: the displacement of point A upward along the Oy direction by the amount of ω (ω ~ 1) and that of point B by the amount of −ω to the left along the

Ox axis. As a result, the square turns counterclockwise by an angle ω (in radians). While we are primarily interested in the strain tensor, ε_{ij}, it is still worth pointing out here that very often angular velocities are represented by vectors and are summed by using vector summation rules. These are permissible operations, but one needs to remember that these variables are in fact antisymmetric second-order tensors, rather than vectors. This distinction often disappears because both a vector and an antisymmetric tensor have three spatial components.

As we have learned earlier, in the case of the 2D deformation model, the square is transformed into a parallelogram and can also undergo rotation. The strain is referred to as *uniform* if it is the same at all points of the body. Let us consider several particular examples of uniform strain utilizing an analogy with stressed-state tensors. The concept of a principal coordinate system in which there are no shear components, as well as the rules for transforming tensors to the principal coordinate system are the same for both ε_{ij} and σ_{ij}.

a. *Uniaxial extension* is an extension along the y axis without any change in dimensions along the x axis. Comprehending the paradigm shift from a "stress without strain" to a "strain without stress" requires that a convenient object be chosen for illustrative purposes. We will use a soap film from Dupré's experiment (Section 1.1) as such an object. The sides of the metal frame are obviously associated with the principal coordinate system in which the strain tensor has a single component $\varepsilon_{yy} = \varepsilon_{22} = \varepsilon_2 = \varepsilon > 0$:

$$\begin{pmatrix} 0 & 0 \\ 0 & \varepsilon \end{pmatrix}$$

The force acting on the frame is directed downward, as indicated by the arrow. However, the same force, that is, the tensor component, is distributed over the top frame as well as over any section of the film. If the side of a film associated with the y axis has the length l, while the "dislocation," that is, the change in length, is Δl, then the deformation is $\varepsilon = \Delta l / l$. Within the framework of this model, the strain is not necessarily small. Turning the coordinate system by 45° orients the axes along the diagonal of the initial square. The new tensor is

$$\begin{pmatrix} \varepsilon/2 & \varepsilon/2 \\ \varepsilon/2 & \varepsilon/2 \end{pmatrix}$$

This tensor is the same, regardless of whether the axes were turned clockwise or counterclockwise.

b. *Uniaxial compression* is the case opposite to the one described earlier: in the principal coordinate system, the strain tensor has a single component, $\varepsilon_{yy} = \varepsilon_{22} = \varepsilon_2 = \varepsilon < 0$.

c. *Pure shear*: In the principal coordinate system, the strain tensor, as well as the stress tensor, has two components with the same absolute value but differing in sign: $\varepsilon_{yy} = \varepsilon_{22} = \varepsilon_2 = \varepsilon$; $\varepsilon_{xx} = \varepsilon_{11} = \varepsilon_1 = -\varepsilon$. When the axes are turned by 45°, the new strain tensor has the components $\varepsilon'_{xy} = \varepsilon'_{21} = \varepsilon'_{xy} = \varepsilon'_{12} = \varepsilon$.

d. *Comprehensive extension* is characterized by $\varepsilon_1 = \varepsilon_2 = \varepsilon > 0$. This is a spherical tensor containing no shear components.

e. *Hydrostatic compression* is the opposite of comprehensive extension, that is, it is characterized by $\varepsilon_1 = \varepsilon_2 = \varepsilon < 0$.

f. *Uniaxial extension (and compression) of a real elastic solid*: Experiments show that an extension ε along the y axis is accompanied by transverse compression along the x axis. This transverse compression is the fraction ν of ε, where ν is the Poisson ratio. In the principal coordinate system, $\varepsilon_1 = -\nu\varepsilon$; $\varepsilon_2 = \varepsilon$.

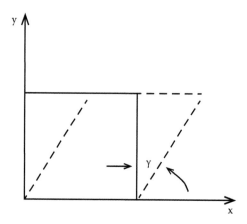

FIGURE 5.6   Schematic illustration of rheological shear.

g. The so-called *rheological shear:* a displacement of the upper side of the cube along the x axis by γ (Figure 5.6). In the case of small strains, γ is both the strain and the angle. In this case, the strain tensor e_{ij} has a single component, $e_{12} = \gamma$. Let us split this tensor into two, one of which corresponds to the strain and the other one to the turn (of the square, not of the axes), that is,

$$\begin{pmatrix} 0 & \gamma \\ 0 & 0 \end{pmatrix} = \begin{pmatrix} 0 & \gamma/2 \\ \gamma/2 & 0 \end{pmatrix} + \begin{pmatrix} 0 & \gamma/2 \\ -\gamma/2 & 0 \end{pmatrix}$$

The left-hand side corresponds to a pure strain $\varepsilon = \gamma/2$, that is, the square has stretched into a rhomb along its diagonal. The right-hand side represents the turn of the square without change in shape. As a result of this turn, point A shifts by a small distance downward, $e_{21} = -\gamma/2$, and point B shifts to the right by $e_{12} = \gamma/2$. The vertical sides of the initial square as a result reveal an inclination to the right by an angle $\gamma = 2\varepsilon$, while the horizontal sides remain horizontal. We have thus arrived at the description of shear strain in terms of the parameter γ present in rheological models.

h. *Strains in a single crystal subjected to an extensional shear:* The crystal was grown with a given orientation out of a wire. We have already pointed out that within such a model, it is customary for one to designate the position of a selected sliding plane by the angle χ measured from the axis of a sample. The rheological shear γ upon the extension of the sample accompanied by a decrease in the angle χ in this case equals to $\gamma = \cot \chi - \cot \chi_0$.

i. Let us again turn our attention to an arbitrary (symmetric) strain tensor, ε_{ij}. Setting the shear component to zero in the principal coordinate system allows one to determine the angle by which the coordinate system is turned, that is, $\theta = \dfrac{1}{2}\arctan\left[\dfrac{\varepsilon_{12}}{1/2\left(\varepsilon_{11} - \varepsilon_{12}\right)}\right]$.

The tensor is then

$$\begin{pmatrix} \varepsilon_1 & 0 \\ 0 & \varepsilon_2 \end{pmatrix}$$

Let us now deviate from the principal coordinate system by turning it by 45°. Such a transformation allows one to subdivide two additive parts of the tensor:

$$\begin{pmatrix} \varepsilon_1 & 0 \\ 0 & \varepsilon_2 \end{pmatrix} \propto \begin{pmatrix} \tfrac{1}{2}(\varepsilon_2 + \varepsilon_1) & \tfrac{1}{2}(\varepsilon_2 - \varepsilon_1) \\ \tfrac{1}{2}(\varepsilon_2 - \varepsilon_1) & \tfrac{1}{2}(\varepsilon_2 + \varepsilon_1) \end{pmatrix} = \begin{pmatrix} \tfrac{1}{2}(\varepsilon_2 + \varepsilon_1) & 0 \\ 0 & \tfrac{1}{2}(\varepsilon_2 + \varepsilon_1) \end{pmatrix} + \begin{pmatrix} 0 & \tfrac{1}{2}(\varepsilon_2 - \varepsilon_1) \\ \tfrac{1}{2}(\varepsilon_2 - \varepsilon_1) & 0 \end{pmatrix}.$$

An arbitrary symmetric strain tensor has been thus subdivided into two additive components: a spherical hydrostatic expansion (or contraction) tensor and a deviator tensor, characterizing shear deformations without any change in volume.

5.1.3 Elastic Deformations

Let us now turn to the relationship between stresses and strains. We have already addressed this in Chapter 3 where we discussed a very broad spectrum of rheological properties found in various systems, namely, elasticity, plasticity, viscosity, and their numerous combinations. Some of the significant limitations that we adapted include a consideration of a single stressed state of a uniform shear and of near steady-state processes. Here, we will limit ourselves to a discussion of a single rheological behavior, that is, elasticity, and will focus on the particular peculiarities and generalizations pertinent to this field.

Let us first address the simplest possible case, that is, the extension of a rod with a cross-sectional area S and a length l_0 (Figure 5.7). Due to the action of the extending force, F, and consequently a single stress component, $\sigma = F/S$, the rod undergoes the elongation Δl. The deformation (strain) in this case is $\varepsilon = \Delta l/l_0$. Let us conduct measurements of the extending force in the course of extending the sample at a constant rate, $d\Delta l/dt$, which corresponds to a constant rate of strain, $d\varepsilon/dt = d\Delta l/l_0 dt$. Depending on the nature of the material, the *strain curve*, $\sigma = \sigma(\varepsilon)$, may exhibit different types of behavior, but in all cases has an initial region corresponding to a linear and reversible behavior where $\sigma \propto \varepsilon$. It is this linear region of *elastic deformation* that is of primary interest here.

The described initial linear (Hookean) behavior is typical for a very large number of solid-like objects. Conversely, objects like soap thin films or modeling clays do not exhibit this type of behavior. A single parameter characterizing the elastic properties of an object is Young's modulus, that is, the dimensional proportionality constant, $E = \sigma/\varepsilon = Fl_0/\Delta lS$. In Chapter 4, we listed some typical values of E. In continuum mechanics, the elasticity moduli are typically described by the letter c and the inverse quantity, *compliance*, by the letter s, that is, $s \propto 1/c$.

In the most general case, the elasticity (or compliance) moduli relate every component of the stress tensor to every component of the strain tensor: there are four components in each tensor in the 2D case and nine components in the 3D case. This corresponds to a total of 81 moduli c and compliances s. However, let us restrict ourselves to a discussion of the principal strains in uniform

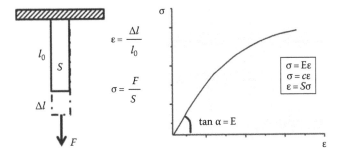

FIGURE 5.7   The extension of a rod.

isotropic material. From symmetry, it is apparent that the same principal axes are also the principal axes of the conjugate stress tensor.

In the simplest case of a uniaxial extension along the Ox axis, the uniaxial stress

$$\begin{pmatrix} \sigma_{11} & 0 & 0 \\ 0 & 0 & 0 \\ 0 & 0 & 0 \end{pmatrix}$$

causes a principal deformation,

$$\varepsilon_{ij} = \begin{pmatrix} \varepsilon_{11} & 0 & 0 \\ 0 & \varepsilon_{22} & 0 \\ 0 & 0 & \varepsilon_{33} \end{pmatrix}$$

Due to symmetry, there should be no shear present within this model, and one can write that

$$\varepsilon_{11} = (1/E)\sigma_{11}$$
$$\varepsilon_{22} = -\nu\varepsilon_{11} = -(\nu/E)\sigma_{11}$$
$$\varepsilon_{33} = -\nu\varepsilon_{11} = -(\nu/E)\sigma_{11}$$

In accordance with the experiment, the transverse compression strains have an inverse sign to the longitudinal extensions, and their contribution, ν, is on the order of 0.2–0.3. These two parameters, that is, the modulus with the units of stress (Pa), $[E] = [\sigma]$, and dimensionless Poisson ratio ν are sufficient for describing arbitrary deformations of an isotropic solid. It is worth pointing out here that *two* parameters (E and ν, G and ν, or G and E) are the smallest "pieces" of fourth-order tensors that are sufficient for the description of strains. It is not possible for one to get away with using a single parameter. For an arbitrary principal stress tensor,

$$\begin{pmatrix} \sigma_{11} & 0 & 0 \\ 0 & \sigma_{22} & 0 \\ 0 & 0 & \sigma_{33} \end{pmatrix}$$

one can write

$$\varepsilon_{11} = (1/E)\sigma_{11} - (\nu/E)\sigma_{22} - (\nu/E)\sigma_{33}$$
$$\varepsilon_{22} = -(\nu/E)\sigma_{11} + (1/E)\sigma_{22} - (\nu/E)\sigma_{33}$$
$$\varepsilon_{33} = -(\nu/E)\sigma_{11} - (\nu/E)\sigma_{22} + (1/E)\sigma_{33}$$

Other characteristics of elasticity, such as the shear modulus G, the volume compression coefficient K, the coefficients μ and λ used in mathematical elasticity theory, the c modulus, and the compliance s used in crystallography can be expressed via E and ν. To illustrate this, let us use the rheological shear modulus G as an example.

Let us start with the principal stresses and strains:

$$\begin{pmatrix} \sigma_{11} & 0 & 0 \\ 0 & \sigma_{22} & 0 \\ 0 & 0 & \sigma_{33} \end{pmatrix} \rightarrow \begin{pmatrix} \varepsilon_{11} & 0 & 0 \\ 0 & \varepsilon_{22} & 0 \\ 0 & 0 & \varepsilon_{33} \end{pmatrix}$$

and then turn the principal axes by 45°:

$$\begin{pmatrix} \sigma_{11}/2 & \sigma_{11}/2 & 0 \\ \sigma_{11}/2 & \sigma_{11}/2 & 0 \\ 0 & 0 & 0 \end{pmatrix} \rightarrow \begin{pmatrix} \dots & \frac{1}{2}(\varepsilon_{11}-\varepsilon_{22}) & \dots \\ \dots & \dots & \dots \\ \dots & \dots & \dots \end{pmatrix}$$

As a result of this transformation, two new components have appeared: $\sigma'_{12} = \sigma_{11}/2$, which corresponds to τ in rheology, and $\varepsilon'_{12} = \frac{1}{2}(\varepsilon_{11} - \varepsilon_{12}) = [(1+\nu)/2E]\sigma_{11}$, which corresponds to $\frac{1}{2}\gamma$ in rheology. By definition, the shear modulus $G = \tau/\gamma = \sigma'_{12}/2\,\varepsilon_{12} = [\sigma_{11}/2]/2[(1+\nu)/2E]\sigma_{11}$, that is,

$$G = \frac{E}{2(1+\nu)} \approx \frac{2}{5}E$$

The bulk compression modulus for an isotropic material can be determined from the relationship between the spherical stress and strain tensors, that is,

$$\begin{pmatrix} p & 0 & 0 \\ 0 & p & 0 \\ 0 & 0 & p \end{pmatrix} \rightarrow \begin{pmatrix} \varepsilon & 0 & 0 \\ 0 & \varepsilon & 0 \\ 0 & 0 & \varepsilon \end{pmatrix}$$

By definition, $K = \Delta p/(\Delta v/v_0)$, and for a solid

$$K = \frac{E}{3(1+2\nu)} \approx \frac{2}{3}E$$

For air, $K = p$ (see comments on entropic elasticity).

In elasticity theory and in the resistance of materials, the moduli μ and λ are often used. The relationship between the stress and strain tensors can be written in this case as follows:

$$\sigma_{11} = (2\mu + \lambda)\varepsilon_{11} + \lambda\varepsilon_{22} + \lambda\varepsilon_{33}$$
$$\sigma_{22} = \lambda\varepsilon_{11} + (2\mu + \lambda)\varepsilon_{22} + \lambda\varepsilon_{33}$$
$$\sigma_{12} = 2\mu\varepsilon_{12}$$

Since $2\varepsilon_{12} = \gamma$, and $\sigma_{12} = \tau$, one immediately finds that $G = \mu$, and $\lambda = E\nu/[(1+\nu)(1-2\nu)]$. By assuming $\nu \approx 0.25$, we get $\lambda \approx \frac{2}{3}E$. Indeed, for common solids with $\nu \approx 0.25$–0.3, all of these coefficients are of a similar order of magnitude. At the same time, $\nu \to 0$ represents either a very loose object (within the limits of uniform isotropic materials), much looser than, for example, cotton or a non-homogeneous anisotropic object ("a mattress") or nonsolid material in a rigid "shell," such as a gas in the cylinder under the piston. Oppositely, the other limit $\nu \to 0.5$ corresponds to the case of complete bulk incompressibility, that is, corresponds to either very large values of E, or, oppositely, to a body comprising a material with a high compliance enclosed in a very soft shell. It is apparent that the values of the elasticity coefficients that one gets on approaching either of these two limits can differ significantly.

5.1.4 ELASTIC ANISOTROPIC BODIES

In this section, we will briefly address the homogeneous, bulk (i.e., 3D), primarily crystalline materials in uniform stressed states within the limits of reversible small deformations.

In the most general case, all nine components of the strain tensor are linearly related to all nine components of the stress tensor. This would yield Hooke's law with 81 parameters: $\varepsilon_{ij} = \sum_{kl}^{1-3} s_{ijkl}\sigma_{kl} = s_{ijkl}\sigma_{kl}$ with the summation taking place over the repeating indices and with the inverse transform matrix of $\sigma_{ij} = c_{ijkl}\varepsilon_{kl}$. Obviously, there is no inverse proportionality between individual s and c. The 81 s and c parameters transform linearly upon turning the coordinate system, that is,

$$c'_{ijkl} = \sum_{mnop}^{1-3} a_{im}a_{jn}a_{ko}a_{pl}c_{mnop} = a_{im}a_{jn}a_{ko}a_{pl}c_{mnop}$$

Here, summation takes place over the four repeating indices. The coefficients a constitute the direction cosines to the fourth power: the parameters c_{ijkl} (or s_{ijkl}) represent a fourth-order tensor, referred to as the *material tensor* describing the physical properties of the crystal.

Thermodynamics yields a particular symmetry of pairwise components c_{ijkl} (or s_{ijkl}), so there are only 36 independent variables. These variables can be written down in the form of a square 6×6 matrix. Further simplification is achieved by a pairwise combination of indices according to the following rule:

$$
\begin{array}{ccccccc}
11 & 12 & 13 & & 1 & 6 & 5 \\
 & 22 & 23 & \to & & 2 & 4 \\
 & & 33 & & & & 3
\end{array}
$$

that is, $\sigma_{22} \to \sigma_2$; $\varepsilon_{12} \to \varepsilon_6$, and so on. Consequently, for fourth-order tensor components, one has $c_{1111} \to c_{11}$; $c_{1122} \to c_{12}$; $c_{1212} \to c_{66}$. It is worth pointing out here that this notation is typically utilized in textbooks and handbooks on crystallography.

Hooke's law can be then written as $\sigma_i = c_{ij}\varepsilon_j$ or $\varepsilon_i = s_{ij}\sigma_j$ with the assumed summation of the *six* variables. This is visualized as the relationship between the two vectors that is established via a second-order tensor, but in fact, we are dealing with two second-order tensors and the material tensor of the fourth order that establishes the connection between them.

Finally, let us introduce another simplification. The matrices of the c_{ij} and s_{ij} tensors are symmetric with respect to their main diagonal, which means that the largest number of independent elasticity or compliance moduli is $6 + (36 - 6)/2 = 21$. This is exactly the case for the "least symmetrical" crystals belonging to the triclinic crystal system. In the opposite extreme case for isotropic media, the number of independent moduli is reduced to two (but never to a single one). All other crystal systems can be arranged by the number of independent elasticity constants in the following series:

Triclinic, 21
Monoclinic, 13
Orthorhombic, 9
Tetragonal and rhombohedral, 7(6)
Hexagonal, 5
Cubic, 3
Isotropic media, 2

For example, hexagonal crystals in the principal coordinate system (system of crystallographic axes) have five independent moduli, $c_{11} = c_{22}$, c_{12}, $c_{13} = c_{23}$; c_{33}; $c_{44} = c_{55}$, and one dependent component, $c_{66} = \frac{1}{2}(c_{11} - c_{22})$. All other components are zero in the principal coordinate system.

It is worth pointing out that the elastic properties of cubic crystals (precisely the elasticity is implied here and not the plasticity or cleavage) reveal anisotropy while being optically isotropic. The reason for this appearing deviance is as follows: the refractive index is a second-order tensor, and its characteristic surface is a second-order surface (an ellipsoid). The principal components of this tensor in the directions, [100], [010], [001], are the same. This means that the ellipsoid is in fact a sphere, and for any other direction, for example, in [111], one gets the same value. In contrast to this, the characteristic surface of the fourth order is not ellipsoidal, and the equality of characteristics in the direction of the principal axes does not in any way imply that the same values would be preserved in other directions. The only monocrystal that is elastically isotropic is that of molybdenum.

For cubic crystals, one has $c_{11} = c_{22} = c_{33}$, $c_{12} = c_{13} = c_{23}$, $c_{44} = c_{55} = c_{66}$. The only components that remain nonzero in an isotropic body are c_{11} and c_{12}. These two moduli can in turn be expressed via E and ν.

Let us now give a few more examples that illustrate the rules of how to interpret the material tensor indices and at the same time explain the meaning of the elastic constants in crystallographic notations. This can be most conveniently done with compliances, s. The parameter $s_{1111} = s_{11}$ shows the relative strain along the Ox axis caused by the unit stress applied along the Ox axis. It is worth emphasizing again that this is the stress applied to the unit cross section of an area perpendicular to this axis. Similarly, the parameter $s_{1122} = s_{12}$ indicates the deformation in the direction of the Ox axis caused by the unit stress applied in the direction of the Oy axis, and so on. All components may either reveal their own role or introduce necessary corrections, but the two parameters in the aforementioned example are closely related to Young's modulus and Poisson's ratio measured in a simple uniaxial extension of a rod, namely, $s_{11} \approx 1/E$ and $s_{12} \approx -\nu/E$. For an isotropic medium, this relationship is rigorous: $s_{11} = 1/E$, and $s_{12} = -\nu/E$.

Let us also say a few words about s_{44}. This parameter represents the relative shear (rheological shear, γ) in a plane y in a direction z caused by a unit shear stress acting on a plane y in a direction of z. For an isotropic medium, it can also be shown that $c_{11} = 2\mu + \lambda$, and $c_{12} = \lambda$.

5.2 HETEROGENEOUS STRESSED STATE: DURABILITY AND FATIGUE

5.2.1 HETEROGENEOUS STRESSED STATE

We now have enough background in elasticity theory to address the specific concepts of the strength of materials. The problem that is of primary interest in material science and engineering and thus in physical-chemical mechanics is that of the estimation of the distribution of stresses and strains and the change in the shape of a body of a given geometry with known elasticity parameters (rheological characteristics). The quantification of the changes in the body's shape is the most difficult problem, which has to do with the complexity of the integration. In addition to known models and methods, one may have to apply an individualized approach to either problem in carrying out the integration. Modern numerical methods make this task much easier, but still can't completely replace the analog model approaches.

In our discussion, we will restrict ourselves to selected characteristic problems that we will address at the basic qualitative and quantitative level. However, at the same time, we will make an attempt to reveal the physical nature of the model. We will use this simplified approach in the discussion of the laws of material failure and the principals of mechanical testing.

To begin, let us address a common and a rather characteristic example:

a. *Twisting of a cylindrical rod*: The rod is mounted at the bottom end, and a pair of forces is applied to the upper end, as shown in Figure 5.8. The geometry of this system is defined by the rod diameter, $2R$, and the length, H. If only elastic strain is present, one needs a single rheological parameter, that is, the shear elasticity modulus, G. The torque (moment), in this case, is given by $M = Fl$, or $2F \times l/2$. The problem is to find the angle of twist ϕ.

The nonhomogeneous nature of this state is manifested in the dependence on the variable radius measured from the axis of the cylinder, ρ. This dependence is the same along the entire length of the rod. In approaching this problem, it is essential to select a region within which the stressed state is uniform. Such a region can be found at a fixed distance from the axis along the entire rod. Consequently, to select a unit volume,

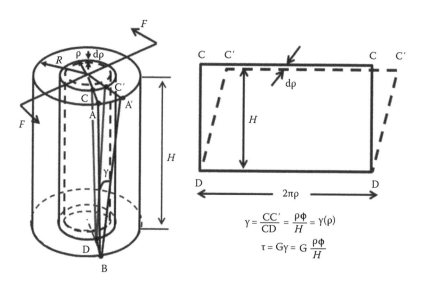

$$\gamma = \frac{CC'}{CD} = \frac{\rho\phi}{H} = \gamma(\rho)$$

$$\tau = G\gamma = G\frac{\rho\phi}{H}$$

FIGURE 5.8   The twisting of a cylindrical elastic rod.

we need to examine a hollow cylinder with a variable radius, ρ, and an infinitely small wall thickness, $d\rho$. This volume can be "rolled out" into a rectangular block of length $2\pi\rho$, height H and thickness $d\rho$ (Figure 5.8). The position of point D remains the same, while point C moves to position C'. The uniform shear, $\gamma = CC'/CD = CC'/H$ (within the approximation of the small strain rheological model).

At the same time, the displacement $CC' = f(\rho) = \rho\phi$, and also $CC' = H\gamma(\rho)$. We have therefore obtained a solution to the first problem, that is, the description of a uniform strain in a given elementary volume and the distribution of the deformation in the sample, that is, $\gamma = \gamma(\rho) = \rho\phi/H$. We have not yet introduced material properties (elasticity, plasticity, etc.) and can't make any conclusions regarding the acting stresses.

Let us first focus on the elastic behavior case (i.e., restrict ourselves to the initial linear portion of the rheological curve). In so doing, we define the "root" of our solution, that is, set the dependence between the stress and the strain in the elementary volume, that is, in our case Hooke's law yields $\tau = G\gamma = G\rho\phi/H$ per unit area. In order to find the stress in the entire ring, that is, within any horizontal cross section of the selected volume element, we need to multiply τ by $2\pi\rho d\rho$, that is,

$$\tau 2\pi\rho d\rho = (G\rho\phi/H)2\pi\rho d\rho = (G\phi/H)2\pi\rho^2 d\rho$$

The applied torque defines the force acting on the body. For this reason, we can use dimensional analysis and use the same dimension for the elementary action. The product of the force and the arm (variable radius, ρ) yields the elementary torque $dM = G\phi/H\, 2\pi\rho^3 d\rho$.

The only thing left to do is to sum the behavior of all the unit volumes over the entire volume of the body. This is in general a complex problem, but in our case, it does have a simple solution, that is, the integration of the elementary torque with respect to radius, ρ, along the axis of the rod between the limits of $\rho = 0$ to $\rho = R$. By doing so, one obtains the total torque $M = Fl = (\pi/2)(G\phi/H)R^4$, from where we can get the angle of twist, namely,

$$\phi = (2\pi)\left(\frac{M}{G}\right)\left(\frac{H}{R^4}\right).$$

We thus have demonstrated a rigorous solution to the classical problem in the strength of materials. In particular, we have determined the distribution of the deformations and stresses in a body and the mean-field change in the shape of a body based on the known geometry, the known mechanical properties, and the known stress.

This reasonably simple problem also demonstrates the power of using *dimensional analysis*. Within the limits of small parameters (strains in the present case), the linear approximation is frequently valid (please refer to the discussion on the "missing equation" in the thermodynamic description, Chapter 1). From Hooke's law, it follows both logically and empirically that the net effect, that is, the twist of the upper base of the rod over the lower base is directly proportional to the applied stress and momentum, is inversely proportional to the elasticity modulus and exhibits a dependence on the cross-sectional area, that is,

$$\phi \propto \left(\frac{M}{G} \right)(H)f(R)$$

In the aforementioned expression, ϕ is dimensionless, and the dimensions of other quantities are as follows: $[M] = $ N m, $[G] = $ N/m^2, and $[H] = $ m. This immediately yields $[f(R)] = $ m^{-4} and $\phi = \text{const}(M/G)(H/R^4)$. A rigorous solution yields the value of the constant as $2/\pi$.

b. *Bend of a beam* represents another frequently encountered nonhomogeneous state. In the strength of materials, a beam represents a body in which both transverse dimensions are much smaller than the longitudinal dimension. In the case of a strip, it is assumed that only the thickness is much smaller than the length.

An example of a bend that has the simplest solution is that of a thin strip of elastic material of length l and thickness h rolled into a ring and fixed in this rolled state (Figure 5.9). In this case, the stressed state is the same in all cross sections. This stress state is referred to as *pure bending*. Between the upper and lower surfaces of the ring there is a so-called *neutral axis*, that is, the surface in which there are no stresses. Let us assume that the radius of the neutral axis can be characterized as $R = l/2\pi$. The inward surface of this ring is compressed, while the outward surface is stretched (extended). For a *thin* strip, one can assume that these stresses increase linearly with the distance from the neutral axis.

The elongation of the outer side of the ring at $dR = +h/2$ is $dl = 2\pi dR = 2\pi h/2 = \pi h$, the relative strain $\varepsilon = dl/l = \pi h/l$, and the stress $\sigma = E\varepsilon = \pi Eh/l > 0$. Similarly, on the inner side the relative compression is $-\pi h/l$, and the stress $\sigma = -\pi Eh/l < 0$. This is essentially all that concerns the magnitude of the residual internal stresses. A more complicated subject is the magnitude of force that one had to apply in order to roll the strip into a ring, that is, the forces that keep the ring in equilibrium until the ends are firmly connected.

c. *Four-point beam bending* is a common test method used in material science to evaluate mechanical properties. In this test, a beam with a rectangular cross section of width b and thickness h is mounted on two point-like sharp supports located at a distance l between them. The beam is then loaded with identical weights, P, located at a distance of $l/4$ moving toward the center from each point of contact. It turns out that in the middle point located at a distance of $l/2$ from the points where the loads are being applied, the stressed state constitutes pure shear, similar to that in the ring described earlier. At $h \sim l/2$, one can assume that longitudinal stresses linearly change between the layers from the maximum compression, $-\sigma_{max}$ on the upper (concave) side of the beam to the maximum extending $+\sigma_{max}$ on the lower (convex) side. This scheme can also be viewed "upside down": there is no principle difference between the concepts of *load* and *support*: they are equivalent on the basis of Newton's third law. The arrangement in which the extended side is facing upward is convenient when testing in the active medium is involved, for example, in the application of the liquid medium to the stressed region in a preformed groove or crack.

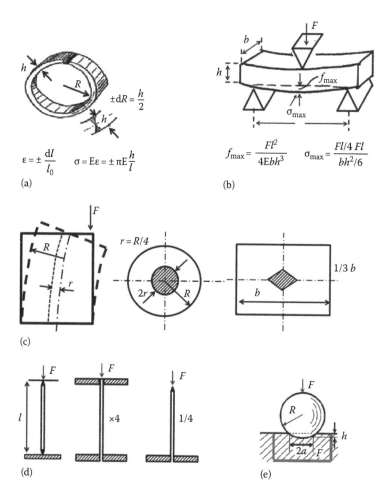

FIGURE 5.9 Schematic representations of nonhomogeneous stressed states: pure bend (a); three-point beam bend (b); eccentric compression (c); the loss of elastic resistance (d); indentation by a rigid sphere (e).

The process of estimating σ_{max} and the change in the shape (deflection from the known geometry) based on the known mechanical properties and the mechanical load closely resembles that used in the example of the twisted rod. In the case of small strains, dimensional analysis produces a correct solution with a precision up to a dimensionless constant. The maximum stress, σ_{max}, is directly proportional to the load force, P, and to the longitudinal linear parameter, l; it is inversely proportional to the transverse linear parameter (thickness), b, and represents some kind of a function of the beam thickness $f(h)$: $\sigma_{max} \sim (P/E)(l/b)f(h)$. Dimensional analysis yields $f(h) = \text{const } h^{-2}$. Consequently, one can write that

$$\sigma_{max} = \text{const } P \frac{l}{(bh^2)}$$

A rigorous quantitative solution with the integration of the stresses (moments of force) over dh in both directions from the neutral axis confirms the same dependence on h and establishes that the constant has a numerical value of 3/2. It is common to write the result obtained as

$$\sigma_{max} = \frac{(Pl/4)}{(bh^2/6)}$$

In the aforementioned expression, $Pl/4 = \sigma_{max}(bh^2/6) = M$ is the *moment of force* in the beam cross section (the integral of the layer-wise distribution). This momentum and σ_{max} are constant along the entire middle portion of the beam of length $l/2$. The quantity $bh^2/6 = W_{r.b.}$ is referred to as the moment of resistance of a beam with a rectangular cross section. For a cylindrical beam (i.e., a rod), one finds that $W_{c.b.} = \pi r^3/4$. As a result, for the maximum extensional stresses on the cambered side in the case of a four-point bent beam, one gets

$$\sigma_{max(beam)} = \frac{(Pl/4)}{(bh^2/6)} = \frac{M}{W_{beam}}; \quad \sigma_{max(rod)} = \frac{(Pl/4)}{(\pi r^3/4)} = \frac{M}{W_{rod}}$$

The same value can also be assigned to the maximum compressive stresses on the opposite (concave) side of the beam.

At every local spot the shear stresses correspond to the uniaxial stressed state. Consequently, upon reaching the right level of stress damage will occur in any cross section along the central part of the beam of length $l/2$. For this reason, an introduced defect, such as a scratch or crack, plays a very significant role in mechanical testing that uses the four-point bend method. Depending on the conditions, the defect may lead to the formation of a crack or to a plastic yield. It is worth mentioning here that the final step in determining the elastic deflection as a characteristic related to the shape change in the case of a bend is more complex than in the case of the rod twist. We will address this in more detail further down.

d. *Simple transverse bending* occurs in the case of a *three-point* beam loading. In this case, the distance between the points of support is l and a load F is applied to the middle of the beam from the side opposite to the points of support. The distribution of both longitudinal and transverse moments is a lot more complex than that in the case of four-point loading. The distribution is obtained using the Zhuravsky formula, which relates the longitudinal distribution of moments M to the shear components in the transverse direction, Q: $Q = -\,dM/dx$ and $M = -\int Q dx$. Without going into details, let us just stress that at the mid-point of the stretched side of the beam the expression for σ_{max} is similar to the one established for a four-point bending, that is,

$$\sigma_{max} = \frac{M_{max}}{W}, \quad \text{and also} \quad \sigma_{max(beam)} = \frac{M}{W_{beam}} = \frac{(Fl/4)}{(bh^2/6)}$$

The result for the deflection f_{max} in the middle of the beam is given by

$$f_{max} = \frac{(Fl^2)}{(4Ebh^3)}$$

Three-point transverse bending is a commonly used method in the experimental evaluation of the mechanical properties of materials. This method is popular for use with materials from which it is either difficult or impossible to prepare suitable samples for uniaxial extension studies. Since this method is especially attractive for short samples, it is worth highlighting a few points here.

First, the distribution of the tangential stresses in three-point transverse bending is principally different from the distribution in the case of pure bending. Point-like loading in the middle of the beam between the supports results in the appearance of new *shear* components. The shorter the beam, the stronger the effect of these components on the entire testing scheme. In the case of a very short beam, the transition from *bending* to *cutting* takes place. In this case, one operates with a completely different strategy for conducting tests, and completely different data analysis methods need to be used.

Second, in the fracture testing of a sample with loads applied using the three-point scheme, one needs to account for the spread in the experimental data that is caused by random deviations in the geometry of a setup, rather than by the sample structure. These include a slight shift, Δ, in the point at which the load is applied, from the middle point between the supports, and the recognition of the fact that the load is not really applied in a point-like manner but is indeed distributed over some area $\pm a$ with respect to the axis (i.e., the point of load application is blunt). It can be shown using the Zhuravsky equation that these two deviations correspond to a correction in the magnitude of the maximum moment, M_{max}. The maximum value of the moment can be expressed as the product of two terms:

$$M_{max} = (Fl/4)(1 - a/l)\left[1 - (\Delta/2l)^2\right]$$

We have examined two characteristic cases of the nonhomogeneous stressed state, those of twisting and bending. They lay a path forward to solving other problems in the area of the strength of materials. Let us briefly examine several examples of other interesting stressed states. We will primarily focus on a qualitative description of these states with the main purpose of understanding the underlying physics behind them.

e. *Eccentric compression*: For illustration purposes, let us consider a cylinder or a block made out of soft rubber (such as that used to make pencil erasers) or out of elastic porous material (e.g., a cleaning sponge). The compression of this material along the narrow side results in an extension on the other side. Such an effect is observed only when the sample is compressed over its edge (Figure 5.9).

Calculation of the elasticity indicates that there is an outer area in the transverse cross section of the body the compression of which results in the appearance of compensating extension stresses on the other side of the sample. At the same time, the border of this region is different from the inner part of the cross section, in which there is no such effect. The border of such a region is referred to as the *cross section core*. For a round cylinder, the cross section core is a circle with a radius equal to ¼ of the cylinder base radius; for a rectangular block it is a rhomb with diagonals parallel to the sides of the cross section and with the length equal to ⅓ of the length of the corresponding side.

f. *The loss of elastic stability*: This problem was originally formulated and solved by Leonard Euler. Let us consider a thin elastic rod with a cross-sectional area that is much smaller than the length. Let a longitudinal compressive load be applied to the rod. Both a tree twig and a razor blade start bending under the applied load, once the load reaches a certain level. From that point on, the resistance to deformation decreases, which causes an elastic-brittle material to break and an elastic–plastic material to become permanently deformed. The level at which the *loss of elastic stability* is achieved depends on how the ends of the rod are mounted.

One can identify three characteristic cases: (1) the sample of length l and Young's modulus E is loosely mounted with both ends touching the walls in grooves; (2) the sample is fixed at one end and the other end is free; (3) both ends are fixed.

In the first case, the critical load, P_{crit}, is given by $P_{crit} = \pi^2 E I_{min}/l^2$, where l_{min} is the minimum moment of inertia of a cross section. For a rectangular cross section, $I_{min} = bh^3/12$; for a round cross section, $I_{min} = \pi r^4/4$. The stability is the lowest in the second case: the value of P_{crit} is four times smaller than in the first case. Conversely, the third case with the firmest rod mounting yields the highest stability: in this case, P_{crit} is four times the P_{crit} value established for case (1). It is worth emphasizing that there is a very strong dependence between the stability and the l/h ratio.

g. *Elastic indentation* is one of the many methods used to determine the mechanical (rheological) properties of various solid materials. The method is based on an analysis of the image formed upon compressing a tip of a given shape (spherical, conical, or pyramidal) into the solid surface with a given force. In this chapter, we will restrict the discussion to the solution of the *Hertz problem*. This problem addresses the geometry of the reversible pressing with a force F of a stainless steel ball with radius $R \sim 1$ cm or of a diamond cone or pyramid with a rounded tip with $R \sim 25$ μm, into the smooth surface of a material having Young's modulus E. Without going into a detailed analysis of the components of the emerging, quite complex stressed state, we will make a rather crude estimate using a dimensional analysis of the variables for the case where a rigid sphere is pressed into the surface.

Let an indentor leave a hemispherical pit hole with a diameter equal to $2a$ and a depth equal to h in the plane of the sample surface. The order of magnitude of the deformations can be estimated as $\varepsilon_{aver} \sim h/a$, which is the only dimensionless combination of the parameters involved. Similarly, the stresses by order of magnitude are $\sigma_{aver} \sim F/\pi a^2$. Within the framework of Hooke's law, we have $\sigma_{aver} \sim F/\pi a^2 \sim E \varepsilon_{aver} \sim h/a \sim Eh/a$. In order to find the final solution, we need one more equation establishing the relationship between the geometric parameters. This relationship is obtained using the Pythagorean theorem, namely, $a^2 = h(2R-h)$, or for $h \ll 2R$, $a^2 \approx 2Rh$, that is, $h \approx a^2/2R$. Substituting this expression into Hooke's law, $F/\pi a^2 \sim Eh/a$, one finds

$$a \sim (FR/E)^{1/3}; \quad \sigma \sim (E^2 F/R^2)^{1/3}; \quad h \sim (F^2/E^2 R)^{1/3}$$

The integration of the elasticity equations (see the simple example of cylinder twisting) yields the values of the dimensionless coefficients. These coefficients are close to 1 and are not precise. For this reason, for a given material and type of indentor, the rate at which the load is applied, etc., one needs to generate a calibration curve, which would then yield the accurate values of these numerical coefficients.

Indentation testing is widely used in many different areas of technology for various materials and disperse systems. Examples include the determination of steel hardness by the Brinell method, using conical plastometers in determining the mechanical properties of grounds and construction materials. The properties of the Moon surface were originally studied using these types of measurements, which indeed confirmed that the surface was solid. Here, it is also worth mentioning an old method of measuring eye pressure, which is critical in the diagnosis of glaucoma: by using the diameter of an imprint caused by of a 15 g weight that has been placed in contact with the eye over several seconds.

5.2.2 CONCEPTS OF STRENGTH THEORIES

The analysis of the processes leading to the degradation of solids and various materials is of interest in many different aspects. These include the analysis of the properties of materials, the nature and character of the stressed state, and the means by which loads are applied. The latter include testing using constant strain rates, creep analysis, measurements of stability with loads applied over extended periods of time, resistance to impact, cyclic fatigue, and many others. Numerous standard test methods were developed for various materials under different conditions. The influence of the specific conditions, such as temperature, humidity, and the presence of surface-active substances also represents the subject of essential studies. Depending on the particular combination of all these factors, a particular type of material degradation can be observed.

The degradation can be brittle or plastic, that is, can be accompanied by the development of large or small residual deformations. Degradation can be of a particular fractographic type, that is, it can occur along the grain boundaries or over the body of the grains. We will use here the phenomenological approach to the concept of strength. This will be done similarly to how we have reduced the entire description of the stressed state to the analysis of the minimal number of clearly defined stress components and the complex behavior of the material to two constants: Young's modulus E and the Poisson ratio ν. In this case, we will establish the principles by means of which one can compare

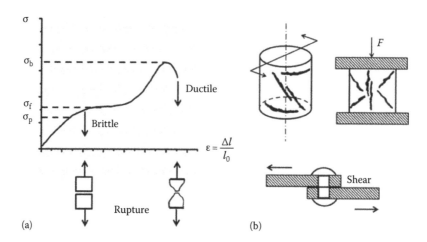

FIGURE 5.10 Schematic deformation curve (a); schematic illustration of types of fracture encountered in different stressed states (b).

the specific means of the degradation of materials under various physical conditions using a single "standard" method of testing. This would allow one to use the minimal number of parameters to interpret the results of such testing. There are numerous methods and algorithms available for use; we typically imply them when we talk about the various *strength theories.*

Let us turn our attention to the *deformation curve* encountered in sample testing in the uniaxial extension at a constant relative strain rate, $d\varepsilon/dt$. Let the curve be plotted in ε–σ coordinates, where σ is the stress applied to the initial section of the sample (Figure 5.10a). If the material is brittle, the failure can occur already within the initial Hookean portion of the curve characterized by a linear dependence between ε and σ. The collapse will then occur before the *proportional limit*, σ_p, has been reached. If the material is plastic, the increase in the stress slows down and the transition to a plastic flow occurs once the yield point, σ_f, has been reached. The increase in the stress then progresses at a slower rate, and the so-called *strain hardening* takes place. A subsequent fast increase in σ in these coordinates corresponds to the formation of a neck. After that, σ reaches a particular maximum and tear takes place. This stress maximum (per initial cross section) is referred to as the *temporary resistance to degradation*, σ_B. This parameter can be determined experimentally under the most simple and reproducible conditions, and for this reason, it is used as the standard characteristic in a number of strength theories. In addition to this characteristic, one also often uses the maximum deformation, $\delta = \varepsilon_{max}$ and the thinning, that is, the decrease in cross-sectional area, ϕ (%).

We refer to the theories of strength in plural because we have to be able to describe principally different types of degradation. In the deformation curve, we can clearly distinguish *brittle* and *plastic* fracture. These two cases are similar to each other in the sense that in extension, both of them are associated with a tear taking place perpendicular to the main extension component of σ. Here, the value of σ_b = (maximum strength/initial cross section) can serve as a single quantitative characteristic describing the sample strength.

The case that is opposite to tensile failure is *shearing failure.* For example, the failure of a rivet takes place along the plane of the highest tangential stresses. Although it may seem that tensile rupture is typical for brittle materials, while shearing failure is typical for plastic materials, it turns out that both of them can be accompanied by larger or smaller residual deformations. A piece of wire breaks upon stretching but undergoes shearing failure if it is used as a rivet. However, the same parameter σ_b characterizing the types of break can't be used to characterize shear failure.

Here, we encounter various types of behavior of materials upon fracture under different stressed states. Let us now address the opposite problem: by assuming the existence of intermediate states between tensile failure and shearing failure, let us compare the results of the testing of bodies of the same shape made out of different materials under similar loading patterns. Let us again refer to the example of twisting a cylinder. Indeed, brittle material undergoes failure mainly under the action of σ_{max} applied at a 45° angle to the axis of the cylinder, while plastic material undergoes shearing failure under the action of τ_{max} in the plane parallel to the base of the cylinder (Figure 5.10b).

Let us now discuss one more example that is essential for the analysis of the strength of concrete and other construction materials. This example relates to the case when a cube of large dimensions is tested in uniaxial compression, that is, tested for crushing. One can apply a number of restrictions here: the upper and lower sides of this cube either cannot slide against the mounts or conversely *can* slide freely due to an applied lubricant, the presence of a gasket, or bulged mounts. In this case, depending on the conditions, the fracture may start with pieces of material falling out on the sides, at a 45° angle, or by longitudinal crack formation.

These examples of various types of material failure allow one to formulate the problem more precisely. Indeed, we are looking for ways to compare various stressed states with a universal one that can be described using the minimal number of parameters, or even better—by a single parameter. *A priori* one can sense that this parameter is primarily the temporary resistance to failure characterized by the maximum on the deformation curve, σ_b. The main task is then to identify methods of reducing complex stressed states to this simplified one. This is addressed by using various strength theories, that is, by employing the hypotheses regarding the most vulnerable factors for a given range of materials and conditions.

a. *The hypothesis of the maximum tensile stress* (the first theory of strength). Let the stressed state of an isotropic solid be defined by the σ_{ij} tensor, which has the following representation in the principal coordinate system:

$$\begin{pmatrix} \sigma_1 & 0 & 0 \\ 0 & \sigma_2 & 0 \\ 0 & 0 & \sigma_3 \end{pmatrix}$$

It is implied that $\sigma_1 > \sigma_2 > \sigma_3$ and $\sigma > 0$ correspond to extension, while $\sigma < 0$ reflects compression. The simplest, but not the most probable, hypothesis states that only the largest extension component, σ_1, plays the determining role in the failure of a solid body, while the other components can be totally neglected. It is therefore assumed that the tensor

$\begin{pmatrix} \sigma_1 & 0 & 0 \\ 0 & \sigma_2 & 0 \\ 0 & 0 & \sigma_3 \end{pmatrix}$ is equivalent to the tensor $\begin{pmatrix} \sigma_1 & 0 & 0 \\ 0 & 0 & 0 \\ 0 & 0 & 0 \end{pmatrix}$ and that material failure

takes place when the stress $\sigma_1 = \sigma_b$. Consequently, the first theory (first hypothesis) of strength can be expressed as $\sigma_1 = \sigma_1 = \sigma_b$, which is equivalent to making *two* assertions: first, the equivalence of two tensors with the same component σ_1 and second, material failure when stress $\sigma_1 = \sigma_b$. The first hypothesis works well for hard stressed states associated with the extension, bending, or twisting of brittle materials. However, this hypothesis is not applicable to cases when tensile stresses are absent, such as the case of the compression of the cube with parallel cracks. In the latter case, one should not neglect the *tensile strain components*. This leads to the second hypothesis.

b. *Hypothesis of the role of the maximum tensile strain* reached in a given point of the material (the second theory of strength). For the same tensor as mentioned earlier, the maximum tensile strength is given by

$$\varepsilon_1 = \left(\frac{1}{E}\right)\left[\sigma_1 - v(\sigma_2 + \sigma_3)\right]$$

This expression includes both the determining role of tensile stress applied along the x axis as well as compressive stresses acting along y and z axes. The essence of the second theory can be formulated as

$$\sigma_{II} = \sigma_1 - \nu(\sigma_2 + \sigma_3) = \sigma_b$$

The left-hand equation states that for the set of failure phenomena under consideration, the initial stress tensor is equivalent to a stress of simple uniaxial extension that causes the maximum tensile strain caused by the single component of the stress tensor to be equivalent to the maximum tensile strain resulting from all the other components of the initial stress tensor. The right-hand equation establishes that failure occurs when the uniaxial extension reaches the value of σ_b.

The second hypothesis is suitable for describing the conditions leading to tensile failures in materials in various "hard" stressed states. The second theory includes all of the cases covered by the first theory but is not suitable for describing shearing failure.

c. *The hypothesis of maximum tangential stresses*: Here, we will follow the same logical scheme as described earlier. In the principal coordinate system for an arbitrary initial stress tensor, one has $\tau_{max} = 1/2(\sigma_1 - \sigma_3)$. We further would like to transform the initial stress tensor so that it applies to the case of uniaxial extension, for which one can write that $\tau_{max} = 1/2\sigma_{III}$. Comparing these two expressions allows one to formulate the third theory of strength, namely,

$$\sigma_{III} = \sigma_1 - \sigma_3 = \sigma_b$$

Here, the left-hand equation reflects the transformation of an arbitrary initial stress tensor to the condition of a uniaxial extension, for which the maximum tangential stresses (one-half of the only extensional stress component) are equal to the maximum tangential stresses of the starting tensor. The right-hand equation states that σ_b is the failure condition for this equivalent uniaxial extension. It is worth emphasizing that here, one is only dealing with the difference $\sigma_1 - \sigma_3$, while the individual components can be either positive or negative. The Lichtman–Shchukin theory, which takes into consideration both normal and tangential stresses, is discussed in Chapter 7 [28,30,31].

d. The so-called *energy strength theory* assumes that neither the stresses σ and τ nor the deformations themselves are of primary importance. The main focus here is on the stored shear energy (the deviator tensor). This theory can be reduced to a theory of maximum tangential stresses (third hypothesis). In the end, the critical elastic energy is realized once some critical value of τ has been reached.

e. *The Davidenkov–Friedman theory*: In all of the strength theories described so far, the properties of a given material are characterized by a single and only parameter, that is, by the extensional strength, σ_b. This is both an advantage and a limitation. A generalized hypothesis that allows one to describe the transition from brittle failure to plastic failure was developed by Davidenkov [1] and Friedman [7] back in the 1940s at the onset of the development of the discipline presently known as the "strength of materials." The Davidenkov–Friedman theory employs multiple parameters for the characterization of material properties: along with the extensional break resistance, σ_b, the shear resistance, τ_s, and the yield stress, τ_y, are used. The latter parameter describes the tangential stress corresponding to the onset of plastic flow, that is, it is equivalent to τ^* in rheology. The stressed state within the framework of the Davidenkov–Friedman model is described in terms of two main parameters: the normal stress used in the second strength theory, $\sigma_{II} = \sigma_1 - \nu(\sigma_2 + \sigma_3)$, and the maximum shear stress, $\tau_{max} = 1/2(\sigma_1 - \sigma_3)$. The stressed state is represented by the point or the region in the coordinate system where σ_{II} is the abscissa and τ_{max} is the ordinate (Figure 5.11).

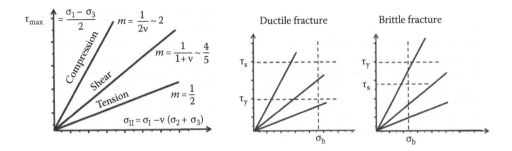

FIGURE 5.11 The parameters of the universal Friedman–Davidenkov theory.

Also, the ratio of these two variables, $m = \tau_{max}/\sigma_{II}$, serves as a very informative parameter, which characterizes the intensity of the given stressed state (absolute value) and its stiffness (slope). While being universal, the Davidenkov–Friedman scheme still has limitations, one of which is the absence of negative values in $\sigma_{II} = \sigma_1 - \nu(\sigma_2 + \sigma_3) < 0$, that is, it assumes the absence of damage under hydrostatic compression. This is not the case for a large number of porous bodies, such as grounds, sponges, various catalysts, and adsorbents.

Overall, the theories described in the textbooks on the strength of materials are primarily oriented toward the description of isotropic materials. At the same time, numerous real objects, such as paper, plywood, particle board, fabrics, and bone and muscle tissue require that specific approaches be utilized. One specific example of a customized approach is discussed in Chapter 7, where we address the determination of the strength of catalysts made in the form of granules, pills, marbles, or rods.

The strength theories described earlier constitute phenomenological schemes. This means that in order to use them for estimating the strength of real systems, one needs to supply the relevant experimental data on the characterization of the given materials under the given set of conditions. Separate and detailed treatment is also necessary for taking into account the effects of macro- and microheterogeneities in real objects and materials, which is the subject addressed in dislocation theory and other areas of *solid state physics* and in particular in *statistical strength theories*. Important factors that influence the strength, strain, and failure of materials are temperature and time, which is manifested by the most fundamental process in nature, that is, by *thermal fluctuations*. The latter is the subject of Zhurkov's *kinetic strength theory* [3].

5.3 MECHANICAL TESTING

5.3.1 PRINCIPLES AND METHODS OF MECHANICAL TESTING

5.3.1.1 Some General Remarks

Strength theories provide the means for describing complex stressed states. However, the strength characteristics themselves, that is, σ_b, τ_y, and others, depend on the conditions under which the tests are being carried out, that is, they depend on the rate of loading, the temperature, the media, etc., and need therefore to be determined under a broad range of conditions. Initially, let us focus on the role of temperature, as this is the factor of primary importance.

The influence of temperature on the mechanical properties of solids manifests itself primarily in the macroscopic dependence of the breaking stress, σ_b, and the yield stress, τ_y. Typically, both parameters decrease with increasing temperature, and the decline in τ_y is typically more rapid than the decline in σ_b. As a result of this "competition," the σ_b and τ_y curves may contain a crossing point, which corresponds to the *brittle-to-plastic behavior transition* temperature, T_c (Figure 5.12). This transition is sometimes referred to as the *cold brittleness*. Because of this, ionic crystals, metals, and

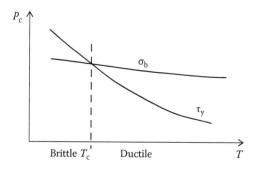

FIGURE 5.12 The effect of temperature on material strength (on breaking stress, σ_b, and the yield stress, τ_y).

other "hard" materials become plastic with an increase in temperature, while "soft" materials that exhibit plastic behavior under normal conditions become brittle when the temperature is lowered. This is the essence of the temperature effect. Other interfering factors include structural changes, the influence of the media, and the presence of impurities.

Consequently, one needs to conduct experimental studies under a broad range of temperatures with these other "factors of influence" taken into consideration. Such experiments can be carried out using commercial and custom-built research instruments that can handle measurements at both very high and extremely low temperatures. Further down, we will address the microscopic aspects of temperature effects. Here, let us briefly describe the measurements focused on investigating the rate of sample deformation, that is, let us closely examine the time factor. One needs to keep in mind that it is the combination of both thermal and time factors that determines the macroscopic and microscopic kinetics of material failure, phase transitions, crystal growth, diffusion deformation, the dissolution and accumulation of various defects, etc. In discussing the time factor in mechanical testing, we can outline three different types of testing regimes.

5.3.1.2 Static Testing Regime

In this regime it is assumed that the inertial forces are negligible under the strain rates and accelerations used (Figure 5.13).

5.3.1.2.1 Testing with constant deformation rate:

Represents one of the most common measurement regimes in which the deformation rate is set to a constant value, $d\varepsilon/dt = \text{const}$, and the force, F, necessary to keep the sample in equilibrium is measured with a dynamometer. If this force is measured in reference to the initial cross section of the

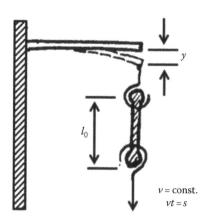

FIGURE 5.13 Schematic illustration of static mechanical testing.

sample, S_0, then the value of $F/S_0 = \sigma$ is referred to as the stress. A common result is the strain curve, $\sigma = \sigma(\varepsilon)$, which is essentially the $\sigma(t)$ dependence at a given strain rate, $d\varepsilon/dt$, and the character of that dependence can strongly depend on the value of $d\varepsilon/dt$.

The principal setup needed to carry out such measurements consists of three elements connected in series: (1) the sample itself, (2) a mechanical device connected to one end of the sample (e.g., a motor) capable of moving the sample end at a preset rate, and (3) a rather stiff spring connected to the other side of the sample.

Here, we need to devote special attention to clarifying the role of the spring. Stress (i.e., the force) is always measured by the reversible deformation of a spring, which is the "measuring body" of the dynamometer. The spring deformation can be determined by many different means, either via a direct observation or with the use of devices capable of transforming and amplifying an initial signal, such as photomultipliers, resistance sensors, capacitance sensors, piezoelectric sensors, as well as digital devices. Regardless of which specific mechanism is used, the measurement of the force is always based on Newton's third law and is related to the ability to measure small displacements with a force sensor of a *finite* stiffness.

Quantitatively, this problem can be defined as follows: let a sample with an initial length l_0 be deformed by extension at a constant rate controlled by a motor, $v = ds/dt = \text{const}$ (in Figure 5.13, $s = vt$ is the displacement of the lower end of the sample). As a result, the force F is being exerted on both the sample and the spring. To be specific, we may assume that the sample is either elastic–plastic or viscoelastic, characterized by an as yet unknown modulus K_{sp} (N/m), compliance, $1/K_{sp}$, and other rheological parameters. The elongation of the sample, $\Delta l = l - l_0$, is a function of all these parameters. Let us also assume that the spring stiffness, that is, the dynamometer modulus, equals K_{dyn} (N/m) and the compliance equals to $1/K_{dyn}$. This implies that the spring will undergo deflection and that the upper end of the sample connected to the spring will move by a distance $y = F/K_{dyn}$. As a result, the true elongation of the sample, $\Delta l = l - l_0$, is not equal to $s = vt$ but instead is given by $\Delta l = x = s - y = vt - y$. Consequently, the relative deformation is given by $\varepsilon = \Delta l/l_0 = (vt - y)/l_0$. For the simplest case of an elastic sample, one can write that

$$x = \frac{F}{K_{sp}} = vt - \frac{F}{K_{dyn}}$$

This means that if the dynamometer (i.e., the entire measuring scheme) has a high stiffness, then one can establish with good precision that the deformation of the sample is $x = vt$. The downside of such a *stiff system* is that the precision of the force measurement is low. If the dynamometer stiffness in not "too high," one can correct for the ratio K_{dyn}/K_{sp}. If, however, the values of K_{dyn} and K_{sp} are comparable, then one should not use this measuring scheme. In the case when the dynamometer is very compliant, that is, the system is "soft," measurements of deformations become impossible. However, it becomes possible to measure forces with very high precision, such devices include those based on magnetoelectric galvanometers that are used in the *contact interaction* studies for measuring the *cohesive forces between* particles. These measurements do not require the measurement of displacements and allow one to measure forces on the order of nN (Sections 1.3 and 2.3).

In contrast to such measurements, popular studies on measuring long-range forces require precise measurements of both the forces and the distances (see Section 4.1), requiring sophisticated instrumentation.

5.3.1.2.2   Testing at constant loading rates

The applied forces and the resulting stresses are, within certain limits, proportional to time, that is, $dF/dt = \text{const}$ and $d\sigma/dt = \text{const}$. When one end of the sample is fixed, the load is applied to the other end, and the deformation is determined by the displacement of this other end. Another example of a still used device is a tensile tester with a heavy "pendulum" that is moved to the side at a constant rate and simultaneously serves as a lever transmitting the amplified load to the sample. This amplified load can be quite high, on the order of several tons.

It is worth pointing out two things here. First, in Figure 5.13, the clamped sample is shown in the form of a dumbbell. The same device can be used for compression, twisting, and bending testing. Second, not all materials allow one to prepare suitable sample for testing. They include numerous porous objects, including catalysts and adsorbents. In Section 3.3, we described and referred to a number of priority studies on the development of the physical-chemical theory of strength. Those studies investigated the vulnerability of such materials to the action of an active medium. Improving the strength and durability of porous catalysts and adsorbents is one of the primary goals of physical-chemical mechanics.

Porous "microspherical" materials consist of granules with typical sizes from a few millimeters to about a centimeter. The actual shape of these granules is determined by their application and in particular by the conditions of the heterogeneous catalytic processes and by the specific method that was used to make them (i.e., caking, pressing, sol–gel method, and hydration hardening with subsequent dehydration). As such, the granules may be perfectly spherical, rodlike, spaghetti-like, ringlike, or sometimes even be of an irregular shape. Unfortunately, it is not possible for making a "good" dumbbell-like or beam-like sample with properties representative of the actual materials. For example, there are crush test methods available for catalysts [33–35]. The only universal static mechanical testing method is the granule *crush test* carried out at a constant rate of loading. At the same time, the results of such crush tests, which reveal the strength (in force units) of *granules* of a certain shape and made by a particular method or manufacturer are not suitable for the adequate characterization of the properties of *the material* that makes up these granules or for a relative comparison of different materials.

The estimation of σ_b or other universal parameters requires the use of the appropriate strength theory. The works of Bessonov and Paransky [26,27] (see Section 7.3) have shown that the basic theories described in the preceding text are not applicable in these cases. Porosity can be viewed as a special "structure state" of the material. It is not possible to reduce the investigation of this state to studies of the defects in isotropic continuous materials. A porous body can undergo a collapse and has a finite and broadly distributed strength under the conditions of full hydrostatic compression, which was the motivation behind developing a theory suitable for the characterization of such materials [10,31,32].

Let us restrict our discussion here to the analysis of a seemingly simple issue: how can one estimate the strength of 5 mm long porous rods? The studies carried out by Bessonov et al. [26,27] with the samples of various porous catalysts showed that many cylindrical granules had the following averaged values of strength, P, expressed as the force needed to crush them in a crush test between two parallel plates:

1. The strength measured for horizontally crushed cylinders, $P_h \sim 25$–75 kg cm^2 (the person conducting the test sees the cylinder end).
2. The strength measured for vertically crushed cylinders (the crushing is done along the cylinder's axis, the person conducting the measurements sees a square), $P_v \sim 80$–240 kg/cm^2.

The (a) and (b) cases do not reduce to each other without the use of some special strength theory. Indeed, case (a) does not conform to the elasticity theories, because crushing starts with the sample collapsing at one end, and the compacted portion of the cylinder further acts like a wedge. Case (b) involves certain difficulties associated with accounting for the friction in the supports and yields principally different results for samples with flat and convex ends. Consequently, one is left with no other choice but to conduct both measurements. The results of the tests have indicated that different materials can produce very broad range of P_v/P_h ratios: from 1:2 to 6:8. The approximate theory of strength for porous materials that has been developed by the author et al. predicts that for porous granules the P_v/P_h ratio should be around 3:4.

The observed discrepancy between P_v and P_h has caused a conflict in setting manufacturing specifications for catalysts: manufacturers prefer to report higher P_v values as their product spec, while end users have focused on using the lower P_h values as a realistic product quality indicator. Since the commercial volumes were very high, this issue and the basic question of "how and what should be measured?" were rather serious ones. One example illustrating the point is that of a calcium phosphate catalyst used in the synthesis of monomer for synthetic rubber production. Instead of the expected service life of 1 year, the catalyst pills failed after 3–4 months of use. Our studies have shown that the P_v/P_h was very high, around 7:8. In conjunction with the established mechanism of the failures (exfoliation), these data have allowed one to establish the reason for the early failure of catalysts [33–35]. It was established that the high pressing pressure allowed one to maintain good values of P_v measured in air but resulted in a high level of residual stresses that caused material failures during catalysis. Lowering the pressure in conjunction by using lubricants made it possible to extend the life of the catalysts to about 2 years. This subject will be discussed in a greater detail in Chapter 7.

5.3.1.2.3 *Mechanical testing with constant deformation*

The testing regime in which ε = const involves the determination of the stress relaxation as a function of time, $\sigma = \sigma(t)$. From our previous discussion, it follows that this regime requires a rather stiff measuring system, higher than the stiffness of the sample. Within the scope of physical-chemical mechanics, our primary interest is in the measurement of the rheological properties of viscoplastic, viscoelastic, and other materials with high compliance (see Section 3.1). A rather simple result of the measurements is the exponential drop of the shear stresses, $\tau(t) = \tau_0 \exp(-t/t_r)$, with a characteristic single relaxation time constant, t_r. The measurements carried out over a rather broad range of times allow one either to identify several characteristic relaxation times, t_r, or to analyze a continuous *time relaxation spectrum*.

5.3.1.2.4 *Mechanical testing under a constant load*

Since it is much more difficult to achieve the condition where σ = const (the sample cross section may change), it is the force that is kept constant in this measurement, that is, F = const. This testing regime is of great importance in engineering, solid-state physics, and in the analysis of geological processes. The F = const regime allows one to observe all the stages of *creep*, that is, the gradual accumulation of irreversible deformations. In particular, there are three main stages in the creep regime: a more or less short stage of *transient* creep, rather rapid at the beginning and slowing down further on; a stage of *steady-state* creep occurring at a constant rate and occupying the longest period in the time domain; and the final stage of accelerated creep. The constant load regime allows one to determine sample *durability*, t_{fr}, that is, the *long-term strength*, which is the time it takes for a material to fail. Consequently, this type of testing allows one to estimate and predict the reliability of both the sample and the material. Furthermore, the influence of active media is best revealed in these "slow" processes. This again shows that everything that we say about mechanical testing is inseparable from the physical-chemical aspects of the mutual action of load and surface-active media. Mechanical testing provides the most information if carried out over a very broad range of timescales: from nano- and microseconds to hours and months with a potential extrapolation to years.

Under the conditions of a gradual failure, such as in creep, the kinetic aspects of the physical-chemical processes are revealed, and the role of thermal fluctuations is emphasized. Thermal fluctuations are the primary reason for the activation of elementary acts of cleavage and rearrangements of the interatomic bonds. They determine the probability (i.e., frequency) of these acts overcoming the potential barrier. Here, we deviate from a macroscopic description of mechanical testing and move to a description at a microscopic and nanolevels.

The rupture of an individual interatomic bond can be described in terms of the need to overcome the energy barrier, U_0, which is defined as the cohesive energy in the body per single interatomic bond. This energy barrier can be overcome as a result of three principal factors: the mechanical work, W, carried out by the external forces (or due to the action of internal stresses) while taking into account the local concentration of stresses originating from structural defects ("the lever"); thermal fluctuations; and the possible lowering of the potential energy barrier, that is, the lowering of the work required due to the interaction with the atoms and molecules of the active medium.

Within the "zeroth approximation," the cohesive energy associated with an individual bond, U_0, is on the order of $\sim e^2/b \sim 10^{-11}$ erg $\sim 10^{-18}$ J, where the elementary charge, $e = 4.8 \times 10^{-10}$ CGSE, and b is the interatomic distance (on the order of a few Å). Within the "first approximation," one also needs to account for the specific properties of a given material, that is, the heat of sublimation, the elasticity modulus, or the surface free energy and the structure of the elementary lattice cell. In the latter case, one obtains similar estimates, that is, $U_0 \sim 10^{-19}$–10^{-18} J. Let us use the symbol U to denote the energy lowered due to adsorption, that is, $U = U_0 - \Delta U$.

The work of a force acting on a lattice of size $\sim b^2$ over a distance $\sim b$ is approximately $W \sim b^2\sigma b$, where σ is the stress. According to the Griffith scheme, the local concentration of stresses can be accounted for by a factor $(L/b)^{1/2}$, where L is the linear parameter characteristic of the average linear dimension of defects. Overall, $W \sim b^3(L/b)^{1/2} \sigma = V\sigma$, where V is the parameter with the dimensions of volume. In a real body, this parameter exceeds the elementary cell volume by one to two orders of magnitude. V can be viewed as the effective activation volume; it is actually Zhurkov's *structural factor*, γ.

Thermal activation, which on average may be expected to take place over the time t, is $kT \ln(t/t_0)$, where t_0 is the period of oscillations in a lattice, and the ratio t/t_0 is the number of oscillations taking place during the time t. The frequency $\nu = 1/t_0$ and the spectrum of these frequencies can be determined from the absorbance spectra. However, within the "quasi-microscopic" approach, one can estimate this time as $t_0 \sim b/\nu_s$, where ν_s is the speed of sound in a given body.

Now, one can express the kinetic condition of rupturing an individual bond as $U_0 - \Delta U_0 = V\sigma + kT \ln(t_{fr}/t_0)$, where the index fr (fracture) implies that the critical average time needed for the bond rupture at a given temperature due to a given stress has been reached. Consequently, t_{fr} describes bond durability. By carrying out a transformation from ln to exponent, one obtains

$$t_{fr} = t_0 \exp\left[\frac{(U - V\sigma)}{kT}\right]$$

This dependence of the bond rupture time on the temperature, the mechanical stresses, and the characteristic energy parameter of a material was experimentally established many years ago by Zhurkov et al. [3] (Figure 5.14). When written using conventional notation, the aforementioned expression can be recognized as the famous Zhurkov formula describing the *long-term strength*:

$$\tau_{fr} = \tau_0 \exp\left[\frac{(U_0 - \gamma p)}{kT}\right]$$

In the discussion that follows, we will retain τ_0 and γ but will not use τ and p in order to avoid confusion with shear and the extensional stresses.

It was established that the dependence given by the Zhurkov formula is a universal one and holds well for very different materials with interatomic bonds of different natures and very different structures. Over the years, the subject of numerous arguments and discussions has been the nature of U_0, that is, whether it should be viewed as the energy of sublimation or the energy of self-diffusion, as well as the physical reasons

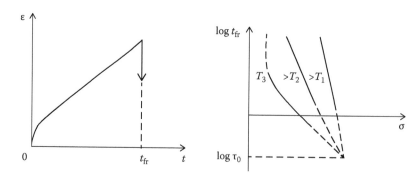

FIGURE 5.14   Time of bond rupture as a function of temperature based on Zhurkov's formula.

for the universal nature of Zhurkov's formula, and the "chicken-and-egg" problem of what precedence in the mechanical failure: the rupture of bonds or their rearrangement (i.e., deformation)? Zhurkov himself was rather skeptical about the explanation of his formula from the standpoint of dislocation theory, which gained common acceptance and recognition in the 1950s. The previously described thermodynamic approach applies to a single bond, while all of the experiments verifying Zhurkov's formula are macroscopic. Nevertheless, the value of the preexponential factor was the same in both approaches ($\sim 10^{-12}$ s). This means that material failure at the final stage of creep takes place at the point when a significant number of bonds in a body conform to the criterion established for each individual bond.

Zhurkov's kinetic theory is similar to other strength theories in the sense that it is also a scheme requiring experimental data comparison. In order to use Zhurkov's theory in specific engineering applications, one needs to conduct material testing over a broad range of parameters, that is, varying times and temperatures with a corresponding variation in the applied loads. The only way to use this scheme is to make certain predictions and extrapolations to very long or very short times.

Typically, creep testing and long-term strength testing are carried out at a constant temperature, $T = $ const. In these measurements, the dependence of the time to failure, $t_{fr} = f(\sigma)$, and of the steady-state creep rate, $(d\varepsilon/dt)_{stead} = f(\sigma)$, on the applied load can be determined. In most cases, σ represents a constant load ($F = $ const) applied to the *initial* cross section of the sample. However, Zhurkov accounted for the possibility of an automatic load tuning, that is, for a decrease in the load in the course of the sample extension, which is especially important in the case when a neck is formed.

Upon the transformation from exponent notation back to decimal logarithm, one observes a linear decay, that is, $\log t_{fr} = a - b\sigma$, where $a = (1/2.3)U_0/kT + \log \tau_0$, and the coefficient b includes the reciprocal of temperature, $b = (1/2.3)\gamma/kT$. At very high values of stress, $\sigma \to U_0/\gamma$, the fracture time reduces to the period of oscillations in the lattice, τ_0. This extrapolation does not realistically apply to the experimental data but helps one in illustrating an important point. In particular, in the $\log t_{fr}$ plots at different temperatures, this dependence corresponds to a family of curves that converge to the pole with coordinates $\sigma = U_0/\gamma$ and $\log \tau_{fr} = \log \tau_0$. The higher the temperature, the lower the position of the curves, that is, the lower the durability.

A more interesting case is that of extrapolation to small loads. Zhurkov's formula predicts a finite fracture time at each particular temperature at $\sigma \to 0$. This extrapolation has also been the subject of numerous debates. The experimental data show a steep increase in the fracture time when the stress is lowered to a particular value, referred as the *long-term breaking point*.

This concludes the discussion on the static methods of mechanical testing.

5.3.1.3 Cyclic Fatigue Testing

In the general case, this testing involves a sinusoidal impact on the sample, which can be superimposed with some constant component: the so-called asymmetric cycle (Figure 5.15). The *harmonic* component of the impact may be represented by both the stress and the strain changing according to a harmonic law. Variable parameters in this type of testing include the amplitude, the frequency, the temperature, and the active media corresponding to the realistic testing conditions (e.g., seawater).

The action imposed on the sample may not necessarily be harmonic but may have the form of *a pulse* with a smaller or larger duty cycle. Various devices can be used to generate a cyclic regime,

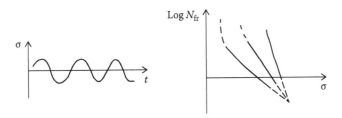

FIGURE 5.15   Schematic illustration of cyclic fatigue testing.

such as a cam, a crank, a drawstring, or an electromagnetic device. This may also involve cantilever bending, twisting, stretching, or compression. One distinguishes between *high-cycle low-amplitude fatigue testing* and *low-cycle high-amplitude fatigue testing*. An example of the first type of testing includes the testing of watch mechanisms, while the second type of testing is typical for such systems as airplane brakes, which need to be replaced after about 100 landings. In fatigue testing, the principal metrics are the residual strength, σ_b, after a given number of cycles, N, and the number of cycles to failure, N_{fr}, measured as a function of the stress amplitude, $N_{fr} = N_{fr}(\sigma)$.

The $N_{fr}(\sigma)$ dependence is typically a linear decaying function when plotted on a log-linear scale, that is, $\log N_{fr} = a - b\sigma$. As one can see, this expression is similar to the one established in the static durability testing. This similarity has a particular physical meaning (in a statistical aspect). Furthermore, the extrapolation to $\sigma \to 0$ is not valid here either: upon the lowering of the applied stress or strain to a particular value, the value of N_{fr} increases rapidly, that is, there is a *cyclic fatigue limit* similar to the static fatigue limit.

We frequently observe the appearance of cyclic fatigue limits in our everyday life, including the combination of cyclic and static fatigue, as in the combinations of high constant loads and vibrations. Common examples include suspension bridges, support beams, trees, and live organism joints. The influence of active media can also have a detrimental role leading to the destruction of the most chemically resistance materials. For example, dangerous cyclic fatigue may lead to the failure of polymeric or even Teflon oil lines in airplane wings.

5.3.1.4 Dynamic (Impact) Testing

This testing regime gives one a different perception of the time factor. Test regimes in which the velocities and accelerations are so high that the forces of inertia start playing a significant role can be recognized as dynamic. The following simple scheme allows one to compare static and dynamic testing. Let a platform be mounted on a rod of length l and cross-sectional area S. A load (weight) of mass m is lowered on to the platform from height H, and the sample undergoes an elongation equal to δl (Figure 5.16). Our task is to compare this elongation under static and dynamic testing conditions. Let us assume that the sample is elastic and can be characterized by Young's modulus, E.

When the loading takes place in a static regime, that is, when it is applied to the platform gradually, the elongation is given by Hooke's law as the ratio of the product of the force and the length to the product of Young's modulus and the cross-sectional area, namely, $\delta l_{stat} = mgl/ES$.

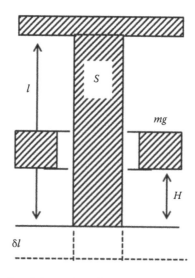

FIGURE 5.16 Schematic illustration of dynamic testing.

In order to determine the elongation under dynamic conditions, δl_{dyn}, one needs to employ a certain model. Let us assume that the kinetic energy of the weight is completely transformed into the elastic energy of the rod. In this case,

$$mg(H + \delta l_{dyn}) = \text{(sample volume)} \times \text{(elastic energy density)} = Sl\tfrac{1}{2}E(\delta l_{dyn}/l)^2.$$

This is a quadratic equation with respect to δl_{dyn}. The positive solution of this equation is

$$(\delta l)_{dyn} = \frac{(\delta l)_{stat}\left[1 + (1 + 2H/(\delta l)_{stat})^{1/2}\right]}{(\delta l)_{stat}\left[1 + (1 + W_{kin}/W_{stat})^{1/2}\right]}$$

where

$W_{kin} = mgH$ is the kinetic energy of the weight at the moment of impact

$W_{stat} = Sl\sigma_2/2E = (mg)^2l/2ES$ is the elastic energy in the case when the same load is applied under static conditions

This indicates that there is a double overload as compared to static loading, even when $H = 0$, that is, when the load is carefully brought into contact with the platform, and then quickly released by removing the support. Doubling the load under the dynamic condition results in doubled elongation and hence doubled stress in the sample. One has to keep this in mind when removing a detent in high-precision analytical balance or when using surface tension instruments.

In the case when $H > 0$, there can be an infinitely large number under the square root sign in the aforementioned equation. However, the square root dependence softens the impact: a 100-fold increase in the energy of the impact causes "only" a 10-fold increase in the overload. Similarly, when a porcelain dish falls on the floor and shatters into pieces, its kinetic energy is many times higher than the work of the elastic forces during a regular p lacing of the same dish on the shelf. This simple example illustrates the danger that the impact has on elastic-brittle materials and the universal importance of impact breaking for solids.

Another characteristic example arises from the comparison of a steel string and a rubber cord. Under static conditions, the cord will fail under loads that are safe for the string. Oppositely, under dynamic loading conditions, the string will fail, while the cord will survive due to the impact amortization. Impact amortization is an important feature stipulating the widespread use of *compliant* materials and various devices.

In the case when a material is not uniform but has a recess, the danger of failure is increased. Under static conditions, the distribution of local stresses and strains is simply inversely proportional to the cross-sectional area, while under dynamic conditions, the stresses and energy are localized within thin areas, which increases the likelihood of sample failure. This is why the unthreaded part of the bolt is typically cut to the lowest diameter of the thread: in this way the impact strength of the bolt can be enhanced.

In the case when material is not ideally elastic but also has some plastic or viscous nature, the effectiveness of impact load is decreased, and the energy conservation equation has a different form, that is,

$$mg(H + (\delta l)_{dyn}) = W_{elast} + A_{plast}$$

The higher the work of plastic deformation, A_{plast}, and the dissipation of energy in general (internal friction, heating, sound waves, etc.), the lesser the detrimental consequences of an impact, and at the same time the lesser the effectiveness of milling or grinding. In order to grind plastic materials, one utilizes liquid nitrogen.

A characteristic that describes the ability of a material to withstand dynamic loads is referred to as the *impact viscosity*. The impact viscosity of glass is very low, and as such glass is very brittle. At the same time, the impact viscosity of steel, and especially steel containing nickel, is very high. Disperse systems can have a very broad range of dynamic viscosities. It is very high in the case of live tissue: we can easily bang our fist against the table and clap our hands without causing any harm. Bones in fact represent the most remarkable material in terms of their ability to self-heal.

A standard device for conducting impact testing is a drop hammer. In a typical test, the sample is mounted horizontally, a heavy pendulum is raised and then released, hitting the sample. The impact viscosity is evaluated as

$$\eta = (H - H_0)\frac{mg}{S}$$

where

$(H - H_0)$ is the difference between the initial pendulum height and the height after the sample failure

m is the pendulum weight

S is the sample cross-sectional area

The invariant characteristic η is obtained when the absorbed energy W_c is normalized by the cross-sectional area S rather than by the volume. For high-quality construction materials, the impact viscosity can be in a range between units and tens of kg m/cm². Various admixtures can have a significant impact on viscosity.

The invariance of the W_c/S ratio for a given material can be confirmed experimentally. Samples of magnesium oxide and hydroxide were chosen as model systems for the experimental studies to be described. A power relationship between the value of η and the linear dimension of a sample D has been assumed, that is, $\eta = W_c/D^n$, which yields a linear relationship on a log–log scale, log $W_c = n$log $D +$ const. When the sample size is varied over a broad range of values, the slope of the log W_c – log D plot, (tan α), yields the value of n. Indeed, the experiments confirmed the linear log–log dependence and yielded tan α = 2, which is indicative of the physical nature of the fracture. Under these conditions, the work of fracture is determined by the development of a new surface area, rather than by the level of energy density needed to disperse the bulk matter.

The studies described require that samples of particular shapes be used. The studies involved catalyst granules, which are important objects for physical-chemical mechanics studies, and utilize material in which the sample shape is dictated by the manufacturing process. In this case, the proportionality (or even a correlation) between the static and dynamic strength is lost. The known strength theories for this system give very crude approximations at best. Granules can't be tested using a standard drop hammer. Bessonov and Shchukin used a rather simple testing method in which the granules were crushed by a drop hammer dropped vertically [26,27,33–35]. In these tests, a "yes–no" scale was utilized, that is, only the fact of crushing was recorded for a given hammer mass, m, and drop height, H. The results were then plotted in the form of the percent of granules that were uncrushed, q, as a function of the impact kinetic energy, W. The mean result corresponding to the critical value of W_c was the W_c value at q = 50%. The width of the transitional interval, $2W_1$, represents the data spread.

These general principles of mechanical testing now allow us to transition to a discussion of the principles for determining the structure–rheological properties of colloidal objects.

5.4   DETERMINATION OF THE STRUCTURE–RHEOLOGICAL CHARACTERISTICS

For many catalysts, the measurements described yielded rather modest values for the mean impact viscosity, $\eta \sim 100$ g cm/cm². This indicates that the static strength of porous granules may be 10–100 smaller than that of common construction materials, and the impact viscosity is 10,000 times worse. This means that reasonably strong porous ceramics can be quite brittle, which needs to be borne in mind in both manufacturing and quality control.

Along with the already discussed standard mechanical testing methods, there are highly precise techniques that are used specifically for the measurements of mechanical properties in colloidal

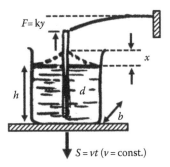

$F = ky$

h

d

b

x

$S = vt \ (v = \text{const.})$

FIGURE 5.17 Schematic illustration of the Weiler–Rehbinder plate.

systems [10,28–32,36–38]. In this section, we will describe several popular methods for the measurement of rheological parameters starting with the ones that utilize the conditions of a *simple uniaxial shear.*

a. *The Weiler–Rehbinder plate:* This simple scheme reveals all the problems and difficulties associated with carrying out rheological studies. In this method, a plate is mounted verti- cally in the middle of a rectangular cuvette, and its upper edge is connected to a dynamome- ter (Figure 5.17). The cuvette is filled with a solid-like material, which can be elastic, plastic, viscoplastic, or viscoelastic (e.g., gelatin). In order to avoid slipping, the walls of the cuvette and the surface of the plate can be corrugated. The cuvette is moved downward with a con- stant rate, v, (the plate is thus moved upward with the same rate). The result of testing is the stress–strain curve, $\tau = \tau(\gamma)$, where τ is the uniform shear stress and γ is the uniform strain ("technical" shear strain, i.e., two times the pure shear strain). The action is characterized by the force F and the deformation by the displacement of the plate, x. Without accounting for the strain of the dynamometer, one can write that $x = vt$.

The classical problem of relating the result of the action to the properties of the medium in the cuvette is solved as follows. Let h and b represent, respectively, the vertical and hori- zontal dimensions of the plate, and let d be the distance between the plate and side walls. The shear strain is $\gamma = x/d$. This picture is symmetrical on both sides of the plate. The shear strain, τ, can be found from the expression for the force, $F = 2bh\tau$, and $\tau = F/2bh$. The plot of $\tau = \tau(\gamma)$ is then generated, from which one can obtain several parameters.

First, the *shear modulus*, which, within the limits of reversible strain, can be obtained as follows:

$G = \tau/\gamma = (d/2bh)(F/x)$ = instrument constant $\times$ (action/displacement). Second, the *strength*: the meaning of the "strength" depends on the type of material and the nature of the breaking process. Here, we can employ the Davidenkov–Friedman theory as the most universal one, since the properties of colloidal disperse systems are very diverse. The uni- form shear is identified by a particular slope, that is, $\tan \alpha = 4/5$. In the case when, within the limits of elastic strain, one can observe a typical decay in the shear stress, characteristic of a fracture, and can see cracks developing at a 45° angle, one can conclude that we are dealing with a *brittle fracture*. In the case when such cracks develop after a certain residual strain, one is dealing with a *ductile tensile fracture*. The case when a vertical fracture has developed, or when the plate simply slipped out, corresponds to a *brittle shear fracture*, if there was no residual strain. If residual strain is present, one is dealing with a *ductile shear fracture*. This case requires the highest energy. Different materials exhibit different behavior in the same testing scheme, so the evaluation of the strength is inseparable from identifying the physical nature of the strength and from the investigation of the type of fracture process involved. Third, the *yield stress* (or yield point: the critical shear stress) for ductile solids is observed when $\tau_{\text{flow}} < \tau_{\text{shear}}$ when material failure does not occur early at $\tau_{\text{shear}} < \tau_{\text{flow}}$. The yield point can't be reached if it is higher than the fracture strength, $\sigma_{\text{fr}} = \sigma_{\text{b}}$.

This is the typical case for brittle and weak materials. Fourth is *the stiffness of the measuring system*. Here, it is worth recalling what has already been said in the discussion of static testing at a constant strain rate. The displacement of a platform, $S = vt$, along with the deformation x of the sample also includes the deformation of the dynamometer, y. The rheological scheme is such that these two are connected in series. Under the action of an applied force F, the deformation of an elastic sample is $x = (d/2bhG)F$ and that of a dynamometer $y = (1/k)F$, where k is the dynamometer spring constant. Since the compliances are additive for springs connected in series, the total displacement can be written as

$$S = vt = \left(\frac{d}{2bhG}\right)F + \left(\frac{1}{k}\right)F$$

If the dynamometer is stiff, the displacement of the platform (cuvette) equals the deformation, but the dynamometer itself is not very sensitive. For dynamometers with moderate stiffness, one can use the correction factor, $x = vt - (1/k)F$. If $2bhG/d \gg k$, that is, the sample is much stiffer than the dynamometer, but the estimation of deformations is no longer possible. Over the years, this had deceived many researchers, who were in fact measuring y instead of x. However, under these conditions, the high compliance of a dynamometer stipulates its high sensitivity. If the goal is to measure the strength, F, without measuring deformation, for example, in the regime where $dF/dt = $ const, one needs exactly this type of scheme. This is also the case for the scheme involving a magnetoelectric galvanometer, which we have referred to on numerous occasions.

Generally, when one needs high precision in measuring both the displacements and their forces, different means for amplifying the displacement signal are used: optical, electromagnetic, photoelectric, those using A/D converters, etc. However, the simple mechanical means of amplification, which has existed since ancient times, such as a lever or a pulley, can still be effectively utilized.

b. A *rotational viscometer* is a commonly used device for measuring rheological properties under uniaxial shear conditions. The viscometer consists of two coaxial cylinders with a small gap between them. Let us describe this scheme quantitatively following the same principles as in the discussion of the plate in a flat-parallel gap (Figure 5.18).

The two cylinders of height h have radii R (essentially the same for both cylinders) and the gap between the cylinders, $d \ll R$. At this point, let M be the torque due to the external action and the resulting displacement is the turn angle around the axis, ϕ. The task here is to determine the distribution of deformations and stresses in the media filling the gap between the cylinders and to characterize them using the geometry of the viscometer as well as to determine the turn angle, ϕ.

While the deformation can be viewed as uniaxial shear, the shear plane, in fact, moves around the axis, while maintaining a constant magnitude. If point A undergoes a transformation into point A′, then the displacement $x = AA'' = R\phi$, and the displacement, $\gamma = R\phi/d$, has a constant magnitude at all points in the gap.

The applied torque, M, is balanced by the shear forces acting in the medium,

$$M = \left(\text{force}\right) \times (\text{distance}) = (\text{stress}) \times (\text{area}) \times (\text{distance}) = \tau \times 2\pi Rh = 2\pi R^2 h$$

from where one gets $\tau = \dfrac{M}{2\pi R^2 h}$.

A more interesting question is what exactly is being measured. The normal regime is not that of $M = $ const but that of $d\phi/dt = $ const, that is, $d\gamma/dt = $ const. One operates therefore at a given strain rate and measures the torque $M = M(t)$ with a rotation dynamometer connected

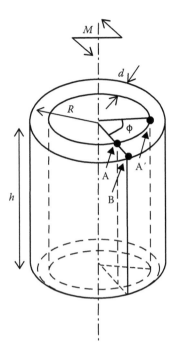

FIGURE 5.18 Schematic illustration of a rotational viscometer.

to the inner cylinder in the direction of its axis. Similar to the Weiler–Rehbinder plate method, the rotational viscometer allows one to determine various rheological character-istics, for example, the shear modulus $G = \tau/\gamma = (d/2\pi R^3 h)(M/\phi)$. This dependence can also be derived using dimensional analysis.

The largest advantage of this method is that it allows one to conduct measurements over a long period of time with the accumulation of large deformations. A plate in a flat-parallel gap does not have this capability: the Weiler–Rehbinder method is suited mostly for studies of strength characteristics, that is, the critical parameters of bodies with a solid-like behav-ior. In contrast to that, the rotational viscometer allows one to study the viscous behavior of liquid-like viscoplastic materials in a steady-state regime. Due to the difficulty in filling the gap properly, this method is of little use in the characterization of solid-like materials.

In the steady-state regime, the viscosity $\eta = \tau(d\gamma/d\tau)$, where $d\gamma/d\tau = R(d\phi/dt)/d$. The steady-state regime is achieved by rotating the outer cylinder with a constant angular velocity, which can vary over a broad range of values. The torque is measured by a small twist of elastic stiff thread connected to the inner cylinder along its axis. The torque M and the turn angle ϕ yield the values of τ and γ and provide a description of the rheological properties of materials in the coordinates $d\gamma/dt = f(\tau)$ and $\eta = f(\tau)$.

It is also possible to employ other regimes in rotational viscometry, such as the creep regime achieved by applying a constant load, $M = const$. In the creep measurements, one determines the $\gamma = \gamma(\tau)$ dependence, and in the steady-state regime, one obtains the $d\gamma/dt = f(\tau)$. It is also possible to conduct measurements at $\phi = const$ and to observe the relaxation of stresses, that is, $M = M(t)$ and $\tau = \tau(t)$. In the latter case, there are special requirements as to the stiffness of the force-measuring device.

One difficulty associated with the use of this method is related to the influence of the cylinder end effects. If the system has a high viscosity, the issue can be to some extent addressed by making the walls corrugated. When the viscosity is not very high, one needs to employ more drastic means for addressing the end effects. As hinted by the geometry itself, one needs to use cylinders with a large side area, $S_{side} \gg S_{base}$, that is, one needs to use tall and narrow cylinders and solve the problem of how to fill the gap. A second option would be to make the gap at the bottom much larger than the side gap, d. Since $\gamma \sim 1/d$, the stresses and deformations at the end are not large. Both of these geometrically based means are utilized. One may also use a "bell," a hollow inner cylinder with a sharp edge.

Another difficulty is associated with the existence of dry friction in the axes of the inner cylinder. Bearings cause a loss of precision due to dry friction at the initial low stresses and deformations. Modern rheological instruments employ advanced air and even magnetic bearings to address the sensitivity issue. In any case, one still faces issues related to the stiffness of the dynamometer and issues related to the loss of stability due to signal amplification.

A radical solution to these problems is to get the stiffness, not from the mechanical stiffness based on employing a spring with the proper spring constant, but rather by establishing a *negative feedback*. The principle of negative feedback is used not only in carrying out precise physical measurements (e.g., in atomic force microscopy) but also in solving engineering problems associated with the stability of various nano- and megasystems and constructions. We will discuss the principle of negative feedback later in this chapter in the section devoted to the discussion of the torque pendulum.

c. *Cone–plate viscometry*: In this method, the gap between a plate and a cone with a "very obtuse" apex is filled with a material to be tested (Figure 5.19). The main characteristics are the cone radius, R, and the angle, β. As in rotational viscometry, one may conduct measurements employing various regimes: creep (i.e., the rotation of a cone with constant torque, M) or the measurement of the torque upon the rotation of a plate with a constant angular velocity, $\omega = d\phi/dt$.

A peculiarity associated with this measurement scheme is the state of the uniform shear in the gap. To illustrate this, let us choose a ring located at the distance ρ from the axis as a unit volume. This volume is essentially a hollow cylinder with a wall of thickness $d\rho$, and a height, $d = \rho\beta$. Let us assume that the cone is turned counterclockwise with respect to the platform by an angle ϕ, which in the figure corresponds to the displacement by $x = \rho\phi$. The deformation due to shear is then $x/d = \gamma = \phi/\beta$, that is, the deformation is independent of ρ and is constant within the entire gap. Within the selected ring, the elementary torque is balanced by the force $dM = \tau 2\pi\rho^2 d\rho$. When one carries out the integration with respect to $d\rho$, one obtains $M = 2/3\pi\tau R^3$ and $\tau = (3/2\pi)MR^{-3}$.

The principal difference between cone–plate viscometry and rotational viscometry is in the materials that are most suitable for investigation by each method. In contrast to the

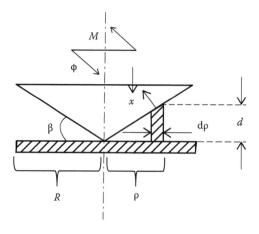

FIGURE 5.19   Schematic illustration of a cone–plate viscometer.

rotational viscometry, cone–plate viscometry is most suitable for studies on concentrated disperse systems and viscoplastic objects, such as pastes, sediments, slurries, and greases.

The advantage of the methods utilizing a uniform uniaxial stressed state is the possibility they offer to characterize the disperse system using standard rheological parameters. This allows one to perform a quantitative comparison of the behavior of a given material to the behavior of other materials or to the behavior of the same material under different conditions. Although in uniform states one is typically dealing with linear dependencies and the simple additivity of the various factors, there can be a broad variety of suitable rheological methods that reveal the nonuniformity of real systems. In fact, the deviations from linearity observed in the measurements can be a source of valuable information regarding the properties of the systems under investigation. As such, the deviations may not necessarily be an obstacle but can serve as means of investigating the behavior of a given system.

d. *Poiseuille's capillary viscometer* (Figure 5.20a) is the most popular tool for conducting rheological measurements under conditions of nonuniform stresses and strains. For a Newtonian fluid that wets capillary walls, the velocity distribution has a reasonably simple appearance. In the absence of slip, the velocity at the walls is zero, while the velocity gradient and the shear stresses are maximal. Along the capillary axis, the velocity reaches the maximum value in the absence of shear forces. Integration of this parabolic profile yields the strain rate (flow of volume, V, over time, t) as a function of the system geometry (capillary length, l, and capillary diameter, $2r$), the stresses applied (hydrostatic pressure differential, Δp), and the properties of the liquid (constant viscosity η), that is, it yields Poiseuille's Law:

$$\eta = (\pi/8)(\Delta p/l)(t/V)r^4$$

The same logic as in the cases described earlier was also employed here. The same expression for V/t can also be derived using dimensional analysis.

In the case of non-Newtonian behavior and especially in the case of viscoplastic behavior, such as that typical for moderately concentrated colloidal dispersions, Poiseuille's law gradually loses its validity. This happens because in the shear force–free central region close to the capillary axis, the structure of the concentrated colloidal system remains intact, so that the viscous shear exists only in the peripheral regions of the capillary. This process causes serious issues in the pumping of cement slurries or crude oil containing crystallizing phases. In laboratory practice, it is beneficial to conduct such measurements in combination with other measurements that utilize uniform states.

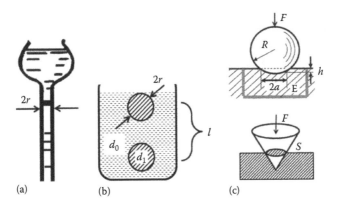

FIGURE 5.20 Rheological studies under conditions of nonuniform stresses and strains: viscosity measurements according to Poiseuille (a); Stokes (b); and the indentation by a sphere and a cone (c).

e. *Stokes viscometry* is another popular method for measuring viscosity (Figure 5.20b). This method is based on the measurement of the steady-state velocity of a ball with radius r falling in a Newtonian fluid. The fluid viscosity is determined via Stokes' Law:

$$\eta = (2/9)r^2(d - d_0)g(t/l)$$

f. *Indentation* is a rather common method for determining the mechanical and rheological characteristics in various kinds of technological testing (Figure 5.20c). The method can be used to obtain crude estimates, as well as for conducting accurate and precise measurements (e.g., for studying dislocations in crystals). We have already referred to indentation on numerous occasions.

The primary use of indentation is in estimating the *hardness* of various solid materials. Indentation using a hard standard (steel, tungsten carbide, or diamond), which can have various shapes, such as a sphere, a pyramid, or a cone, with a force F leaves an impression of size D. The area of the impression is then estimated as $S = kD^2$, where $k \leq 1$. This area may be the entire area of an impression or the area of a cross section. The hardness is then determined as $H = F/S$.

The hardness determined by the indentation test represents an effective mean-field parameter. However, it is also a very useful, simple, and universal characteristic if the type of mechanical behavior is known. The effective hardness can serve as an estimate when dealing with *elastic–plastic* objects. The value of the effective hardness H in this case is close to the yield stress. This applies to the well-known Brinell hardness test, which yields a hardness value, HB, using a steel ball to indent the tested object.

A very gentle indentation may yield information on the residual strains in very brittle bodies, such as glasses.

For plastic materials, a *conical plastometer* can be used. The immersion of a cone with a defined apex angle into the body down to a depth h due to the action of force F can yield an area of contact, S_t and the tangential component of the force, F_t. If the ratio $F_t/S_t = \tau^*$ is invariant with respect to the force applied, then the rheological behavior can be interpreted in terms of the Coulomb model and $\tau^* = \mathrm{Inv}$ is a critical shear stress. This is a common way to study the rheology of cement, flour, soils, etc. A conical plastometer was used in the evaluation of the mechanical properties of the surface of the Moon.

If in the addition to the Coulomb's element there is also a Maxwell's element present, that is, the irreversible *viscous flow* of the medium is possible, then such an invariant is not present. The applied force is distributed between the resistance of dry friction and the viscosity. However, varying the forces and measuring times may allow one to qualitatively differentiate those two components.

The case of a perfectly elastic contact between the solid surface and the absolute solid ball is known as the *Hertz Problem* of contact mechanics. The *Hertz Problem* has a rather cumbersome solution. With the application of dimensional analysis (Section 5.2), one can get a characteristic nonlinear dependence of the size of the impression on the indenting ball's diameter, the applied force and the Young's modulus of the material. In its reverse version, that is, for the case of a contact between a compliant sphere and a solid surface (bottom of a 15 g weight), this method was used for a long time to measure the internal eye pressure of the eye.

g. The *torque pendulum* represents a good example that illustrates both the specifics of solving a problem involving a nonuniform stressed state, as well as the general methodology of carrying out precise measurements in colloid science. This method was been originally developed by Trapeznikov [39] and was successfully used for the investigation of the rheological properties of interfacial adsorption layers [40,41].

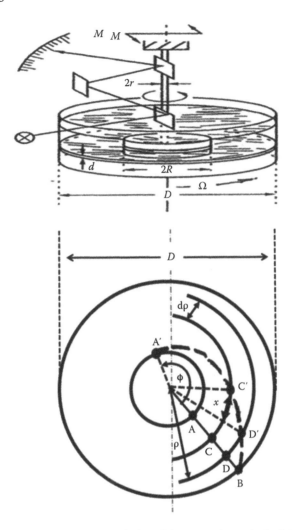

FIGURE 5.21   The torque pendulum. The description of the stressed state in the interfacial film.

The principal scheme of the torque pendulum is shown in Figure 5.21. A disk of radius R is placed at the surface of the liquid or at the interface between the two liquids (a polar aqueous surfactant solution phase and a nonpolar phase, that is, hydrocarbon or fluorocarbon). The disk is suspended on a thin wire that serves as a dynamometer. Turning this wire by an angle ψ produces a torque M, while the turn of a cuvette by an angle ϕ with respect to the disk leads to the total shear deformation of the adsorption film. The principle of operation of this instrument is similar to that of a rotation viscometer, with one principal difference: in the torque pendulum it is not possible to utilize a thin gap between a disk and a cuvette. Because of this the stresses and the deformations in the film are not uniform.

For a wire of diameter $2r$, length H_w and a shear modulus G_w (e.g., tungsten wire 50–100 μm in diameter and about 50 cm long), the torque is related to the angle ψ_w via the following relationship (see Section 5.2):

$$M = (\pi/2)(G_w \psi_w / H_w) r^4$$

This torque is balanced at the disk contour by the shear stresses τ acting within a film of thickness δ:$M = \tau(R)2\pi R\delta R$. The same is true for any other ring of radius ρ, that is, $M = \tau(\rho)2\pi\rho\delta\rho$. We can then express the distribution of the shear stresses along the radius as

$$\tau = \frac{M}{(2\pi\rho^2\delta)} = \tau(\rho)$$

This decay in the stresses from the center of the disk to the periphery results in a heterogeneity that makes a solution difficult. The simplest case is to assume that the film is elastic and has a shear modulus G. In this case, one can write the expression for the displacement of point A belonging to a radius ρ with respect to point B belonging to a radius $\rho + d\rho$: $dx = CC'-DD' = d\rho\gamma = (\tau/G)d\rho$. Substituting the expression for $\tau(\rho)$, one gets $dx = M/(2\pi\rho^2\delta G)d\rho = \rho d\phi$, where $d\phi$ is the angle element corresponding to the turn around the axis. By choosing an elementary ring, where the stressed state is homogeneous, one can relate the distribution of the deformations in the studied object (the film) to its properties and the loading conditions.

Since the film reveals Hookean behavior over its entire area, one can carry out integration with respect to $d\phi$, that is,

$$d\phi = dx/\rho = M/(2\pi\delta G)(1/\rho^3)d\rho$$

The integration is carried out from R to $D/2$. One thus gets a total turn angle ϕ corresponding to the transition of point A located at the edge of the disk to point A′ relative to point B located at the wall of the cuvette:

$$\phi = M/(4\pi\delta G)\left[R^{-2} - (D/2)^{-2}\right]$$

Let us now compare the net film deformation ϕ with the results of the torque measurements, M, that is, after a small twist of the dynamometer wire by an angle ψ_w. This angle can be measured, for instance, by the deflection of a reflected light beam. If the film thickness δ is unknown, it is included in the product with the unknown modulus G, that is, the net result of the measurements is the modulus, that is, the elasticity *per unit length* (perimeter), and *not* per unit area. One can thus write

$$\delta G = (1/4\pi)\left[R^{-2} - (D/2)^{-2}\right](1/\phi)\left[(\pi/2)(G_w r^4/H_w)r^4\right]\psi_w.$$

The aforementioned expression is valid to the extent that the film behavior remains linear along its entire area and primarily along the edge of the disk, where the stresses are the highest. At the same time, this expression provides an important estimate pertinent to the experimental method. Using common values for all parameters on the right-hand side, one finds that $\delta G \sim 10\psi_w$. This means that for a rather rigid film with $\delta \sim 10^{-6}$ cm and $G \sim 10^5$ dyn/cm^2, one needs to be able to determine the angles $\psi_w \sim 10^{-2}$ radians with the precision of a few angular minutes. In the case when the film is very thin and elastic, that is, $\delta \sim 10^{-7}$ cm and $G \sim 10^4$ dyn/cm^2, the angles will need to be measured with a precision up to angular seconds.

Up to this point the linear dependence between M and ϕ has been assumed. This dependence corresponds to linear elastic behavior, that is, $M = $ const ϕ. This means that in the construction of the deformation curve other rheological factors may start to play a role. These factors include the viscous flow and failure. When conducting measurements in nonhomogeneous stressed states, we typically do not know the relative contribution of all these factors; thus, one needs to remember that under such conditions, the obtained estimates may lose both quantitative and quantitative significance.

h. *The estimation of the strength of the interfacial adsorption layers* by observing the behavior of individual droplets in surfactant solutions [42,43]. This method presents a difficult case for carrying out rigorous calculations but yields a clearly defined effective characteristic, that is, the critical force of mutual compression resulting in drop coalescence. We have discussed this method and some results obtained with it for fluorinated systems in Section 4.2.

REFERENCES

1. Davidenkov, N. N. 1936. *Dynamic Testing of Metals*, 2nd edn. Moscow, Russia: Oborongiz.
2. Davidenkov, N. N. 1949. *Fatigue of Metals*. Kiev, Ukraine: Izd. AN USSR.
3. Zhurkov, S. N. and T. P. Sanfirova. 1955. Temperature and time dependence of the tensile strength of pure metals. *Doklady AN SSSR*. 101: 237–240.
4. Belyaev, N. M. 1965. *Strength of Materials*, 14th edn. Moscow, Russia: Nauka.
5. Rabotnov, Y. N. 1966. *Creep of Structural Elements*. Moscow, Russia: Nauka.
6. Timoshenko, S. P. 1972. *Strength of Materials*. Kiev, Ukraine: Naukova Dumka.
7. Fridman, Y. B. 1974. *Mechanical Properties of Metals*. Moscow, Russia: Mashinostroenie.
8. Cherepanov, G. P. 1974. *Mechanics of Brittle Fracture*. Moscow, Russia: Nauka.
9. Smirnov, A. F. 1975. *Strength of Materials*. Moscow, Russia: Vysshaya Shkola.
10. Shchukin, E. D. 1985. Physical-chemical theory of the strength of disperse structures and materials. In *Physical-Chemical Mechanics of Natural Disperse Systems*, E. D. Shchukin, N. V. Pertsov, V. I. Osipov, and R. I. Zlochevskaya (Eds.), pp. 72–90. Moscow, Russia: Izd. MGU.
11. Birger, I. A. and R. R. Mavlyutov. 1986. *Strength of Materials*. Moscow, Russia: Nauka.
12. Feodosyev, V. I. 1986. *Strength of Materials*. Moscow, Russia: Nauka.
13. Lemaitre, J. and J.-L. Chaboche. 1994. *Mechanics of Solid Materials*. Cambridge, U.K.: Cambridge University Press.
14. Hertzberg, R. W. 1995. *Deformation and Fracture of Engineering Materials*. New York: Wiley.
15. Hudson, J. B. 1996. *Thermodynamics of Materials: Classical and Statistical Synthesis*. New York: Wiley-Interscience.
16. Shames, I. H. and J. M. Pitarresi. 1999. *Introduction to Solid Mechanics*, 3rd edn. Upper Saddle, NJ: Prentice Hall.
17. Schijve, J. 2001. *Fatigue of Structures and Materials*. Berlin, Germany: Springer-Verlag.
18. Ambrose's J. 2002. *Simplified Mechanics and Strength of Materials for Architects and Builders*. New York: Wiley.
19. Girifalco, L. A. 2003. *Statistical Mechanics of Solids* (Monographs on the Physics and Chemistry of Materials). Oxford, U.K.: Oxford University Press.
20. Alexandrov, A. V., Potapov, V. D., and B. P. Derzhavin. 2003. *Strength of Materials*, 3rd edn. Moscow, Russia: Vysshaya Shkola.
21. Christensen, R. M. 2005. *Mechanics of Composite Materials*. Mineola, NY: Dover Publications.
22. Riley, W. F., Sturges, L. D., and D. H. Morris. 2006. *Mechanics of Materials*. New York: Wiley.
23. Beer, F., Johnston, E. R., DeWolf, J., and D. Mazurek. 2008. *Mechanics of Materials*, 5th edn. New York: McGraw-Hill.
24. Wei, R. P. 2010. *Fracture Mechanics: Integration of Mechanics, Materials Science and Chemistry*. Cambridge, U.K.: Cambridge University Press.
25. Shchukin, E. D. 2006. The influence of surface-active media on the mechanical properties of materials. *Adv. Colloid Interface Sci.* 123–126: 33–47.
26. Shchukin, E. D. 1965. On some problems of physical and chemical theory of the strength of fine porous materials—Catalysts and sorbents. *Kinet. Catal.* 6: 641–650.
27. Shchukin, E. D., Bessonov, A. I., and S. A. Paranskiy. 1971. *Mechanical Testing of Catalysts and Sorbents*. Moscow, Russia: Nauka.
28. Rehbinder, P. A. and E. D. Shchukin. 1972. The surface phenomena in solids in the process of their deformation and fracture. *Uspehi Fiz. Nauk.* 108: 1–42.

29. Shchukin, E. D. 2002. Surfactant effects on the cohesive strength of particle contacts: Measurements by the cohesive force apparatus. *J. Colloid Interface Sci.* 256: 159–167.
30. Shchukin, E. D. and V. I. Lichtman. 1959. On brittle fracture of zinc single crystals. *Doklady AN SSSR.* 124: 307–310.
31. Shchukin, E. D. 1981. Some problems of physical and chemical theory of strength of disperse structures. In *Physical-Chemical Mechanics and Lyophilic Properties of Disperse Systems*, Ovcharenko F. (Ed.), pp. 46–53. Kiev, Ukraine: Naukova Dumka.
32. Shchukin, E. D. (Ed.). 1992. *The Successes of Colloid Chemistry and Physical-Chemical Mechanics.* Moscow, Russia: Nauka.
33. Romanovskiy, B. V., Shchukin, E. D., Burenkova, L. N., and L. N. Sokolova. 2002. Influence of catalysis on the strength of porous materials with a globular structure. *Zh. Fiz. Him.* 76: 1044–1047.
34. Shchukin, E. D., Bessonov, A. I., Kontorovich, S. I. et al. 2003. Physical and chemical mechanics of catalysts in active media. *Fiz.-Him. Meh. Mater.* 39: 28–43.
35. Abukais, A., Burenkova, L. N., Zhilinskaya, E. A. et al. 2003. The influence of the catalytic conversion of the alcohol on the strength of porous materials ZrO_2 and $ZrO_2 + Y_2O_3$. *Neorg. Mater.* 39: 602–608.
36. Somasundaran, P., Lee, H. K., Shchukin, E. D., and J. Wang. 2005. Cohesive force apparatus for interactions between particles in surfactant and polymer solutions. *Colloids Surf.* 266: 32–37.
37. Steblin, V. N., Shchukin, E. D., Yaminskiy, V. V., and I. V. Yaminskiy. 1991. Hydrodynamic interaction of surfaces in electrolyte solutions. A new method of surface forces studies using capacitor ultradynamometer. *Kolloidnyi Zh.* 53: 684–687.
38. Yaminskiy, V. V., Steblin, V. N., and E. D. Shchukin. 1992. Viscous interaction between surfaces. Studies by means of capacitor ultradynamometer. *Pure Appl. Chem.* 64: 1725–1730.
39. Trapeznikov, A. A. 1941. Viscosity of monolayers and adsorption layers in solutions. In *The Viscosity of Liquids and Colloids*, pp. 87–115. Moscow, Russia: Izd. AN SSSR.
40. Izmaylova, V. N., Alexeeva, I. G., Shchukin, E. D., and P. A. Rehbinder. 1972. Rheological properties of interfacial adsorption layers of proteins and surfactants. *Doklady AN SSSR.* 206: 1150–1153.
41. Izmaylova, V. N., Alexeeva, I. G., Shchukin, E. D., and P. A. Rehbinder. 1973. Rheological properties of interfacial adsorption layers of gelatin on the boundary with benzene. *Kolloidnyi Zh.* 35: 860–866.
42. Parfenova, A. M., Amelina, E. A., Vitvitskiy, V. M. et al. 1990. The effect of various Pluronics on the resistance to coalescence of perfluorodecaline droplets. *Kolloidnyi Zh.* 52: 800–803.
43. Shchukin, E. D., Amelina, E. A., and A. M. Parfenova. 2008. Stability of fluorinated systems: Structure-mechanical barrier as a factor of strong stabilization. In *Highlights in Colloid Science*, D. Platikanov and D. Exerova (Eds.), pp. 41–53. New York: Wiley-VCH.

6 Structures with Phase Contacts

6.1 PHASE CONTACTS BETWEEN PARTICLES IN DISPERSE STRUCTURES

In the first section of this book, we have extensively discussed structures having *coagulation contacts* in which the particle–particle interactions are limited to simple "touching," either directly or via the equilibrium gaps containing the dispersion medium (Figure 6.1a and b). Low strength and mechanical reversibility are the main characteristics associated with such structures. Mechanical reversibility implies that such structures are thixotropic, that is, they can be spontaneously restored after undergoing mechanical degradation.

In this chapter, we will address structures with phase contacts in which the particles are bound via short-range cohesive forces acting over an area with dimensions exceeding those of an elementary cell. That is, we will focus on the particle cohesion resulting from at least 10^2 to 10^3 interatomic bonds. In this case, the contact surface is similar to the grain boundary in a polycrystalline solid, and the transition from the bulk volume of one particle to the bulk volume of another particle takes place continuously within *the same phase* (Figure 6.1c), which is where the term "phase contacts" originates from. In this chapter, we will heavily reference a number of early works in the area dealing with phase contacts [1–8]. The primary objects of interest here are silicates and cement (i.e., mineral binders). A number of essential publications devoted to these materials are covered in references [9–50].

The minimum value of the strength of the phase contacts can be estimated as

$$p_1 \approx \frac{10^2 e^2}{(b^2 4\pi\varepsilon_0)} \sim 10^{-7} \text{ N}$$

where
  b is the lattice parameter
  e is the elementary charge

Accounting for specific types of chemical bonding allows one to come up with more precise estimates suitable for particular materials. Since a phase contact with an area $s_c \sim (10^2 - 10^3)b^2 \sim 10^{-16} \text{ m}^2$ can be considered defect-free, such a contact has the theoretical strength of an *ideal solid* (see Section 7.1). Using this approach, we conclude that the minimal values $p_1 \sim P_{id}s_c$ are on the order of $\sim 10^{-8}$ N for fusible low-strength materials, $\sim 10^{-7}$ N for ionic crystals and medium-strength metals, and 10^{-6} N or more for refractory high-strength materials. The strength of the phase contact increases with an increase in the area s_c and can reach even higher values that are on the order of 10^{-4}–10^{-3} N. In the limiting case of a continuous polycrystalline material (e.g., a metal), we are dealing with the cohesive strength at the grain boundaries.

Estimates of the cohesive strength in structures with phase contacts allow one to conclude that, depending on the degree of dispersion (i.e., on the number of contacts per unit area) and on the mean strength of an individual contact (i.e., on the chemical nature of particles and all physical-chemical factors corresponding to the formation of a given structure), the values of $P_c \sim \chi p_1$ cover a very broad range from about 10^4 to about 10^8 N/m² or higher. In contrast to coagulation contacts, phase contacts undergo irreversible destruction. Since the contacts between the particles are the main carriers of the strength, an investigation of the mechanisms of their formation and rupture under various physical-chemical conditions provides one with the basis for developing effective methods for controlling and tuning the properties of disperse structures and of materials.

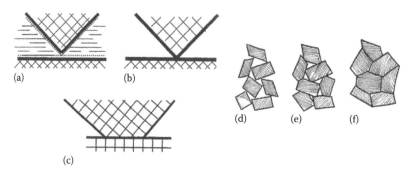

FIGURE 6.1 Coagulation (a,b) and phase (c) contacts and the corresponding structures with coagulation (d) and phase (e) contacts and full bridging of grains (f).

In a certain sense, the formation of phase contacts can be viewed as a result of partial coalescence between solid particles. Such coalescence occurs due to an increase in the area of contact between the particles and the transition from point-like contact to cohesive interaction over an area that is substantial relative to atomic dimensions. Such a transition may take place gradually, due to the diffusional transport of substance into the contact zone, as in the course of sintering (Figure 6.2), or due to the isothermal transfer of substance via the menisci of residual fluid. The latter can be frequently observed during the storage of fertilizers, which are known to form hard clumps.

Although gradual transition is possible, the experiments show that in most cases this process is abrupt. This is the case when the formation of phase contacts is associated with the need to overcome the energy barrier defined by the work of formation of a nucleus of a contact, that is, of a primary bridge connecting two particles.

In agreement with the concepts developed by Polak [10], the appearance of a contact nucleus may take place when the *new crystalline phase* forms in the contact zone between the newly formed crystals in the course of crystallization from metastable solutions. The bridging of crystals results in the formation of fine disperse polycrystalline products, such as those formed during hydration hardening of mineral binders and cements.

Similarly, phase contacts can also form when the new *amorphous* phase (organic and inorganic) is precipitated out of a metastable solution in sol–gel transitions, which are common in technology and in nature.

The formation and further growth of the primary hydration bridge can be the result of mutual particle deformation taking place at the point of contact due to mechanical stresses that exceed the yield stress of the material making up the particles. This is common in all friction and wear processes.

The processes leading to the formation of phase contacts can be studied experimentally by direct measurement of the cohesive forces between the particles. Such studies were described in studies carried out by Amelina [25–27] and Kontorovich [34–36] and will be discussed in detail further in this chapter. Throughout the book, we have described techniques and instrumentation for measuring small cohesive forces. These measurements also yield the energetic and geometric parameters of the process, such as the size of the critical nucleus of contact and its work of formation.

Typical experiments can be carried out in the following manner: Two crystals are brought into contact in a given medium and are kept under the conditions necessary to form a contact. Such conditions include contact time, compression, solution supersaturation, temperature, and the presence of surfactants. After the contact has been formed, the crystals are forced to separate from each other, and the contact strength, p_1, is measured.

The results of such experiments typically fall on a rather broad distribution curve that reflects the formation of microcontacts between geometrically and energetically heterogeneous surfaces of the two particles. For this reason, the results are typically presented as differential distribution histograms, which typically contain two maxima: one corresponding to the "weak" (coagulation)

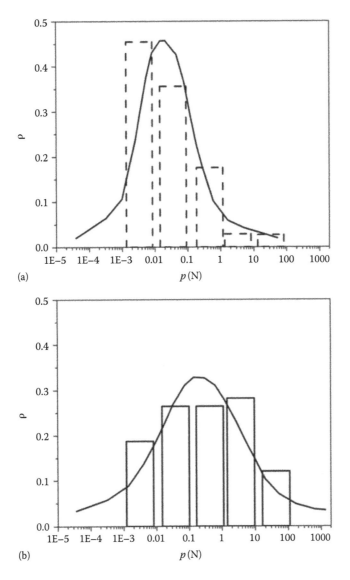

FIGURE 6.2 The growth of phase contacts in the course of sintering: the differential distribution of the strength of contacts, ρ, formed in the process of the sintering of two crystals of ammonium nitrate at 100°C (a) and 140°C (b).

contacts and the other one corresponding to the "strong" (phase) contacts. The difference between these two types of distributions is characteristic of the barrier-like nature of this transition.

Histograms summarizing the results of cohesive force measurements between two silver chloride crystals are shown in Figure 6.3. The mechanical behavior of these crystals is similar to that of plastic metals. The abscissa shows the logarithm of the contact strength, while the ordinate shows the fraction of contacts, ρ, having a strength in the given interval p_1. This figure shows that there are only two types of contacts between the crystals: those with strength p_1 around or lower than 10^{-7} N (coagulation contacts) and those with strength p_1 on the order of or higher than 10^{-6} N (phase contacts). This barrier-like transition from "weak" coagulation contacts to substantially stronger phase contacts indicates the absence of contacts with an intermediate strength. Changing the conditions of this experiment does not result in a gradual increase of the strength of the coagulation contacts and their gradual transition into phase contacts, as would have been seen on the histogram by a simple

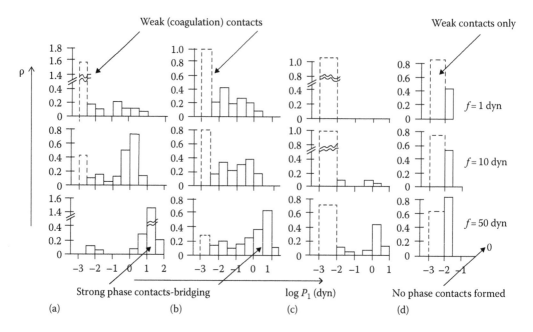

FIGURE 6.3 The example of a barrier-like transition between coagulation and phase contacts. Contacts between AgCl crystals in different media: in air (a) and octadecyl amine-coated samples in air (b), in heptane solution of octadecyl amine (c), and in water in the presence of traces of octadecyl amine for different compressions, $f = 1$ dyn, $f = 10$ dyn, and $f = 50$ dyn (d). Strong protective action of surfactant monolayer is clearly visible in (d): there are no phase contacts.

shift of the maximum. Instead, the increase in contact strength follows a barrier-like pattern: the histogram becomes bimodal with the maximum of the second mode being separated from the first one by several orders of magnitude of p_1. This second maximum corresponds to the appearance of qualitatively new contacts, the strength of which and the fraction of which increase with an increase in the applied compressive force. At the same time, the data shown in Figure 6.3 illustrate the possibility of preventing cohesion between the surfaces in contact by using surfactants (octadecyl amine in the present case) (Figure 6.4). This example illustrates the essence of using surfactants for controlling the processes of friction and wear as well as the processes of the mechanical treatment of solids, such as pressing or cutting. We will address the subject of surface damage and protection in detail in Chapter 7.

Similarly, one can also observe the barrier-like nature of the phase contacts in the course of the formation of particles of a new amorphous or crystalline phase in metastable solutions.

It is worth noting here that the case of sintering is principally different: this is a nonbarrier diffusion process, and the histograms of the contact strength have a single maximum that shifts with an increase in temperature as shown in Figure 6.2.

One can thus conclude that any given structure may contain both coagulation (residual) and phase (newly formed) contacts. Depending on the prevailing type of contacts, the structures can be subdivided into two large classes: coagulation structures and structures with phase contacts.

Disperse structures with phase contacts can form in a variety of physical-chemical processes, such as in the sintering of ceramic powders and in the course of pressing powders into pellets. Disperse structures with phase contacts that are formed in the process of generating a new phase in metastable solutions or melts are referred to as *condensation structures*. If the particles forming structures are crystalline, then the structures are referred to as *condensation–crystallization structures*. These structures can be viewed as "opposite" to the condensation structures composed of amorphous particles.

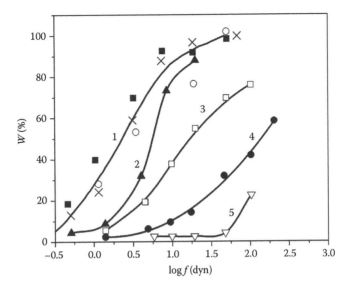

FIGURE 6.4   Fraction of phase contacts between AgCl crystals, W (%), with the strength p_1 on the order of or greater than 0.1 dyn (probability of tearing the adsorption layer between the particles) as a function of applied compressive force, f_1, in air, water, and heptane (crosses, circles, squares) (1); in heptane solutions of various substances such as decyl alcohol (2), cetyl alcohol (3), and octadecylamine (ODA) (4); and in aqueous solutions containing traces of ODA (5).

Structures with phase contacts represent the foundation for the formation of most materials. The formation of phase contacts can be the principal stage in the formation of continuous bodies or the final stage in the formation of porous bodies.

Indeed, the formation of cast metal is the result of nucleation in a melt, the growth of solid phase nuclei, and further bridging leading to the solid phase, which occupies the entire available volume. Similar processes taking place in the Earth's crust lead to the formation of mineral deposits. One can also name here numerous materials that are based on sintered particles, such as brick, various types of a cement, porous minerals, and catalysts with crystalline phases. The critical role of condensation–crystallization in the formation of structures in the course of the hardening of mineral binders and cements and in the course of the formation of artificial rocks has been described in a number of studies by Polak et al. [9–11]. Although the structure formation in glasses and organic polymers is beyond the scope of this book, we will discuss one essential example of structure formation in systems with amorphous phases, namely, the formation of silica gels and aluminosilicate gels.

Structures with phase contacts also form in the course of the caking of powdered materials. This is especially pronounced if the materials are hygroscopic. The caking process is in fact a type of recrystallization accompanied by the formation and proliferation of phase contacts between individual particles under variable humidity. Caking is an issue in many processes that are associated with the handling of powders: feeding powdered chemicals, transporting and handling fertilizers, dosing drugs, transporting crude oil with precipitated paraffins, etc.

One can say that in some cases the formation of structures with phase contacts is favorable, as it carries with it an increase in strength, while in other cases it should be either minimized or prevented.

The entire spectrum of natural and technological processes involving the formation of structures with phase contacts is very broad. Some principal publications on this matter are listed in References 11–24. In addition to essential studies conducted by the author [1–8], other significant publications relevant to the subject matter discussed in this chapter are listed in References 25–37.

A reasonably simple and yet "classical" example illustrating the formation of a crystallization disperse structure is the hardening of gypsum plaster upon its reaction with water: $CaSO_4 \times \frac{1}{2}H_2O + 1\frac{1}{2}H_2O = CaSO_4 \times 2H_2O$.

Over a broad temperature range, gypsum dihydrate, $CaSO_4 \times 2H_2O$, is a thermodynamically more stable substance than calcium sulfate hemihydrate, $CaSO_4 \times \frac{1}{2}H_2O$. At 20°C, the solubility of dihydrate is approximately 2 g/L, while that of hemihydrate is 6–8 g/L, depending on the type of calcium sulfate modification. Because of this, in sufficiently concentrated aqueous suspensions of $CaSO_4 \times \frac{1}{2}H_2O$, the liquid phase is saturated with respect to hemihydrate but supersaturated with respect to dihydrate. Supersaturation results in the formation of a new colloidal disperse phase under these conditions. This new phase consists of crystals of $CaSO_4 \times 2H_2O$, which, together with the particles of initial mineral binder, that is, $CaSO_4 \times \frac{1}{2}H_2O$, form a primary coagulation structure.

A decrease in supersaturation due to the separation of a new phase is compensated by the dissolution of new portions of hemihydrate. This allows one to maintain a continuous supersaturated condition and hence ensures further growth of dihydrate crystals. The preservation of supersaturation as well as the time during which the supersaturated state is maintained depends on the ratio between the rates at which the dissolution of hemihydrate and crystallization of dihydrate take place. The existence of sufficiently high supersaturation allows for the formation of seed nuclei for crystallization contacts between crystals of gypsum dihydrate at their points of contact.

A rapid increase in the number of primary crystallization bridges and their further growth result in qualitative structural changes. The initially formed plastic, thixotropic, and reversible coagulation structures turn into strong, rigid, and brittle structures that undergo irreversible fracture upon crushing. The formation of the new phase contacts and the expansion of their area results in a further increase in the strength of the formed structure. As the hydration of the gypsum hemihydrate progresses further, the degree of solution supersaturation in the system decreases. Consequently, the probability of the formation of phase contacts becomes lower. For this reason, at later stages of the process, hydration does not result in the formation of new contacts any longer, but leads rather to the slower growth of crystals and to an increase in the strength of previously formed contacts. The formation of a compact solid eventually takes place. If the formed structure is destroyed at later stages, the crystals will not bridge together, resulting in the formation of a reversible coagulation structure.

A typical peculiarity associated with the formation of crystallization disperse structures is the development of *internal stresses*. These stresses are the result of pressure originated during the directional growth of crystals that are bound to each other via a spatial network. X-ray data reported in References 44–46 have shown that these stresses are on the order of 10^7 N/m or even higher. If the internal stresses that build up in the structure reach the level of the structure strength, further crystallization in the course of the hydration of the mineral binder results in the development of fractures at the weakest points. This action on the part of the internal stresses may be facilitated as the strength of structure declines in the process of hydration. In the case when the internal stresses are lower than the strength, structure failure resulting in stress relaxation may not necessarily occur. Instead, these stresses remain present inside the material in the form of the elastic deformation of crystals and hence in the form of excessive energy associated with it. Further use of a material with the internal stresses preserved may result in a decrease in strength and durability, especially upon exposure to a surface active medium. This will be discussed in detail in Chapter 7. One therefore needs to decrease the internal stresses in the course of the formation of a disperse structure in order to improve performance characteristics and the reliability of materials. It has been shown in the works of Rehbinder and Mihaylov [49] and Uriev and Potanin [50] that the internal stresses can be effectively lowered at early stages of hydration by using a proper combination of vibration and surfactant additives.

Internal stresses can also be generated in other industrial processes associated with the formation of a structure, that is, in the processes used to pelletize powders.

Crystallization disperse structures can also form in aqueous suspensions of other single-component mineral binders. For instance, the hydration hardening of MgO results in the formation of a fine disperse structure of $Mg(OH)_2$. This process is used to make catalysts granules of a high strength. The same processes also take place in the hydration hardening of calcium aluminosilicate and of other frequently utilized cements.

Silicates and aluminosilicates, including hydrated and dehydrated silica and aluminosilicate gels, represent other examples of noncrystalline condensation structures. Silica gels form in the course of a *sol–gel* transition in which the new amorphous phase is formed due to the reaction of sodium silicate with an acid:

$$Na_2SiO_3 + 2HCl + H_2O \rightarrow 2NaCl + Si(OH)_4$$

$$(HO)_3 - \underset{|}{Si} - OH + OH - \underset{|}{Si} - (OH)_3 \rightarrow (HO)_3 - \underset{|}{Si} - (OSi(OH)_2-)_nO - \underset{|}{Si} - (OH)_3 + (n-1)H_2O$$

This polycondensation reaction may also involve the participation of aluminum salts. In this reaction, first, a sol of silicic acid or aluminosilicic acid is formed. This sol then undergoes coagulation and forms a gel, which serves as the primary *coagulation* structure. In the presence of supersaturation, the gel particles undergo bridging with the formation of phase contacts, and the *condensation* structure develops. This process is the basis for the synthesis of numerous catalysts and adsorbents.

In the past few decades, this so-called sol–gel technology has been widely used. The concept of this technology is the controlled hydrolysis of various metal and nonmetal alcoxides, such as of tetraethyl orthosilicate. In contrast to the polymerization reaction of sodium silicate, hydrolysis of alcoxides yields sols containing essentially monodispersed particles ranging in size from nanometers to micrometers. Monodispersed sols of alumina, zirconia, titania, and of course silica can be obtained by this method [51]. The coagulation and subsequent drying of these sols result in their conversion first into *hydrogels* and then into *xerogels*. Catalysts, adsorbents, and ceramics are produced by the thermal treatment of xerogels.

The formation of condensation structures is the main cause for the gelling of solutions of various natural and synthetic polymers. Gel formation may be accompanied by conformational changes of the polymer molecules, as is the case with gelatin and other biopolymers, or by various chemical interactions. Such is the acid-catalyzed synthesis of synthetic leather by the partial acetylation of polyvinyl alcohol with formaldehyde. Under supersaturation conditions, the fibers of the polyvinyl formal form in this system and develop into the network structure of synthetic leather.

The described individual examples of disperse structures and of various materials composed on the basis of these structures indicate the essential role that structured disperse systems play in various areas of our daily life. It is thus imperative to make use of the principles of colloid science and physical-chemical mechanics to find the means to fine-tune the properties (primarily mechanical ones) of various nanostructures. Depending on the particular application, one needs to identify ways to decrease or increase the strength of structures and thus to alter the properties of materials based on such structures.

The dependence $P = \chi p_1$ of the strength, P, on the number of contacts, χ, and the contact strength, p_1, provides a path forward that has led to several means to control the mechanical properties, namely, (1) changing the number of contacts by varying the particle size and the packing density and (2) altering the strength of individual contacts by influencing the conditions under which the contacts form and grow. These principal methods allow one to vary the strength over a very broad range from $P_c \sim 10^4$ N/m^2 for coarse structures with coagulation contacts to $P_c \sim 10^7$–10^8 N/m^2 for fine disperse structures with phase contacts. This indicates that the high strength of the material is primarily achieved by fine dispersion, which is the case in both porous and continuous materials. In the latter case, "fine dispersion" means the absence of large structural nonuniformities (defects). Simple dispersion methods typically do not yield particles with $r < 1$ μm. Reducing the size further down is facilitated by the presence of surface active media. Finely dispersed systems with particles on the order of 10^{-8}–10^{-9} m are formed by condensation methods involving the nucleation and growth of particles of the new phase.

Due to the current high interest in nanomaterials and nanotechnology, it is worthwhile to briefly address here the principles of the formation and the unique properties of ultrafine particles with sizes on the order of single nanometers or tens of nanometers [1,8,14,38–42]. The mechanical milling does not enable one to produce materials with particles that small. This limitation can, on the one hand, be explained by the growth in the specific surface area, which is related to the increase in the number of interparticle contacts that need to be broken, which consumes milling energy. On the other hand, as the particles become smaller, they approach a "defect-free state" and thus become stronger and more resistant to further breakdown. Finely dispersed colloidal particles that have sizes in the nanodomain are formed solely by condensation processes, that is, by the formation of a new phase in a homogeneous system under conditions of high supersaturation.

The governing Gibbs–Volmer principles of condensation leading to the formation of nanoparticles are based on a description of the nucleation process in the presence of a potential barrier. The latter can be overcome when the critical work of nucleus formation can be reached by fluctuations. The work required to form a spherical particle of the new phase, $W(r)$, within the initial phase includes the work needed to create a new surface area, $4\pi r^2\sigma$, and the "work gain" of $(4/3)\pi r^3|\Delta\mu|/V_m$ associated with the phase transition, that is,

$$W(r) = 4\pi r^2\sigma - \frac{(4/3)\pi r^3\left|\Delta\mu\right|}{V_m}$$

where V_m is the molar volume of the new phase. This work is maximal when

$$dW/dr = 8\pi r\sigma - \frac{4\pi r^2\left|\Delta\mu\right|}{V_m} = 0$$

This indicates that the critical size of the nucleus is $r_c = 2\sigma V_m/|\Delta\mu|$. Depending on the particular system, the value of $|\Delta\mu|$ is determined either by the pressure excess (in the case of vapor condensation), by supercooling (in the case of crystallization from a melt), or by the level of supersaturation achieved in the liquid phase. Since in the condensed phase $\Delta\mu \sim V_m\Delta p$ and Δp is the Laplace capillary pressure, p_σ, we conclude that the "peculiarity of a nanophase" is simply dictated by the high value of $p_{\sigma(c)} = |\Delta\mu|/V_m = 2\sigma V_m / (r_c V_m) = 2\sigma/r_c$. The elevated chemical potential (i.e., metastability) of the nanoparticles is manifested as a change in solubility, a change in the vapor pressure, or a change in other thermodynamic parameters. That is, the nanophase can be characterized by an *elevated activity*.

By all means, the most important and at the same time the *oldest* nanotechnology in the world is the hydration hardening of cements. Of special interest here is the fact that the nucleation taking place in the supersaturated phase controls both the formation of new phase particles and the formation of bridges between these particles.

Another nanotechnology that we have already mentioned, that is, the sol–gel transition taking place in the formation of fine porous aluminosilicates and of finely dispersed oxide powders, is also based on the nucleation concepts of Gibbs and Volmer.

In order to avoid misunderstanding, one needs to say here that the formation of the new phase particles can be facilitated by lowering the work of the process through the introduction of seed particles.

It is also worth mentioning the special role that fine colloidal sols with low concentrations of the disperse phase have played in the development of science. Such sols were utilized by the ancient Greeks to make colored pigments for mosaics. Michael Faraday used the famous red gold sols in his studies, while Svedberg was awarded the Nobel Prize for proving the existence of molecules by observing the motion of particles in such sols.

Supersaturation (metastability) in the mother liquor is the necessary condition for obtaining fine disperse structures in the processes of new phase formation by condensation. Since this excess in the chemical potential is in one way or another transferred into the newly formed phase, the particles reveal properties that differ from those of a macroscopic phase, such as increased solubility, higher vapor pressure (Ostwald–Kelvin), and thermodynamic parameters. One can say that the "transfer" of chemical potential results in the elevated activity. Some interesting changes may take place in semiconductors, as evidenced by the change in the zone structure. At the same time, fine disperse structures may maintain their nonequilibrium nature in the form of residual internal stresses.

While the fundamental principles that were developed by Gibbs and Volmer, Kaishev, and Stransky are well known, there are a number of aspects that have not been fully reflected in the abundant literature on nanotechnology.

First, there is the formation and preservation of nanosize systems under conditions when kinetic and structural factors prevent the transition of the new phases into a macroscopic phase. This includes the processes of the hydration hardening of various mineral binders.

Second, there is the formation of nanosystems that are preserved in their "nanostate" due to thermodynamic reasons. This is possible because of the strong interatomic bonds between small polyvalent atoms. Examples are the C–C and Si–O–Si bonds, which give special properties to the natural and synthetic materials containing them. The unique features of C–C bonds are manifested in a macroscopic phase, that is, in a diamond, as well as in numerous products of nanotechnology, such as fullerenes, nanotubes, and porous carbon nanofibers. The structures with Si–O–Si bonds are not as unique and are rather abundant. These include colloidal silica and aluminosilicates, as well as crystalline aluminosilicates, that is, zeolites, and quartz and numerous silicate minerals. In these structures, the siloxane bonds play a determining role as the carriers of remarkable strength. Here, it is worth addressing the question of what do the processes of drilling for oil and gas and crude oil cracking have in common. The answer to this question is most remarkable. In drilling, one is interested in breaking siloxane bonds in the rock while preserving them in the diamond bit. Conversely, in the process of crude oil cracking, the breaking of C–C bonds takes place, and the problem that one needs to address is how to minimize the mechanical wear of a catalyst containing siloxane bonds. One therefore faces the need to optimize *both* processes, primarily by controlling the conditions, such as choosing the proper pH and surface active media. Specific studies along these lines are described further in this chapter.

Following Rehbinder, we will state here one more time that high strength in materials requires that they be composed of small tightly packed particles with the maximal number of strong phase contacts between them. However, in fine disperse systems, the process of achieving tight particle packing becomes complicated, since even relatively weak individual coagulation contacts, when combined, give rise to substantial resistance. This is often encountered in the molding of powders and pastes. The influence of coagulation contacts can be overcome at elevated pressures, but this approach introduces additional difficulties, namely, that high pressure leads to the formation of residual stresses that prevent the formation of phase contacts under optimal conditions and result in a decrease in the strength of materials. This means that, at the preparation stages and during molding itself, the high viscoplastic resistance of the disperse system needs to be overcome by the liquefaction of the system, that is, by lowering η_{eff} and τ^* (see Chapter 3). However, the simplest approach to increase liquefaction by an excess in the dispersion medium is completely unsuitable in many instances. For example, an excess of water content in cement paste results in poor resistance of the concrete to cold weather: it simply bursts once the water unbound to hydrate freezes. It is thus evident that liquefaction of a system needs to be carried out by selecting an optimum combination of mechanical and physical-chemical factors. In order to achieve the best possible particle packing density (yielding the maximum number of contacts in the structure) and at the same time avoiding the development and accumulation of internal stresses, vibration is often used. At the same time, in order to weaken interparticle cohesion (e.g., in making dry and moist catalyst and ceramic masses), one would use various surfactants that weaken contacts in coagulation structures due to adsorption and prevent the formation of phase contacts.

In order to control the formation and development of structure in the processes of the hydration hardening of mineral binders, electrolytes are used in addition to surfactants. This allows one to have directional control over supersaturation as well as change the conditions of crystallization and bridging in newly formed formations, so that control over hardening process can be maintained. In fabric and yarn manufacturing, surfactants are used to prevent strong cohesion between the fibers due to the formation of adsorption layers. Similar problems are encountered in papermaking, where there is a need for the fine control over the adhesion forces acting between the fibers.

The investigation of the physical-chemical means needed to control the structure and the mechanical (rheological) properties of disperse systems and materials by using an optimal combination of mechanical action and physical-chemical conditions at interfaces is thus the main subject of physical-chemical mechanics.

In the next section of this chapter, we will present the results of one essential study on the formation of a fine disperse phase in the course of sol–gel transitions in amorphous system conducted by Kontorovich et al. [34–36]. To a large extent, these works were inspired by the interest to the aluminosilicate systems. These systems serve as catalysts in crude oil cracking when they are promoted with transition metals and filled with zeolites. Elsewhere in this book, we will also address the issues associated with the strength of these catalysts in their granulated form, specifically the difficulties associated with introducing zeolite (poor mechanical strength) into an aluminosilicate matrix. One key feature of the essay that follows is that the size of the globular structures in the sol–gel transition was measured for the first time by small-angle x-ray scattering (SAXS) in an aqueous dispersion, and not after drying.

6.1.1 Effect of the Duration and Temperature of Ageing on the Size of Aluminosilicate Hydrogel Globules

A number of works have dealt with studies on structure formation in aluminosilicate hydrogels obtained by coprecipitation. An early basic study carried out by Planck (see in References 1 and 34–36) compared the properties of silica gels and aluminosilicate gels synthesized under the same conditions and concluded that there was a substantial difference in their properties. The same change in the synthesis conditions within the defined interval of parameters (increase in pH, concentration of the hydrogel solid phase, duration of syneresis, etc.) resulted in the opposite trends in the change of the specific surface area of xerogels. In particular, the specific surface area, S_1 (m²/g), of the aluminosilicate gels increased, while the specific surface area of the silica gels decreased.

Planck interpreted the increase in the specific surface area of aluminosilicate xerogels in syneresis as the evidence for a decrease in the size of the globules and suggested that the latter was caused by the dispersion of the globules due to the cleavage of the –Si–O–Al= bonds formed in the process of sol formation.

The idea that the specific surface area of xerogels might be a characteristic of the particle size is rather common. This indeed was confirmed by adsorption and electron microscopy studies, which showed comparable particle sizes obtained by these two different methods. However, this agreement is not universal; there are cases when adsorption and microscopy studies yield different results. One such example is the treatment of silicic acid hydrogel with an acid at pH 1.9. Under these conditions, the adsorption experiments revealed a substantial decrease in the xerogel surface area, while the electron microscopy indicated a decrease in the particle size. Since the observed decrease in the surface area was accompanied by a significant decrease in porosity, the observed discrepancies were explained by the inaccessibility of the surface in the vicinity of particle–particle contact to the adsorbate molecules.

The apparent density (d_v, g/cm³) of aluminosilicate gels is typically much higher than the apparent density of silicic acid gels. For gels obtained under comparable experimental conditions, the apparent density of aluminosilicate gels is nearly 1.5 times higher than the apparent density of silica gels. The apparent density of silica gels approaches that of aluminosilicate gels only when the former are treated with a very strong acid (pH = 0.6). Under these conditions, the apparent density of silica gels was 1.23 g/cm³, which is close to the 1.39 g/cm³ reported for aluminosilicate gels [35].

Since in this case where there was no correlation between the change of specific surface area and the particle size was observed with a silica gel of high apparent density, $d_v = 1.3$ g/cm³ (porosity of 0.33 cm³/g), one may assume that there is no correlation in the region of high apparent densities. This suggests that the "anomalous" trend in the specific surface area of aluminosilicate gels may not be much related to the differences in the mechanisms by which globules form in the hydrogel, but is due to the inaccessibility of a large fraction of the internal surface of these xerogels to the adsorbate molecules. The latter, obviously, increases with an increase in the particle packing density. To verify this, the change in the size of globules in aluminosilicate gels was studied in hydrogels with varying depths of syneresis. The particle size, R, was assessed from the SAXS, which made it possible to determine the particle size directly without exposing the hydrogel to heat, which might lead to a change in the material morphology and particle size [52]. It is worth pointing out that the analysis of xerogels by SAXS is complicated by the inability to distinguish the size of particles from the size of pores, due to the identical diffraction from the particles and the pores. For this reason, the analysis of xerogels by SAXS is only possible when reference data on particle size obtained by other means are available. In the hydrogels with a very low volume concentration of the solid phase ($P_v \sim 2.5\%$ to 3.5%), significantly smaller than that in xerogels, there is no such issue because the size of the pores is six to seven times larger than the size of the particles. For this reason, the particle size of 7–30 Å determined in fine disperse aluminosilicate hydrogels must be related to the size of the particles and not the size of the pores (assuming the opposite would yield an unrealistic particle size of 1–4 Å). A complete dehydration of the hydrogel yields a decrease in pore radius by about 25%, which corresponds to a decrease in the pore volume by a factor of 2.5, while the volume of the sample itself shrinks by a factor of 10.

In the experiments described, the aluminosilicate hydrogels had a pH of 8.0–8.3, $P_v = 3.2\%$, and a weight ratio of $Al_2O_3:SiO_2 = 6$. The hydrogel was prepared by mixing cold 0.6 N solution of aluminum sulfate acidified to pH ~ 0.7 using sulfuric acid with a 1.25 N solution of sodium silicate in a ratio of 2.8:1. The syneresis was conducted in the prepared mother liquor for 1, 48, and 96 h at 22°C and for 3, 12, and 27 h at 70°C. After the completion of the syneresis, the gel was thoroughly washed for 2 days at 22°C–25°C in order to remove sodium sulfate. The completeness of the removal was tested qualitatively with the addition of $BaCl_2$ solution (lack of turbidity).

A rectangular 1 mm thick specimen of hydrogel was used for the SAXS analysis. In order to prevent drying, the specimen was placed into a special cuvette made of a thin film transparent to x-rays. The wash solution from the final stage of the two-day wash cycle was used as a background to correct for scattering from the dispersion medium. The SAXS experiments were conducted in a 4-slit chamber with an ionization detector using 18–20 kV x-ray source. The experimental data were presented as differential distribution curves showing the inhomogeneity dimensions.

The results of the particle size analysis obtained from the SAXS are summarized in Table 6.1, which shows the sizes corresponding to the particle size distribution maxima, R_{max} (Figure 6.5), and the two types of mean radii: those obtained from the distribution curves and those calculated using the Hosemann approach [52] R_m and $R_{m,H}$, respectively.

TABLE 6.1

Inhomogeneity Sizes Obtained at Various Syneresis Times

Globules Radii, Å	Time, h	At 20°C			At 70°C		
		1	48	96	3	12	27
R_{max}		7.2	8.2	7.8	16.2	22.0	25.0
$R_{m,H}$		17.2	15.2	16.0	23.5	35.4	34.2
R_m		22.4	20.4	22.0	27.7	32.9	32.9

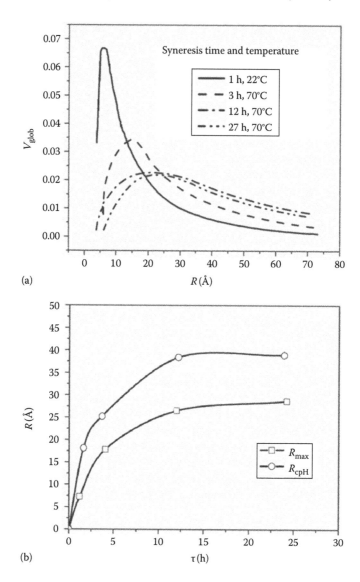

FIGURE 6.5 Differential distribution of volume of aluminosilicate globules as a function of time for different times of syneresis (a). The size of globules as a function of syneresis time (b).

The data in Table 6.1 clearly show that the values of R_m, $R_{m,H}$, and R_{max} do not exceed 30–35 Å. Apparently, these values are related to the size of the globules rather than to the size of the pores, and the change in the particle size only in the range from 7–17 Å to 25–35 Å due to changes in the syneresis conditions is indicative of the change in the size of the hydrogel globules. The interpretation of the microheteregeneities with sizes >40 Å is ambiguous, as these dimensions can be assigned to both particles and pores: in the studied gels ($P_v = 3.2$) with $R_{max} = 7$ Å, one may anticipate to find a large number of pores with sizes $R \sim 50$ Å. As seen in Figure 6.5, the differential particle size distribution curves for different syneresis durations at room temperature are nearly identical, which indicates a very low rate of hydrogel ageing under these conditions (Table 6.1). An increase of the syneresis temperature to 70°C resulted in a sharp increase in globule size: within 3 h at this temperature, the percentage of the smallest globules (7 Å < R < 15 Å) decreased almost twofold (from 50% to 24%), as compared to hydrogels that were aged at room temperature. The mean radius, $R_{m,H}$, increased approximately 1.4 times, while R_{max} more than doubled. When the syneresis duration was increased to 12 h at 70°C, the processes that result in the growth in the globules continued to take place: the percentage of the smallest globules (7 Å < R < 15 Å) decreased to 7%, and $R_{m,H}$ and R_{max} increased by a factor of 1.5. Further ageing of the gel resulted in slower growth in the globules—the particle size distribution curve obtained for gel aged at 70°C for 27 h is practically the same as the curve obtained after 12 h of ageing at the same temperature. These data are shown in Figure 6.5b.

The results of this study indicate that the radius of the globules in the course of the ageing of the aluminosilicate hydrogel increases as the hydrogel ages. This behavior is in fact *the same* as that previously observed for polysilicic acid hydrogel. These results confirm that there is no correlation between the trends in the change in the specific surface area and particle size in the corresponding hydrogels. This is especially well seen in the aluminosilicate hydrogels that underwent acid treatment at pH ~ 3. Ageing in acid resulted in a large decrease in the surface area, while the SAXS data indicated no change in the size of globules. These results indicate that one needs to be extremely careful when using adsorption data in estimating the particle size in dried hydrogels.

6.2 MECHANISMS OF THE FORMATION OF CONTACTS AND OF THE DEVELOPMENT OF INTERNAL STRESSES

Here, we will address the mechanism of the formation of phase contacts in the process of the bridging of amorphous (SiO_2) and crystalline ($CaSO_4 \times 2H_2O$) particles in solutions supersaturated with respect to one or more of the constituents. We will also discuss the results of the x-ray analysis of the internal stresses developing in the hydration products of single-component mineral binders, such as $CaSO_4 \times 0.5H_2O$, CaO, MgO. These studies allowed one to answer the principal question as to why the bridging takes place. The results discussed are of principal importance both for the understanding of the modern nucleation theory of nanophase and for solving practical problems associated with the improvement in the quality of modern construction materials and the manufacturing of catalysts and adsorbents.

In this section, we will present an overview of essential studies that focus on understanding the mechanisms of particle bridging in the course of the formation of solid structures (rocks) in crystalline and amorphous systems. In this discussion, we would like to emphasize the importance of and acknowledge the principal works by Kontorovich, Amelina, Shchukin, and their coauthors, collaborators, and graduate students [1,25–27,34–37]; we would like to specially acknowledge the contribution by Yusupov, Vaganov, and Lankin [26,27,36].

The formation of concrete involves the bridging of particles in the course of the hydration hardening of cement and is in fact the oldest nanotechnology in the world: the concrete dome of the Pantheon in Rome has been around for more than 2000 years. This masterpiece of ancient nanotechnology reflects the level of technological perfection that can be achieved. At the same time the answer to the question as to why the hydration hardening yields a stone is a very difficult one. The search for an answer to this principal question has revealed the need to consider a particular combination of supersaturation and ageing duration, along with the mutual mechanical compression of the newly formed solid particles in the analysis of the thermodynamics and kinetics of these processes. While the mechanical compression stipulates particle bridging, it is also associated with the generation of residual stresses in the system. In the works by Rybakova et al. [44–46], the influences of the two opposing factors, the microstresses of the second kind and the dispersity, were separated. This was achieved by the analysis of the line broadening in the x-ray diffraction patterns of these substantially amorphized systems. It has turned out that in certain cases, the internal stresses achieved levels comparable to the material strength. The role of various physical-chemical factors and conditions in the development of these internal stresses and the influence of these factors on their magnitude were thoroughly investigated. This provides a means for controlling the development of the internal stresses and hence for fine-tuning the properties and stability of materials. These studies have made an essential and critical contribution to the understanding and advancement of the technology of making strong and long-lasting construction materials, as well as of porous adsorbents and catalysts.

6.2.1 FORMATION OF CONTACTS BETWEEN THE PARTICLES AND THE DEVELOPMENT OF INTERNAL STRESSES DURING THE HYDRATION HARDENING OF MINERAL BINDERS

The processes of particle bridging and specifically the formation of strong contacts are responsible for the transition from a free-disperse system (i.e., a sol) to a connected-disperse system, such as a solidifying paste with a particular mechanical strength. Consequently, one needs to achieve complete understanding of the reasons, conditions, and mechanisms of particle bridging (agglutination). One needs to pay special attention to the origination and development of the internal stresses in the forming structure. We have already pointed out several times that there are two principal types of contacts: weak coagulation contacts with a strength of $\sim 10^{-9}$–10^{-7} N and strong phase contacts with a strength of $\sim 10^{-6}$ N and higher. The formation of phase contacts requires the displacement of the dispersion medium from the area exceeding the dimensions of the elementary cell. Phase contacts are mechanically irreversible, and the structures based on such contacts typically exhibit elastic–brittle and elastic–plastic behavior. Nearly all construction materials belong to the class of structures with phase contacts ranging from point like (contacts are small relative to the particle size, such as in catalysts and carriers) to continuous (in the case of a complete bridging of particles, such as in metals, alloys, melts, and various composites). Many ceramic materials as well as materials obtained by means of various powder processing technologies and the structures obtained in the course of hydration hardening can be viewed as intermediate ones. The strength of the contacts between the particles in heterogeneous systems is the principal factor controlling the strength of the resulting structures and hence the durability and properties of the material. Nevertheless, coagulation contacts play an equally important role at various stages of the formation of materials, stages at which one needs to control the mobility of the system by controlling the adhesion forces.

Let us now discuss the transition from coagulation contacts to phase contacts, that is, from the touching of the particles to their bridging, using two characteristic examples. The first example is the formation of contacts between the crystals of gypsum (calcium sulfate dehydrate) in the supersaturated solution of calcium sulfate, and the second one is the formation of the contacts between the amorphous particles of polysilicic acid (i.e., silica) in the supersaturated solution of silicic acid. The first case represents the process of bridging taking place in hydration hardening, while the second one corresponds to the formation of a silica xerogel due to particle bridging in the course of a sol–gel transition. These two basic examples are essential for a better understanding of the bridging processes taking place in more complex multicomponent systems.

A comparative study of the conditions under which particle bridging takes place in the individual contacts and the principal laws governing the development of the corresponding macrophase allows one to understand the dependence of these processes on various factors, such as the degree of supersaturation in the starting system, time, the crystallographic orientation of crystals with respect to each other, and the presence of various additives (electrolytes or surfactants). These studies have also allowed one to emphasize the special role of mechanical factors, such as the mutual compression of particles. The example with gypsum indicates that bridging does not take place without compression. However, in the absence of external forces, the compression can only originate from the action of the *internal stresses* (crystallization pressure, phase transitions, etc.). To crystallographers, these are known as stresses of the second kind, or microstresses. These stresses take active part in the processes of particle bridging and in the development of structure strength, but at the same time may manifest themselves as *residual stresses* that may be the cause of a significant lowering of the strength, especially in an active medium.

All of these processes indicate the need to examine the conditions of particle bridging as they relate to the origination of residual stresses in the products of hydration hardening. This approach is essential for achieving a more complete understanding of the physical-chemical concepts of hydration hardening and for the optimization of various industrial processes.

The formation of contacts can be studied experimentally using the magnetoelectric galvanometer–based device described in detail earlier in this book (see Chapter 5). Since there is no need to measure distance, the measuring scheme allows one to measure forces on the order of 10^{-8} N.

In a typical measurement, two gypsum crystals of size $5 \times 5 \times 0.5$ mm, cleaved from a gypsum single crystal, were mounted in the measuring device—one on the manipulator and the second one attached to the galvanometer arm (Figures 1.28 and 4.15). The crystals were then brought into contact with their edges normal to each other by varying the crystallographic planes forming these edges. The samples were placed in calcium sulfate solutions supersaturated with respect to gypsum. The supersaturated solutions were prepared by dissolving calcium sulfate hemihydrate, filtering the resulting solutions, and diluting them to the necessary concentration, which was maintained at a constant level (the solubility of hemihydrate and dihydrate in water is 6.2 and 2 g/L, respectively). In the study, the following parameters were varied: supersaturation, α; crystal contact time, t; compression force f; and the crystallographic orientation of the gypsum crystal contact edges.

The bridging of gypsum crystals with fluorite, calcite, and quartz, as well as the bridging of these crystals with each other in supersaturated calcium sulfate solutions, was also studied. In some cases, the calcium sulfite solutions contained electrolytes and surfactants.

In the studies with silica, it was not possible to use individual particles because the silica globules forming silica sols are very small. For this reason, the bridging in this system was studied using two crossed glass threads coated with silica particles. First, silica sols of pH 7 and containing 0.6% SiO_2 by mass were prepared by mixing a solution of sodium silicate with hydrochloric acid [6,31]. The threads were then immersed into these silica sols, which resulted in the deposition of single-silica particles (2–3 nm) and silica particle aggregates (tens and hundreds of nm) on the surface. While the surface of the particle coating was mainly covered with silanol groups, in the bulk of the deposited layer, these groups were converted into the siloxane groups formed by condensation:

$$-\overset{|}{\underset{|}{Si}}-OH+OH-\overset{|}{\underset{|}{Si}} \rightarrow -\overset{|}{\underset{|}{Si}}-O-\overset{|}{\underset{|}{Si}}-+ H_2O$$

The silica-coated threads were oriented perpendicular to each other and were placed into contact in freshly prepared silicic acid sols supersaturated with respect to amorphous silica. The degree of supersaturation, α, was determined as the ratio of the amount of silica dissolved in the dispersion medium of the silica sols to the solubility of silica at a given pH and temperature. The value of the supersaturation was varied by changing the concentration of sodium silicate from 0.06% to 0.012% by mass. The concentration of the dissolved silica decreased with time, and consequently the value of α decreased as well. The sols were used while the change in the value of α was less than 6%. Temperature, pH, α, t, and f were all varied within a broad range, while the temperature was maintained constant at 20–22°C.

The bridging of silica particles with the crystals of quartz, fluorite, periclase, bruscite, white asbestos, anthophyllite, and amorphous quartz (uncoated threads) was also studied. The specimens used in these studies were crystals with size $2 \times 5 \times 5$ mm, which were mounted on the manipulator (Figure 1.28) and were cross-contacted with the silica-coated quartz thread. The area of contact was a plane with the dimensions 2×5 mm. The thread was positioned parallel to the short edge of the specimen.

6.2.1.1 Formation of Phase Contacts between Crystals of Gypsum

The measurements of the cohesive forces carried out with the individual crystals yielded widely scattered data for contacts measured under the same conditions. This indicates the need to use statistical analysis and to employ suitable distribution functions. The data can be represented in the form of histograms showing the differential distribution function $\rho = dn/(n_0 d \log p_1)$ as a function of

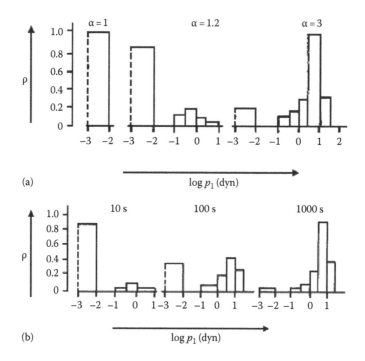

FIGURE 6.6 The differential distribution of the contact strength for the contact between two gypsum dehydrate crystals brought into edge-to-edge contact with a force, $f = 10$ dyn, in solutions of calcium sulfate: the various degrees of supersaturation, α, for the contact time $t = 100$ s (a); the various contact times, t, for supersaturation, $\alpha = 1.8$ (b). The ordinate depicts the density of the distribution, ρ.

the logarithm of the strength, p_1, n as the current measurement number, and n_0 as the total number of measurements. To generate each histogram, at least 100 measurements were carried out. The histograms showing the distribution of the strength of edge-to-edge contacts between two gypsum crystals are shown in Figure 6.6.

Results shown in Figure 6.6 indicate that, based on the strength, there are two types of contacts: "weak" coagulation contacts with $p_1 \sim 10^{-2}$ dyn and "strong" phase contacts with $p_1 \sim 10^{-1}$ dyn, corresponding to particle bridging. The "conversion" of weak contacts into strong ones involves a barrier-like process. In Figure 6.6, this is seen as the shift of the second (right) maximum in the histogram to the right from the position of the first maximum by several orders of magnitude. There are no contacts with intermediate strength between 10^{-2} and 10^{-1} dyn. This very sharp transition corresponds to bridging and provides one with the possibility of studying the mechanism of the bridging process. For this purpose, the dependence of the probability of bridging, W_b, on various parameters has been studied. The probability of bridging is defined as the ratio of the number of "strong" contacts, w_t, to the total number of contacts, w_0, that is, $W_b = w_t/w_0$. As seen in Table 6.2, the bridging probability increases with an increasing degree of supersaturation and an increasing time of contact.

The dependence of the bridging probability on the contact time and the degree of supersaturation allows one to relate the initial step of bridging to the fluctuation process of the formation of the primary contact nucleus. This process is critical for a given supersaturation. Mathematical analysis of the data in Table 6.2 reveals the exponential dependence of the probability of bridging, W_b, on the time, t:

$$\frac{w_t}{w_0} = W_b = 1 - \exp(It) \tag{6.1}$$

TABLE 6.2

Probabilities of Bridging, W_b, for Gypsum Crystals and Amorphous Silica Particles as a Function of the Contact Time, t, and the Degree of Supersaturation, α

				W_b		
	Gypsum Crystals[a]			Silica Particles (pH 7)		
t (s)	$\alpha = 1.2$	$\alpha = 1.8$	$\alpha = 3.0$	$\alpha = 1.2$	$\alpha = 1.4$	$\alpha = 3.0$
10	0	7	38	3	16	32
100	17	49	83	16	54	81
1000	33	96	100	31	64	92

[a] The gypsum crystals were brought into edge-to-edge contact in the [001] direction.

where I is the reciprocal characteristic time, sufficient to cause bridging in 63% of the cases out of the total number of experiments, w_0. Consequently, I may be viewed as the "rate of bridging." In accordance with the fluctuation theory of nucleation [2,6,8–10],

$$I = I_0 \exp\left(-\frac{A_c}{kT}\right) \tag{6.2}$$

where

I_0 is a prefactor with the units of reciprocal time, s^{-1}
A_c is the work of formation of the critical nucleus of contact

The magnitude of A_c depends on the degree of supersaturation in the solution, α, and on the value of the specific surface energy, σ, of the interfaces that either form or disappear during the process of the contact nucleus generation. This dependence may be expressed quantitatively, provided that the model describing the geometry of the forming nucleus is specified. For example, one may use a model similar to Polak's model describing "quasi 2D" nuclei [10], but with one principal difference: the height of the nucleus, h, is a finite value with distinctive upper and lower limits (Figure 6.7a).

The contact nucleus is assumed to have the shape of a rectangular prism with a square base of dimension $a \times a$ formed in a gap of width h. If the bridging crystals have random orientation, we assume that the contact nucleus is cooriented with the crystal lattice of only one of the crystals in contact, as shown in Figure 6.7b. Figure 6.7c corresponds to the case when the contact nucleus is formed between crystals that have a chemical composition different from that of the nucleus.

The energy required to form a nucleus with the volume a^2h consists of two portions—the energy of a phase transition, $(-a^2h\, kT/V_1)$, where V_1 is the molecular volume, and the energy released due to the disappearance of two crystal/solution interfaces $(-2a^2\sigma_3)$. At the same time, free energy equal to $4ah\sigma_1$ is spent in the formation of the lateral surface of the contact nucleus, and the energy $a^2\sigma_2$ is spent in the formation of a nucleus–crystal boundary similar to the grain boundary in a polycrystalline material. The terms σ_1, σ_2, and σ_3 are the specific surface free energies of the boundaries between the nucleus and the solution, the nucleus and the crystal, and the crystal and the solution, respectively. The size of the contact critical nucleus is obtained from the expression of the extremum of the free energy:

$$a_c = \frac{2h\sigma_1}{[(hkT/V_1)\ln\alpha + \sigma^*]} \tag{6.3}$$

where $\sigma^* = 2\sigma_3 - \sigma_2$, and, respectively,

$$A_c = \frac{4h^2\sigma_1^2}{[(hkT/V_1)\ln\alpha + \sigma^*]} \tag{6.4}$$

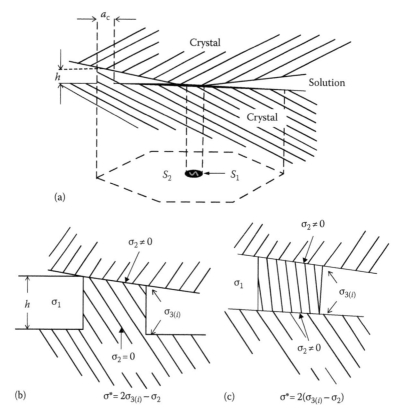

FIGURE 6.7   The model of the contact zone (a); models of contact nuclei-co-oriented with the crystal lattice of one of the crystals (b) and having an orientation differing from both surfaces (c).

Comparison of the expressions for the work of formation of various nuclei formed in crystallization processes indicates that the contact nucleus is substantially different from both a 2D and 3D nuclei. The work of formation of a contact nucleus in a saturated solution ($\alpha = 1$) has a finite value. Indeed, the works of formation of a 2D nucleus, A_2, and the work of formation of a 3D nucleus, A_3, depend on the degree of supersaturation as $A_2 \sim 1/(\ln \alpha)$ and $A_3 \sim 1/(\ln^2 \alpha)$ [15]. For the contact nucleus, $A_c = b/(\ln \alpha + c)$, where b and c are constants independent of the degree of supersaturation: $b = 4h\sigma_1^2 V_1\ (kT)^2$; $c = \sigma^* V_1/hkT$. Consequently, for the case of $\alpha = 1$, one obtains $A_c = b/c$. Furthermore, one may expect that there exists critical supersaturation that depends on the sign of σ^* ($\alpha_c > 1$ when $\sigma^* < 0$ and $\alpha_c \sim 1$ when $\sigma_3 \sim \sigma_2$). This conclusion is indeed supported by the experimental results indicating that phase contacts also appear in the saturated solution, albeit with a very low probability, not exceeding 6% at $t = 1000$ s.

The experimental data provide a means for evaluating A_c and establishing the dependence of A_c on α. First, let us rewrite Equation 6.2 as follows:

$$I = I_0 \exp\left[-\frac{b}{(\ln \alpha + c)} \right] \tag{6.5}$$

By substituting the experimental values of W_b into Equation 6.1, we obtain the values of I corresponding to various degrees of supersaturation. The values of I may be used to estimate the values of b, c, and I_0. The latter can be accomplished by solving a system of equations similar to Equation 6.5 graphically. These results are shown in Table 6.3, which also contains the values for the work of formation of 2D (A_2) gypsum nuclei, calculated using Polak's expression $A_2 \sim 9\ kT/\ln \alpha$. The data in Table 6.3 indicate that the A_c values are consistent with the predictions of the fluctuation theory of nucleation and that these values are consistently lower than the corresponding A_2 values.

TABLE 6.3

Work of Formation of Contact Nuclei, A_c, and of 2D Nuclei, A_2, for Gypsum and Silica at Various Supersaturations, α, of Calcium Sulfate and Silicic Acid Solutions ($f = 10$ dyn)

Particles	Gypsum[a]				Silica (pH 7)		
	$\alpha = 1.2$	$\alpha = 1.5$	$\alpha = 1.8$	$\alpha = 3.0$	$\alpha = 1.2$	$\alpha = 1.4$	$\alpha = 3.0$
A_c/kT	9.2	8.2	7.3	5.8	3	1.5	0.5
A_2/kT	45	22.5	15	8.2	25	10	3

[a] The gypsum crystals were brought into edge-to-edge contact in the [001] direction.

Our model implies that the gap has a constant width, h. In order to assess the value of this width rigorously, one would need to analyze the thermodynamic factors, such as the concentration, and the kinetic factors, such as the rate of diffusion in the gap. One can, however, get an estimate of h on the basis of the available experimental data. For a crude estimation, we can neglect particular values of σ_2, just assuming that $\sigma_1 = \sigma_3 = \sigma \gg \sigma_2$. From the expressions for b and c, and $V_1 = 1.2 \times 10^{-22}$ cm^3, we find $\sigma \sim 20$ erg/cm^2 for gypsum dehydrate at room temperature. Substitution into the equation for b yields $h \sim 1$ nm (10 Å). This value of h may be viewed as the mean width of the gap between the crystals at the point where the contact bridge is most likely to form. Near the contact zone, the surfaces of the crystals may not necessarily be parallel to each other. The values of $\sigma \approx \sigma_1$ at the nucleus/solution interface are in reasonable agreement with the experimental data reported for the nucleation in solutions, particularly for the nuclei of gypsum [16,54].

The value of the size of the critical nucleus, a_c, may be found using Equation 6.3. If $h \approx 1$ nm, a_c appears to be within the range of values between 0.6 and 0.9 for all achievable supersaturations, $\alpha = 1$–3. This estimate for a_c appears to be reasonable, which supports the hypothesis that the formation of crystallization contacts in the process of the bridging of crystals has a fluctuation nature. It also confirms that the dependence of A_c on α is described by $A_c \sim b/(\ln \alpha + c)$.

6.2.1.2 Effect of the Mutual Orientation of Crystals

In the process of the formation of crystallization structures, the isolated small crystallites, which are the primary elements of the emerging structure, appear to be randomly oriented toward each other. It was therefore of interest to investigate the effect of the crystallographic orientation of the gypsum crystals in contact on the probability of bridging.

In these experiments, the gypsum crystals were brought into contact along edges of various crystallographic directions, belonging to particular faces, as shown in Figure 6.8. The edge along the [100] direction is formed by the (011) and the (010) faces; the edge along the [001] direction is formed by the (120) and (010) faces. The corresponding bridging probabilities are summarized in Table 6.4.

Data in Table 6.4 show that the value of W_b strongly depends on the type of faces between which the crystallization bridges are formed. The maximum value of W_b is observed when the crystals were brought into contact along the edges following the [100] and [100] directions; the minimum values are observed when the crystals were brought into contact along the edges following the [001] and [001] directions. This difference may be related to the difference in the specific surface free energies of the surfaces to which the edges belong. The latter can be arranged into series $\sigma_{(011)} > \sigma_{(120)} > \sigma_{(010)}$. This inequality should still hold valid despite the fact that at the interface of a solid with its own solution, the specific surface free energy values should be reduced. Indeed, in the process of gypsum crystal growth, the (120) and the (010) faces are the most stable ones.

According to Equation 6.4, the difference in the specific surface free energy of the faces between which a crystallization bridge forms is the reason for the difference in the work associated with the formation of the contact nucleus, A_c. The calculated values of A_c are summarized in Table 6.5.

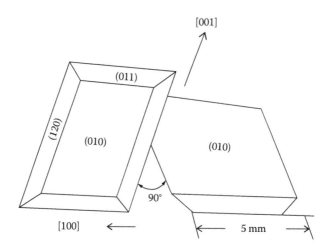

FIGURE 6.8 Schematic illustration of the mutual orientation of crystals in contact.

TABLE 6.4

Effect of the Mutual Crystallographic Orientation of CaSO$_4$ × 2H$_2$O Crystals on the Probability of Their Bridging (W_b) in Supersaturated Calcium Sulfate Solutions of Supersaturation α and the Time of Contact, t

Directions of the Edges in Contact	t (s)	W_b (%)		
		$\alpha = 1.5$	$\alpha = 1.8$	$\alpha = 3$
[100] and [100]	100	24	49	88
	1000	68	96	100
[100] and [001]	100	6	20	62
	1000	36	67	100
[001] and [001]	100	0	0	29
	1000	15	21	87

TABLE 6.5

Work of Formation, A_c, of the Contact Nucleus for Different Mutual Crystallographic Orientations of CaSO$_4$ × 2H$_2$O Crystals in Contact in Supersaturated Calcium Sulfate Solutions for Four Values of the Supersaturation, α

Directions of Edges in Contact	A_c			
	$\alpha = 1.5$	$\alpha = 1.8$	$\alpha = 2.2$	$\alpha = 3$
[100] and [100]	8.2	7.3	6.6	5.8
[100] and [001]	9.0	8.0	4.2	6.3
[001] and [001]	10.7	9.4	8.4	7.2

6.2.1.3 Influence of the Compression Force

Mechanical stresses are necessary for the crystallization contacts to be established in hydration structures in the course of hydration hardening. These stresses are either applied externally or originate from the directional growth of the crystals inside the already formed network. Mechanical stresses may also be caused by the weight of powders or result from the action of capillary forces acting in the drying menisci.

TABLE 6.6

Probability of Bridging, W_b, of $CaSO_4 \times 2H_2O$ Crystals in Supersaturated $CaSO_4$ Solutions as a Function of the Compressing Force, f, for a Given Degree of Supersaturation, α, and Time of Contact, t

	W_b (%)											
	$\alpha = 1.2f$ (dyn)			$\alpha = 1.5f$ (dyn)			$\alpha = 1.8f$ (dyn)			$\alpha = 3f$ (dyn)		
t (s)	0.1	1	10	0.1	1	10	0.1	1	10	0.1	1	10
10	0	0	0	0	1	5	0	2	7	0	7	38
100	0	4	17	0	8	24	1	14	49	4	53	83
1000	4	22	33	7	4	68	10	60	96	33	91	100

Mechanical stresses may cause a decrease in the degree of supersaturation in the contact zone because the stressed regions of the crystals have higher equilibrium solubility. The resulting supersaturation would then be defined as

$$\alpha^* = \alpha \cdot \exp\left(-\frac{\Pi V_1}{kT}\right)$$

where Π is the stress. Generally speaking, this decrease of supersaturation could result in a lower probability of bridging due to the increase in the work of formation of the contact nucleus, but the data in Table 6.6 show that the probability of bridging increases when the compression force, f, acts on the crystals.

The increase in the probability of bridging of crystals with the increase in the applied mechanical stress should thus be related to the increase in the factor I_0 in Equation 6.5. This factor I_0 should account for the area of the "active zone" where the formation of the primary crystallization bridge, that is, of the contact nucleus, takes place.

Experiments showed that for gypsum dihydrate there is a linear dependence between I_0 and f. This can be seen in Figure 6.9, in which the experimental data are presented in the form of $D + 0.43\,A_c/kT$ as a function of $\log f$. In this figure, A_c is the energy of formation of the contact nucleus for the corresponding values of f, and $D = \log 2.3 \times [2 - \log(100 - W_b)]$. The dependence between $D + 0.43\,A_c/kT$ and $\log f$ is essentially linear with the slope approximately equal to I, which corresponds to the proportionality between I_0 and f.

Gypsum is a ductile material and as such may, due to the action of force f, undergo plastic deformation in the vicinity of the contact zone. The area of such "plastic contact," shown in Figure 6.7a, may be estimated as $S_1 \sim f/\tau^*$, where τ^* is the yield stress of the material. In order to explain the observed results, one may assume that the formation of the contact nucleus most likely does not take place in the zone of "plastic" contact, S_1, but rather occurs in the periphery of this zone, S_2. Indeed, the high stresses present within the contact zone S_1 cause the solution to be displaced from the gap, which makes bridge formation difficult. At the same time, in the S_2 zone, the stresses are lower, some solution is present there, and the gap has a width adequate for bridging to take place (Figure 6.7a). If we assume that $S_2 > S_1$, we may write that $S_2 \sim f/\tau^*$.

In the case of solids that undergo only elastic deformation in the contact zone, the dependence of I_0 on f is different: the size of the "active zone" can in this case be estimated from the Hertz equation (see Section 5.2).

We can thus use experimental data to obtain the probability of the appearance of crystallization contacts between crystals during the formation of a new phase as a function of various parameters, namely,

$$W_b = 1 - \exp\left\{-\left(\frac{\kappa f}{\tau^*}\right)\exp\left[-\frac{4h^2\sigma^2}{kT\left\{(hkT/V_1)(\ln\alpha - \Pi V_1/kT) + \sigma^*\right\}}\right]\right\} t \qquad (6.6)$$

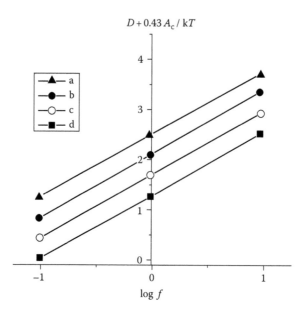

FIGURE 6.9 The value of $D + 0.43 A_c/kT$ as a function of the compression force, f, for gypsum dehydrate crystals in solutions of $CaSO_4$ with various supersaturations, α, and various times of contact, t: (a) and (b) for $t = 1000$ s with $\alpha = 1.8$ and $\alpha = 1.5$, respectively, (c) and (d) for $t = 100$ s with $\alpha = 3$ and $\alpha = 1.8$, respectively.

where the parameter $\kappa \sim 10^{10}$ cm^{-2}s^{-1}. This parameter characterizes the activity of the contact zone S_2 and is the only parameter that cannot be directly estimated from the experimental data.

It turns out that the stresses generated in the course of crystal growth do not have solely a negative impact, as was suggested earlier, but also promote the formation of crystallization contacts and thus serve as the basis for the strength of the structure forming in the process of hydration hardening. The "destructive" role of internal stresses may manifest itself differently in different mineral binders, such as calcium sulfate hemihydrate, calcium oxide, or magnesium oxide. The net result strongly depends on the ratio between the total density of the cohesive energy in the contacts and the elastic deformation energy density accumulated in the structure.

6.2.1.4 Bridging Crystals of Different Types

In order to reduce the cost of mineral binders and improve the properties of the resulting materials, one typically uses various filling additives, such as sand or limestone. Such fillers can significantly influence the kinetics of formation and the properties of a structure formed in the course of hydration hardening. Here, we will address the studies on the bridging of gypsum crystals with quartz, calcite, and fluorite as well as the bridging of quartz with quartz and of calcite with calcite in supersaturated calcium sulfate solutions [53]. In these experiments, various crystals were brought into contact along the following edges:

- In the [2113] direction formed by the (1010) and (1121) faces of quartz
- In the [0111] direction formed by rhombohedral (1010) and scalenohedral (1231) faces of calcite
- In the [101] direction formed by a hexahedral (010) face and the cleavage (111) plane of fluorite

The experiments were carried out using a supersaturated calcium sulfate solution $\alpha = 1.8$ and the time of contact $t = 1000$ s and with the applied compressive force, f, of either 1 or 10 dyn. The results of the experimental studies showed in all cases a sharp transition from contacts with a strength

TABLE 6.7

Probability of Bridging (W_b) of Different Crystals in Supersaturated Calcium Sulfate Solutions ($\alpha = 1.8$; $t = 1000$ s)

N, N	Crystals and Directions of Edges in Contact	W_b (%)	
		$f = 1$ dyn	$f = 10$ dyn
1	Gypsum [100]—gypsum [100]	60	96
2	Quartz [2113]—gypsum [100]	25	68
3	Calcite [0111]—gypsum [100]	30	60
4	Fluorite [101]—gypsum [100]	23	45
5	Calcite [0111]—gypsum [001]	0	14
6[a]	Quartz [2113]—quartz [2113]	N/A	10
7	Calcite [0111]—calcite [0111]	N/A	2

[a] $t = 3000$ s.

of ~10^{-2} dyn or lower to those with a strength of ~ 0.1 dyn or higher. Such a transition agrees with the formation of crystallization contacts between the various crystals. These contacts appear as a result of the fluctuation-based formation of the contact nucleus and the subsequent bridging.

Results summarized in Table 6.7 indicate that the formation of crystallization contacts was facilitated by the applied compressive force, f. In the case of the same applied force, the probability of contact formation depended on the crystallographic orientation of the gypsum crystals. The effect of compressive force on the probability of bridging must be primarily related to the properties of the gypsum crystals. Among all materials used in the study, gypsum was the softest with stress-dependent solubility; its concentration therefore increased in the contact zone.

The probability of bridging between gypsum and other crystals underwent a small change in the following series: quartz > calcite > fluorite. At the same time, the probability of bridging upon the contact of the [100] edge of the gypsum crystal with quartz, calcite, and fluorite was similar to the probability of bridging between two gypsum crystals that were brought in contact along different edges, for example, in the [100] and [001] direction. The most likely reason for this is the similarity in the surface energy, σ_2, of the grain boundaries formed in bridging. The experimental results indicate that upon the bridging of two gypsum crystals, a grain boundary emerged between one of the crystals and the contact nucleus that was completely cooriented with the other crystal (Figure 6.7b). This confirms the validity of the assumption that was used earlier in the analysis of the thermodynamics of contact formation.

Bridging between crystals with a chemical composition differing from that of the solution in which they are immersed, such as quartz–quartz or calcite–calcite, should apparently involve the formation of two grain boundaries, that is, $\sigma^* = 2(\sigma_3 - \sigma_2)$ (Figure 6.7c). Such bridging is thermodynamically less favorable (and hence less probable) than bridging under conditions when the chemical composition of the solution is similar to that of one or both crystals. Therefore, in this case, the values of W_b are low.

The studies conducted indicate that the fine disperse filler added to mineral binders did not act solely as a substrate support for the hydration nuclei, but played an important role as an active structure-forming element and took direct part in the contact formation process.

6.2.1.5 Effect of Chemical Modifiers and the Solution Composition on Crystallization Contacts

The kinetics of mineral binder hydration and the strength of the resulting structures may be significantly impacted by the addition of surfactants or electrolytes. These additives cause chemical modification of the surfaces of the particles making up the structure. A common viewpoint is that electrolytes and surfactants impact the dissolution kinetics of the starting mineral binder and thus affect the kinetics (and the degree) of supersaturation in the liquid phase. Because of this, surfactants and electrolytes affect the rate of formation of new crystals, their morphology, and, hence, the strength of the resulting material. With respect to the formation of crystallization contacts, these additives may change the conditions of the bridging, influencing both the surface properties of the crystals (due to chemical modification of the surface) and the properties of the solution by affecting the supersaturation or mobility of ions.

Here, we will review the results of specific experimental studies on the bridging of crystals with methylated surfaces modified with adsorption layers of octadecylamine (ODA) and gelatin. In these studies, the surface of the crystals was chemically modified in such a way so that the properties of the solution were either impacted insignificantly or not impacted at all. In order to achieve this, quartz crystals were first exposed to aqueous solutions of dimethyldichlorosilane (DMS), ODA, and gelatin. In the case of gelatin and ODA, gypsum and quartz crystals were immersed into saturated calcium sulfate solutions containing small amounts of ODA and gelatin. The concentration of these additives was 0.01% by weight. Crystals were aged in these solutions for 3–5 days, and further experiments were performed in calcium sulfate supersaturated solutions containing 0.01% of ODA or gelatin.

Some of the results on the bridging of the modified crystals are presented in Table 6.8, which also contains the values of the supersaturation, α; the time of contact, t; and the compressive force, f. The same table also contains the bridging probabilities of the unmodified crystals. Data in Table 6.8 clearly show that the chemical modification of the crystals resulted in lower bridging probabilities in all cases, which, in agreement with Equation 6.6, is related to a decrease in the nucleus formation activity, κ, in the zone where the contact nucleus is most likely to appear, as well as to the difference in the specific surface free energy at the crystal/solution and the nucleus/crystal interfaces. The results also indicate that in addition to a decrease in the bridging probability, the maximum possible strength of the contacts formed was reduced.

The decrease in the bridging probability for two gypsum crystals with their surfaces modified by adsorption layers of ODA or gelatin depends on the nature of the crystallographic faces that were brought into contact. This may be related to the difference in the adsorption capacity of the gypsum crystal faces and the history of the face growth. The surface concentration of an impurity present on a given specific face depends on the rate of growth of that face [54]. When crystals

TABLE 6.8
Probability of Bridging, W_b, between Crystals with Chemically Modified Surfaces in Supersaturated CaSO$_4$ Solutions

N, N	Crystals and the Directions of the Edges in Contact	Surface Modifier	t (s)	W_b (%)	α
1	Quartz—quartz	DMS	3000	0(10)	1.8
2	Gypsum [100]—quartz	DMS(quartz)	3000	32(69)	1.8
3	Gypsum [100]—gypsum [100]	ODA	1000	85(96)	1.8
			100	40(49)	1.8
4	Gypsum [100]—gypsum [001]	ODA	1000	36(67)	1.8
5	Gypsum [100]—quartz	ODA	1000	27(68)	1.8
6	Gypsum [100]—gypsum [001]	ODA	1000	12(47)	2.2
7	Gypsum [100]—gypsum [100]	Gelatin	100	42(49)	1.8
8	Gypsum [100]—gypsum [001]	Gelatin	1000	9(47)	2.2
9	Gypsum [100]—quartz	Gelatin	1000	36(68)	1.8

Note: The compressive force $f = 10$ dyn. The values in parenthesis are the W_b values for the original (unmodified) crystal samples.

undergo growth in supersaturated solutions, the adjoining molecules or ions of the crystallizing substance hinder the adsorption of the admixture or impurity (a surfactant molecule in this case). These adjoining molecules can even replace admixture molecules that have already been adsorbed at the surface. It was observed that the effect of the impurities becomes smaller with an increase in the rate of crystallization. The (120) faces of gypsum crystals grew slowly and hence were capable of adsorbing higher amounts of an admixture, so that both the probability of bridging and the contact strength decreased. The same mechanism is likely responsible for the close probabilities of bridging observed between gypsum crystals in contact over their (001) and (120) faces (the edges of the corresponding directions were brought into contact) in the presence of adsorbing substances and the bridging of a gypsum crystal, the (011) face of which was brought into contact with quartz under the same conditions.

Since the faces of quartz crystals do not undergo growth in the supersaturated calcium sulfate solution, the adsorbing substance can form dense adsorption layers on their surfaces and thus decrease the probability of the formation of crystallization contacts. In this respect, the effect of the chemical modification of the surfaces is similar to that of the vibration treatment of crystallization structures applied at the initial stages of mineral binder hydration. The primary goal of such treatment is the destruction of the primary framework of the future crystallization structure in order to prevent the possible accumulation of elastic stresses. The internal stresses that develop in the course of structure formation may lead to irreversible degradation of the material, especially at the later stages of formation. At the same time, the vibration treatment may cause a decrease in strength. The destruction of the primary framework and the prevention of crystal bridging at the early stages of hydration results in the plastification of the system and in the greater strength of the end material.

Electrolytes, both weak and strong, such as KCl, K_2SO_4, or H_3BO_3, were shown to affect the probability of bridging between gypsum crystals in $CaSO_4$ supersaturated solutions due to their effect on the supersaturation. Specifically, the addition of KCl reduced supersaturation and hence decreased the probability of bridging, while the addition of K_2SO_4 increased supersaturation and increased the probability of bridging. The addition of orthoboric acid, H_3BO_3, which is a weak electrolyte, impacted the probability of bridging due to the adsorption of ions at the gypsum crystal surface.

6.2.1.6   Bridging of Silica Particles

The same procedure as used with gypsum crystals was also employed to study bridging taking place in freshly prepared silica sols. In this system, particle bridging leads to the formation of a silica gel. Soluble forms of silicic acid are formed upon the acidification of sodium silicate solution [13]. The concentration of the silicic acids formed depends on the sodium silicate concentration. The monomeric silicic acid species undergo polymerization, resulting in the formation of fine disperse polysilicic acid particles, that is, the formation of a sol takes place. Sol formation causes a decrease in the silicic acid concentration. However, over an extended period of time, this concentration is rather high and exceeds the solubility of amorphous silica, so that the resulting solution becomes supersaturated. Although the silicic acid system is chemically more complex than the $CaSO_4 \times 0.5H_2O$–H_2O system [13], one can assume that the bridging of silica particles in a supersaturated solution of silicic acid also involves the formation of contact nuclei. The hydration of gypsum can be regarded as the "addition of water," while the polycondensation of silicic acids can be regarded as the "subtraction of water" with a siloxane bond substituting for the two silanol groups.

The polysilicic acid globules in silica sols are very small, which does not allow one to conduct the experiments described with single particles. Hence, the experiments were conducted with glass threads (rods) modified with silica particles. These threads were brought into cross-contact in supersaturated silicic acid solutions.

As in the case of gypsum, the values of the contact strength, p_1, measured under a fixed set of conditions (pH, f, α, and t) are spread over a very broad range, as can be seen in the histograms shown in Figure 6.10. In the case of the silica-coated glass rods, the histograms also show a sharp

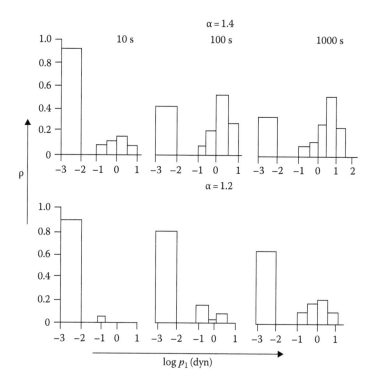

FIGURE 6.10   Histograms showing the differential distribution of the contact strength, p_1, between silica-coated rods in supersaturated silicic acid solutions at pH 7, with the compressive strength $f = 10$ dyn, and for supersaturations $\alpha = 1.2$ and $\alpha = 1.4$.

transition from "weak" coagulation contacts to "strong" phase contacts ($p_1 > 10^{-1}$ dyn). Again, the absence of contacts of intermediate strength, $p_1 \sim 10^{-2}$–10^{-1} dyn, allows one to distinguish this type of bridging and to study the effect of various conditions on this process.

The results at pH = 7 (Figure 6.10 and Table 6.2) indicate that the probability of bridging between silica particles increased with the increase in the degree of supersaturation, the duration of the contact, and the magnitude of the compressive force. These results are very similar to those observed with gypsum (Table 6.2), which supports the hypothesis regarding the role of the fluctuation-induced formation of the critical contact nucleus at the initial stage of particle bridging in silica sols.

Calculations using the same equations as for gypsum can be also carried out for silica using the experimental data on the dependence of W_b on the supersaturation, α, and the time of contact, t. Such calculations indicate that the estimated parameters for the contact nucleus in a silica sol system ($h = 1.2$ nm, $A_c = 0.3 - 0.8$ nm at $\alpha = 3.0$–1.2) are similar to the parameters obtained for the contact of nucleus of gypsum and that the work of their formation, A_c, is significantly lower than the work of the formation of the 2D nucleus, A_2 (see Table 6.3). The surface free energy at the contact–solution interface estimated in this way is around 6 erg/cm^2, which is significantly lower than that of the thermally treated silica particles (46 erg/cm^2).

There are a number of peculiarities associated with the bridging in silica sols. Namely, the bridging probability depends on the polymerization rate of the silanol groups, as indicated by the dependence of W_b values on the pH.

The rate of the silicic acid polycondensation, resulting in the conversion of silanol groups into siloxane groups, did not reveal a monotonous change when the pH was changed. The residual concentration of the dissolved silica, C_1, measured 1 h after the mixing of the sodium silicate solution with the perchloric acid solution was a strong function of the pH, as shown in Figure 6.11a. The trend in the concentration correlates well with the known trend in the rate of polycondensation [13,24].

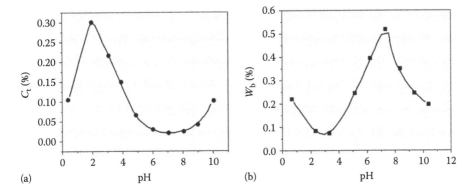

FIGURE 6.11   The residual concentration of the dissolved silica, C_t, in a silica sol (a) and the probability of phase contact formation in silica sol, W_b, as a function of pH (b).

The lower is the residual silica concentration, the higher is the reaction rate. This dependence can be explained by the ratio between the molecular and the ionic forms of silicic acid in a solution. The highest rate corresponds to unequal concentrations of these two forms, which occurs at pH 7. A deviation in pH to either side from pH 7 resulted in a decrease in the reaction rate. At pH 2–3, the degree of ionization of the silicic acid molecules is low and hence the reaction rate was very low as well.

The values of W_b for solutions of different pH are shown in Figure 6.11b. A comparison of the W_b data to the residual concentration data at the same pH indicates that the maximum in W_b occurs between pH 7 and 8 and corresponds to the highest rate of polymerization (and low final value of supersaturation, α). The lowest values of W_b are observed around pH 2–3, where the supersaturation is the highest and the polymerization rate is the lowest. At the same time, at each given pH, the probability W_b increases with an increase in α, as demonstrated in Tables 6.2 and 6.9. These data show that the formation of molecular associates that reach the size of critical nuclei is not sufficient for the formation of contact nuclei in silica sols. For bridging to be initiated, the associates must contain dissolved silica capable of quick siloxane bond formation.

The same experimental technique was also used to investigate bridging between silica particles and crystalline and amorphous particles of other chemical composition [54–56]. The studies on contact formation were carried out using the same "standard" conditions, that is, $\alpha = 1.4$, pH = 7, $f = 10$ dyn, and $t = 100$ s. The bridging of silica with other solid species is not only of theoretical interest but also of practical significance, since many of them are used as fillers in catalysts and adsorbents and in building materials. Table 6.10 contains the values of experimentally determined bridging probabilities between silica particles and other solid materials. The results with original materials and materials modified by coating their surface with silica nanoglobules are reported. The data indicate a considerable decrease in the W_b values for contacts formed between a thread covered with silica and uncoated thread, a similar thread made of unmodified quartz or quartz monocrystal,

TABLE 6.9

Probability of Silica Particle Bridging, W_b, as a Function of the Supersaturation, α, at Different pH Values

	pH								
	0.5		**5.0**		**7.0**				**8.5**	
α	3.5	7.0	2.5	5.0	1.2	1.4	1.8	3.0	1.1	2.2
$W_b(\%)$	3.0	22.0	6.0	25.0	25.0	16.0	54.0	68.0	81.0	25.0

Note: The compressive force and contact time were kept constant at $f = 10$ dyn and $t = 100$ s.

TABLE 6.10

Bridging Probability, W_b, between Silica Samples and Other Particles in a Supersaturated Solution of Silicic Acid

	W_b (%)	W_b' (%) after Particle Modification	$(W_b' - W_b)$, (%)
Silica–silica	54	N/A	N/A
Silica–quartz glass	13	54	41
Silica–quartz	7	37	30
Silica–leucosapphire	4	30	26
Silica–brucite	1	9	8
Silica–fluorite	3	11	8
Silica–lithium fluoride	1	10	9
Silica–antophillite	1	9	8
Silica–chrysotile	5	7	2
Silica–natrolite	5	25	20
Silica–heulandite	4	30	26
Silica–periclase	3	19	16

Note: The supersaturation $\alpha = 1.4$, pH = 7, the contact time $t = 100$ s, and the compression force $f = 10$ dyn.

as compared to contacts formed between two identical silica-modified threads. The probability of bridging between a silica-modified thread and other crystals of "foreign" materials was even lower than in the case of glass and quartz.

The decrease in the values of W_b can be explained by the increase in the work of nucleus formation due to an increase in the interfacial free energy at the boundary between the nucleus and another particle. Indeed, particles of amorphous silica are chemically most similar to the nuclei of contact. The values of W_b for the contact between silica-modified threads and quartz glass (a highly condensed polysilicic acid) or quartz (crystalline silica) were higher than those for contacts between silica-modified threads and other materials because of a higher degree of chemical similarity. This further emphasizes the role of chemical composition and structure of the substance in the process.

The main conclusion that one can draw is that bridging between silica particles and particles of other chemical composition or morphology is less probable than bridging between the silica particles themselves. This indicates that the boundaries between nonporous fillers and silica hydrogel in catalysts, adsorbents, and hardened cement are weakened. At the same time, the pretreatment of fillers with silica sols can increase the strength of these boundaries. This is illustrated in Table 6.10 showing an increase in all W_b values after the exposure of the materials to silica sols. The difference between the bridging probabilities between silica-coated threads and silica-modified materials, W_b', and the bridging probabilities prior to modification, $W_b' - W_b$, can apparently serve as characteristics of the differences in the adhesion between various materials and silica particles and is indicative of the effectiveness of coating these materials with silica under given ageing conditions.

6.2.1.7 Investigation of the Microstresses in Disperse Structures Using X-Ray Diffraction

In order to cause bridging between crystals and particles, one needs to apply external compressive forces. In the laboratory studies, these forces were introduced by applying suitable compression. However, in the course of hydration hardening, such forces are originated in the system internally. One is thus faced with the need to answer a number of questions. Specifically, what is the origin of those forces in the process of structure formation? What are the nature, magnitude, and distribution of these forces? How are these forces preserved in the residual stressed state? How do these forces depend upon the conditions, and what is their further fate? How can one control and influence these forces?

The well-known phenomenon of "strength drop" during hydration hardening, along with the known impact of other hydration conditions on the structure strength, prompted Chernov and Lyubov [54] to consider the role that the *crystallization pressure* between particles in the forming structure plays in the possibility of generating high internal stresses. This phenomenon can be described as follows: At the beginning of the hydration hardening process, the initial supersaturation is high, and nucleation takes place, resulting in the formation of numerous nuclei of the new fine disperse phase. At low volume ratios of water to the solid phase and at high degrees of supersaturation, bridging takes place with a substantial probability, even for short-term collisions between particles. This in turn results in the formation of the primary structure, the framework of which is supported by the vessel walls and the particles that have already bridged via phase contacts. As the growth process continues, the newly formed crystals collide with other particles that make up the framework. At this stage, two competing thermodynamic factors come into play. On the one hand, everything in the system favors the formation of phase contacts, including the compression force, f, resulting from the crystallization pressure associated with crystal growth. On the other hand, as a direct consequence of supersaturation, the crystallization pressure may transfer tensile stresses to the adjacent links of the framework. In certain cases, these stresses can be sufficiently high to cause the destruction of the already formed contacts, which comprises the elementary act of "strength lowering." In some other cases, the destruction of contacts does not take place and the bridging proceeds, but the tensile stresses that compensate each other in the adjacent links of the framework may be retained in the structure as residual internal stresses. These residual internal stresses can be effectively studied by x-ray diffraction, as described in the following.

6.2.1.8 General Methodology of X-Ray Diffraction Studies

X-ray diffraction (XRD) studies on the distortions in crystal lattices enable one to distinguish three main types of deformation, or "stresses," as evidenced by changes in the XRD patterns of polycrystalline samples. *Line displacement* in the patterns corresponds to a more or less homogeneous deformation, such as an increase or decrease in the interatomic spacing d in the lattice (change in the lattice constant), at least in the range of the beam cross section. Such elastic macroscopic deformations are referred to as stresses of the first kind. *Line broadening* indicates that in the area covered by the x-ray beam, numerous microcrystals (grains) have undergone various types of deformation—some are compressed, while others are stretched or bent. In these cases, the heterogeneity of the lattice elastic deformations is referred to as stresses of the second kind, or elastic microstresses. The *attenuation of line intensity* and the increase in the background signal points to the presence of short-range distortions, for instance, the accumulation of point defects or the so-called body "amorphization." The latter are referred to as stresses of the third kind. A discussion of the details of the analysis of XRD patterns is beyond the scope of this chapter; here, we will only address the experimental measurements of stresses of the second kind, that is, heterogeneity in the lattice elastic deformations, $\Delta d/d$. These are the deformations that are behind the generation of internal stresses during hydration/hardening.

The principal issue in the analysis of line broadening in XRD patterns is related to the fact that it is caused not only by elastic deformations (of both signs) in microcrystals but also by the disperse nature of the sample itself, that is, by the diffraction on the fine particles, both crystalline and amorphous. To separate these two contributions, classical metallography employs quantitative line shape analysis methods. In the cases discussed, these methods could not be used because of the high background signal. Also, one has to keep in mind that the dispersion leading to line broadening not only is due to the size of particles but is also associated with the features of their internal structure—blocks, fragments, and dislocation structures, including those formed as a result of the plastic deformation of particles.

Shchukin et al. developed the procedure that enables one to isolate the part of the line width, β_d, associated with the microstresses themselves (i.e., with the heterogeneity of the elastic deformation) and part β_b that is associated with the structure dispersion (i.e., with the presence of blocks) [43,44]. The method is based on the fact that the first factor is reversible, while the second is not. The procedure involves comparing the line widths, β, in the XRD patterns of samples that underwent hydration hardening and in the XRD patterns of powders obtained by gently crushing the samples in an inert liquid, such as isopropanol (the "true" line width, β, is assumed to be that determined from Equations 6.8 and 6.9). The gentle crushing of samples into powders allows one to destroy the contacts between the particles and to isolate crystallites from each other so that the elastic deformations vanish, while the "dispersion" (i.e., the block structure) is preserved. The difference in the line width before and after sample crushing characterizes the reversible contribution, $\beta_d = (\Delta d/d) \tan \Theta$, that is, the microstresses themselves:

$$P = \left(\frac{\Delta d}{d} \right) E \tag{6.7}$$

where
d is the spacing in the undeformed lattice
Δd is the mean deviation (of both signs)
Θ is the angle of reflection from a given system of planes in the lattice
E is Young's modulus

One can also conduct comparative studies using a specimen of a hydrated powder with very high water-to-solid (W/S) ratio that has been pounded to prevent bridging. The same methodology can also be used to study the residual stresses in samples obtained by powder pressing. In the latter case, the reference material is the original uncompressed (unstressed) powder.

The experimental XRD studies were conducted in backscattering chambers positioned for exposure from the back side of the cylindrical samples of 10 mm in diameter and 10 mm in height. An x-ray tube with a cobalt anode was used. The half width of the interference line ($\Delta \Theta$, radians) was determined geometrically from the intensity curve, using the integral intensity method.

The calculation of β values was carried out using the following expressions representing the two extremes for the approximation of the line shapes:

$$\left(\Delta \Theta \right)^2 = b^2 + \beta^2; \quad \beta^2 = \beta_b^2 + \beta_d^2 \tag{6.8}$$

$$\left(\Delta \Theta \right)^2 = b^2 + \beta^2; \quad \beta = \beta_b + \beta_d \tag{6.9}$$

where b is the part of the width defined by the experimental conditions. The studies were carried out using a specimen obtained by the hydration hardening of single mineral binders, specifically magnesium oxide, calcium oxide, and gypsum. In the experiments, the main variables were the W/S ratio, the degree of dispersion of the binder, the content and degree of dispersion of a filler (amorphous silica), and the concentration and type of electrolyte additives (e.g., $CaSO_4$). The coarseness (degree of dispersion, as defined by the mean particle size, a) was measured by microscopy and obtained from the surface area (s, m^2/g) measured by the Brunauer–Emmett–Teller isotherm method. The angles of reflection, Θ, registered in the x-ray patterns were 80° and 62° for Mg(OH)$_2$ for the scattering from the (3030) and (1231) plane systems, respectively; 69° for Ca(OH)$_2$ for scattering from the (3032) and (2133) plane systems, and 74° for gypsum, in which the specific planes of scattering were not identified. The relative error in the measured $\Delta \Theta$ values was between 5% and 15% for different lines.

6.2.1.9 Internal Stresses in Structures Obtained in the Course of the Hydration Hardening of Various Binders

The data on the interference line width for magnesium hydroxide samples ($s \sim 50$ m²/g), obtained in the course of the hydration of magnesium oxide under various conditions, are summarized in Table 6.11. In one case, the hardening process was carried out in a sample with high water content (W/S = 50) with a continuous abrasion to prevent bridging. In the other case, the process was carried out in samples with low water content (W/S = 1.0 and 0.8) without any abrasion, so that hardened samples could be obtained.

The data in Table 6.11 show that the true line width, β, of the hardened "stone," obtained at both W/S = 1.0 and 0.8, was approximately 20%–30% larger than that of powder consisting of crystals of similar size but not bridged during hydration. The same value of β was obtained for unbridged crystals (W/S = 50) and for the crushed "hydration stone" obtained with W/S = 0.8. This finding indicates the similarity in the substructure of magnesium hydroxide crystals obtained at different W/S ratios. The heterogeneity of the lattice elastic deformation, $\Delta d/d$, determined in these experiments reaches $d \approx 4 \times 10^{-3}$, which according to Equation 6.7 corresponds to internal stress values of several kg/mm². For comparison, the modulus of elasticity of the bulk material is around 400 kg/mm². The strength of the magnesium hydroxide specimen was also on the order of several kg/mm². Consequently, the residual stresses that had built up in the crystalline $Mg(OH)_2$ structure by the end of the hydration process were close to its strength.

In the fine disperse samples of $Mg(OH)_2$ having $s \sim 20$–50 m²/g, the true width of the interference lines decreased after sample crushing, which can serve as confirmation of the presence of elastic deformations (microstresses). Conversely, in the case of a coarse sample with $s = 1$ m²/g and in the case of a ductile gypsum sample, the line width did not decrease after crushing of the structure formed during hardening, but appeared to be larger than that observed with unbridged (unstressed) gypsum crystals obtained at high W/S ratios. In each case, internal stresses are generated in the course of the gypsum hydration, but they undergo a complete relaxation due to the plastic deformation of the "soft" gypsum crystals (for smaller regions of coherent scattering, for the splitting of blocks).

It is noteworthy that both elastic and plastic deformations were identified in the $Ca(OH)_2$ structures ($s = 12$ m²/g), which may be due to the polydispersity of the specimen. Such a dependence on the degree of dispersion can be explained as follows: The range of dispersion lies between single units and tens of m²/g, which corresponds to a range of crystal sizes from micrometers to tenths and hundredths of a micrometer. Based on the characteristic values of the density of dislocations in crystals, this range corresponds to a transition from crystals with defects (similar to macroscopic "real" crystals) to microcrystals with properties close to those of "ideal," defect-free crystals. This is the reason why the effective microhardness (yield point) can increase considerably in the range of dispersions corresponding to the transition to submicron crystals. In the latter case, the forces that

TABLE 6.11

Effect of the Hydration Conditions of the MgO Specimen on the True Line Width of Interference Patterns, β, and on $\Delta d/d$ Values in $Mg(OH)_2$ Structures

Water-to-Solid Ratio	$\Delta\Theta \times 10^3$ (rad)	$\beta \times 10^3$ (rad)	$\beta_d \times 10^3$ (rad)	$(\Delta d/d) \times 10^3$	$\beta_d \times 10^3$ (rad)	$(\Delta d/d) \times 10^3$
50 (abrasion)	32.0	29.5	0	0	0	0
1.0 ("stone")	38.0	36.0	21.0	3.5	6.5	1.1
0.8 ("stone")	41.5	40.0	27.0	4.6	10.5	1.8
0.8 (after "stone" crushing)	32.0	29.5	0	0	0	0
		Equation 6.8			Equation 6.9	

are responsible for plastic deformations should increase as well. The latter was indeed confirmed experimentally in studies on compressing $Mg(OH)_2$ and gypsum powders with different degrees of dispersion.

The studies conducted clearly showed that internal stresses developed in the course of the hardening of all studied mineral binders. However, they could show up in both elastic and plastic deformations depending on the crystal hardness and the degree of dispersion.

6.2.1.10 Stresses Formed in Structures Obtained by Pressing Powders

Pressed magnesium hydroxide powders containing rather coarse particles ($s \sim 3$ m^2/g, ground brucite) reveal plastic deformations of the crystals in samples compressed at a pressure on the order of several metric tons per cm^2 (at lower pressures the elastic deformations prevailed). However, in the case of a $CaSO_4 \times 2H_2O$ specimen of the same grain size, plastic deformations were observed at a lower pressure of 1 ton/cm^2. This difference agrees well with the difference in the microhardness, which is 80 kg/mm^2 for brucite, and 35 kg/mm^2 for gypsum. When the degree of dispersion is higher, the pressures promoting the formation of a fine block structure are considerably higher. This means that different manifestations of internal stresses in materials with different hardness and degrees of dispersion are related to differences in the yield points of crystallites.

6.2.1.11 Dependence of Internal Stresses on the Conditions of the Hydration Hardening

The effect of hardening conditions on internal stresses ($\Delta d/d$) was studied using specimens of finely disperse $Mg(OH)_2$ with a surface area on the order of tens of m^2/g. The results indicated a considerable dependence of the internal stresses on the degree of supersaturation, α ($\alpha = c/c_0$; c, concentration; c_0, equilibrium solubility), reached in the course of the hydration of a mineral binder in an aqueous slurry. The $\Delta d/d$ values increased with both the increase in the degree of dispersion and the decrease in the W/S ratio, as shown in Table 6.12.

The opposite situation, corresponding to a decrease in the $\Delta d/d$ value, was observed when there was a decrease in the supersaturation, for instance, in cases when one used seed crystals or when electrolytes such as small amounts of Na_2SO_4 were present in the CaO suspension. The effect of the presence of the seed crystal is illustrated in Table 6.13.

A significant decrease in the internal stresses also occurred upon the introduction of a filler. In the example summarized in Table 6.14, the filler is a sample of amorphous silica introduced into the MgO suspension. One can assume that in this case, the decrease in the $\Delta d/d$ value was also caused by the decrease in the supersaturation because of the dissolution of the $Mg(OH)_2$ in the course of magnesium silicate formation.

These results indicate that the internal stresses that build up due to recrystallization of a new hydrate phase are the results of the crystallization pressure, Π. The latter is known to develop in the

---

TABLE 6.12

Effect of the Water-to-Solid Ratio and the Degree of Dispersion of MgO on Values of the True Width of the Interference Line, β, and on the Value of Δd/d in the Mg(OH)₂ Structures Formed in the Hardening Process

	MgO Dispersity (s, m^2/g)			
	150		45	
Water-to-solid ratio	0.8	1.0	0.8	1.0
$\beta \times 10^3$ (rad)	73.5	68.5	42.3	40.2
$(\Delta d/d) \times 10^3$ (Equation 6.9)	4.0	3.2	1.8	1.5
Degree of dispersion of the "stone"(s, m^2/g)	45.5	54.0	22.0	24.0

TABLE 6.13

Effect of the Addition of Seed Crystals of Mg(OH)₂ on the Values of the True Width of the Interference Line, β, and on the Δd/d Values in Structures Formed in the Course of Mg(OH)₂ Hardening

Mg(OH)₂ Content (%)	Water-to-Solid Ratio	"Stone" Dispersity (s, m²/g)	β × 10³ (rad)	(Δd/d) × 10³ (Equation 6.2.9)
0	0.8	19	45.9	2.0
15	0.8	17	40.4	1.1

TABLE 6.14

Effect of the Addition of a Filler (Amorphous Silica) on the Values of the True Width of the Interference Line, β, and on the Value of Δd/d in the Structures Formed in the Hydration Hardening of the Mg(OH)₂ Samples

MgO Dispersity (s, m²/g)	SiO₂ Content (%)	"Stone" Dispersity (s, m²/g)	β × 10³ (rad)	(Δd/d) × 10³ (Equation 6.2.9)
150	0	54	68.5	3.5
	5	87	57.3	0.7
	15	84	55.7	0.4
45	0	23	40.2	1.5

course of directional crystal growth in supersaturated solutions. The dependence of Π on $\alpha = c/c_0$ can be obtained from the thermodynamics of crystal growth under the action of an external pressure:

$$\Pi = \frac{kT \ln(c/c_0)}{V_1} \tag{6.10}$$

where V_1 is the volume of a molecule of the solid phase. The maximum supersaturation values observed in suspensions of calcium sulfate hemihydrate ($\alpha \sim 3$–4) and in suspensions of magnesium oxide ($\alpha \sim 100$) correspond to the maximum crystallization pressures of 4 and even 40 kg/mm², respectively. These crystallization pressures not only are capable of producing elastic and plastic crystal deformations but can also result in the partial destruction of the structure formed during crystallization. The latter can halt the buildup of strength or result in the lowering of the latter in the process of hydration hardening.

The residual stresses in filled xerogels were studied in a similar way by comparing x-ray patterns of strained and crushed samples [35]. The results of these tests showed that internal stresses building up in the course of the drying of the filled hydrogel appeared in the patterns as the shift of the filler interference lines, rather than as line broadening. The direction of the shift (increasing Θ) corresponds to the compression of crystals distributed in the volume of a gel (here, CaF₂ crystals can be used as sensors). It is apparent that the gel was extended in this case, as the compression of each crystal was compensated by the extension existing in the surrounding drying gel. The data on the line shift and on the Δd/d value for samples with various fillers showed that in these materials, the residual stresses were also on the order of the material strength. The stresses, however, could be effectively suppressed by increasing the degree of dispersion of the filler or by introducing anisometric particles such as fibers.

In summary, one can say that x-ray diffraction studies have proven the presence of residual stresses in the structures formed in the course of hydration hardening and have made it possible to quantitatively estimate their magnitude under various physical-chemical conditions. The stresses that originate in the course of crystal bridging can reach hundreds of kg/cm² and manifest themselves in many different ways depending on the hydration hardening conditions. In particular, the stresses can undergo partial or complete relaxation in the course of the plastic deformation of the particles, disappear as a result of the destruction of the contacts formed, or be retained in the material as residual stresses. The residual stresses can become comparable to the entire strength of the material and can significantly affect the material properties by causing the formation of cracks and a decrease in durability, especially upon the exposure of the materials to active media.

The combination of the physical-chemical conditions existing in the hydration hardening process may significantly impact the root cause and level of the internal stresses. The factors impacting the degree of supersaturation and the duration ("life") of the supersaturated state's existence are of primary importance. Supersaturation is the main cause for the appearance of residual stresses, but it is also the necessary condition for the formation of new phases and for the bridging of crystals. In the latter case, these internal stresses stipulate the necessary compressive forces. For this reason, control of the supersaturation and the changes in the supersaturation over time is essential in optimizing all of these factors. Therefore, one needs to choose the most appropriate and beneficial methods and conditions, for example, vibration and the use of surfactants. Enhanced knowledge and understanding of the mechanisms behind the formation of the individual contacts and internal stresses that form and build up in the course of the hydration hardening of mineral binders can provide one with the means for better controlling and further improving various technological processes.

It is worth pointing out here that the manifestations of internal stresses in single-phase crystalline materials (such as MgO, CaO, and gypsum) do not in any way limit their variety in more complex structures, for example, in such two-phase materials as amorphous silica or aluminum-modified silica filled with various crystals, or in even more complex materials, such as cement compositions.

6.2.1.12 Concluding Remarks on Hydration Hardening

The bridging of particles in the formation of structures generated in hydration hardening, which we have now discussed, is believed to be the key factor responsible for the material strength. The studies in which single-component (monomineral) cements, gypsum, and brucite were used as model systems allowed one, as we have described, to develop a general approach and extend it to the investigation of more complex systems.

A systematic investigation of the bridging of individual gypsum crystals indicated that the principal step in the hardening process is the fluctuation-based formation of a critical contact nucleus in the gap between the two crystals. Once such a contact is established, it spreads rapidly over the entire contact zone. The probability of the formation of a contact nucleus is dictated by the degree and duration of the supersaturation (metastability) in the solution, the time of contact, and the compressive force acting on the particles. This probability varies for crystals with different orientations and for contacts between particles of different nature and morphology (e.g., fillers) and can be influenced by the addition of electrolytes, surfactants, etc. Similar observations were made for systems of amorphous nature, such as for the bridging of particles of polysilicic acids with varying degrees of supersaturation, times of contact, compression forces, pH, etc.

The bridging of particles is closely related to the development and buildup of internal microstresses in the forming structures. Direct x-ray diffraction studies made it possible to estimate the magnitude of these stresses and indicated that in some cases these stresses can be comparable to the strength of the structure.

Internal stresses comprise the necessary precondition for the bridging of crystals, because they are the source of the necessary compressive forces that must be present for the bridging to take place. The internal stresses may undergo relaxation in the course of plastic deformation or can be preserved in the material in the form of residual stresses, which results in a decrease in the strength and durability of the material.

These phenomena, observed with monomineral binders as model systems, are the essential elements in the hydration hardening process of all complex multicomponent systems, such as cements and concretes. Achieving better understanding and control of the physical-chemical processes of hydration hardening would make it possible for one to predict and microengineer particle bridging and take control over the development of the internal stresses and their subsequent relaxation. This provides one with a key to making materials with superior strength and durability.

REFERENCES

1. Shchukin, E. D., Amelina, E. A., and S. I. Kontorovich. 1992. Formation of contacts between particles and development of internal stresses during hydration processes. In *Materials Sciences of Concrete*, J. Skalny (Ed.), pp. 1–35. Westerville, OH: American Ceramic Society.
2. Shchukin, E. D. (Ed.). 1992. *The Successes of Colloid Chemistry and Physical-Chemical Mechanics*. Moscow, Russia: Nauka.
3. Shchukin, E. D. 1996. Some colloid-chemical aspects of the small particles contact interactions. In *Fine Particles Science and Technology*, E. Pelizzetti (Ed.), pp. 239–253. Amsterdam, the Netherlands: Kluwer Academic Publishers.
4. Sokolova, L. N., Shchukin, E. D., Burenkova, L. N., and B. V. Romanovskiy. 1998. Sintering of the porous zirconium oxide in the course of a catalytic reaction. *Doklady AN Khimia*. 360: 782–783.
5. Sokolova, L. N., Shchukin, E. D., Burenkova, L. N., and B. V. Romanovskiy. 2000. Effect of strengthening of porous materials during catalytic reaction. *Doklady AN Khimia*. 373: 491–492.
6. Shchukin, E. D. and E. A. Amelina. 2003. Surface modification and contact interaction of particles. *J. Dispers. Sci. Technol*. 24: 377–395.
7. Shchukin, E. D., Bessonov, A. I., Kontorovich S. I. et al. 2006. Effects of adsorption-active media on the mechanical properties of catalysts and adsorbents. *Colloids Surf*. 282–283: 287–297.
8. Shchukin, E. D., Pertsov, A. V., Amelina, E. A., and A. S. Zelenev. 2001. *Colloid and Surface Chemistry*. Amsterdam, the Netherlands: Elsevier.
9. Segalova, E. E. and P. A. Rehbinder. 1962. The formation of crystallizational structures of curing, and conditions for the development of their strength. In *New in Chemistry and Technology of Cement*, P.A. Rehbinder (Ed.), pp. 202–212. Moscow, Russia: Gosstroyizdat.
10. Polak, A. F. 1966. *Hardening of Mineral Binders*. Moscow, Russia: Promstroyizdat.
11. Rehbinder, P. A. 1979. *Selected Works. Surface Phenomena in Disperse Systems: Physical-Chemical Mechanics*. Moscow, Russia: Nauka.
12. Adamson, A. 1979. *Physical Chemistry of Surfaces*. Moscow, Russia: Mir.
13. Iler, R. K. 1979. *The Chemistry of Silica: Solubility, Polymerization, Colloid and Surface Properties, and Biochemistry of Silica*. New York: Wiley.
14. Gibbs, J. W. 1982. *Thermodynamics: Statistical Mechanics*. Moscow, Russia: Nauka.
15. *Very High Strength Cement-Based Materials*. 1985. Ed. J.F. Young. MRS Proc.
16. Skalny, J. (Ed.). 1992. *Materials Sciences of Concrete*. Westerville, OH: American Ceramic Society.
17. Nonat, A. and J. C. Mutin. (Eds.). 1992. *Hydration and Setting of Cements*. London, U.K.: E & FN Spon.
18. Taylor, H. F. W. 1997. *Cement Chemistry*, 2nd edn. London, U.K.: Thomas Telford Ltd.
19. Johnston, C. D. 2001. *Fiber-Reinforced Cements and Concretes*. London, U.K.: Taylor & Francis.
20. Lea, F. M. 2004. *Lea's Chemistry of Cement and Concrete*, 4th edn. Oxford, U.K.: Butterworth.
21. Allcock, H. R. 2008. *Introduction to Materials Chemistry*. New York: Wiley.
22. Eckel, E. C. 2009. *Cements, Limes, and Plasters: Their Materials, Manufacture, and Properties*. Ithaca, NY: Cornell University Library Publication.
23. Gjørv, O. E. 2009. *Durability Design of Concrete Structures in the Severe Environments*. London, U.K.: Taylor & Francis.
24. Auerbach, S. M., Carrado, K. A., and P. K. Dutta. (Eds.). 2003. *Handbook of Zeolite Science and Technology*. Marcel Dekker, New York: CRC Press.
25. Shchukin, E. D., Yusupov, R. K., Amelina, E. A., and P. A. Rehbinder. 1969. Experimental study of cohesive forces in individual microscopic contacts between crystals when compressing and sintering. *Kolloidnyi Zh*. 31: 913–918.
26. Shchukin, E. D., Amelina, E. A., Yusupov, R. K., Vaganov, V. P., and P. A. Rehbinder. 1973. Experimental study of the effect of supersaturation and the contact time on the fusion of individual crystals. *Doklady AN SSSR*. 213: 155–158.
27. Shchukin, E. D., Amelina, E. A., Yusupov, R. K., Vaganov, V. P., and P. A. Rehbinder. 1973. Experimental study of the effect of mechanical stresses on the formation of crystallization contacts at accretion of crystals. *Doklady AN SSSR*. 213: 398–401.
28. Shchukin, E. D. and E. A. Amelina. 1979. Contact interactions in disperse systems. *Adv. Colloid Interface Sci*. 11: 235–287.
29. Shchukin, E. D. and V. S. Yushchenko. 1981. Molecular dynamics simulation of mechanical behavior. *J. Mater. Sci*. 16: 313–330.
30. Gani, M. S. J. 1997. *Cement and Concrete*. London, U.K.: Chapman & Hall.

31. Shchukin, E. D. 2002. Surfactant effects on the cohesive strength of particle contacts: Measurements by the cohesive force apparatus. *J. Colloid Interface Sci.* 25: 159–167.
32. Roque-Malherbe, R. 2009. *The Physical Chemistry of Materials: Energy and Environmental Applications.* CRC Press/Taylor & Francis Group.
33. Kiselev, A. V., Lukyanovich, V. M., and E. A. Poray-Koshits. 1958. Research methods of structure of highly disperse and porous materials. *Izvestiya AN SSSR.* 11: 161.
34. Kontorovich, S. I., Lavrova, K. A., Plavnik, G. M., Shchukin, E. D., and P. A. Rehbinder. 1971. Effect of temperature and aging on the size of the globules in the aluminosilicate hydrogel. *Doklady AN SSSR.* 196: 633–635.
35. Kontorovich, S. I., Lavrova, K. A., Plavnik, G. M. et al. 1971. Investigation of internal stresses in the filled aluminosilicate xerogels. *Doklady AN SSSR.* 199: 1360–1363.
36. Lankin, Ya. I., Kontorovich, S. I., Amelina, E. A., and E. D. Shchukin. 1982. Experimental study of the formation of accretion contacts between silica particles in supersaturated solutions of silicic acid. *Kolloidnyi Zh.* 42: 649–652.
37. Romanovskiy, B. V., Shchukin, E. D., Burenkova, L. N., and L. N. Sokolova. 2002. Influence of catalysis on the strength of porous materials with a globular structure. *Zh. Fizicheskoy Himii.* 76: 1044–1047.
38. Schwarz, J. A. and C. L. Contescu. (Eds.). 1999. *Surfaces of Nanoparticles and Porous Materials.* New York: Marcel Dekker.
39. Feldheim, D. L. and C. A. Foss. (Eds.). 2001. *Metal Nanoparticles: Synthesis, Characterization, and Applications.* New York: Marcel Dekker.
40. Rosoff, M. (Ed.). 2001. *Nano-Surface Chemistry.* New York - Basel: Marcel Dekker.
41. Starov, V. M. 2009. *Nano-Science: Colloidal and Interfacial Aspects.* Boca Raton, FL: CRC Press.
42. Chatterjee, A. 2010. *Structure Property Correlations for Nanoporous Materials.* CRC Press/Taylor & Francis Group.
43. Shchukin, E. D. and E. A. Amelina. 1992. Bridging of crystals in process of hydration hardening of gypsum. In *Hydration and Setting of Cements*, A. Nonat and J. C. Mutin (Eds.), pp. 219–234. London, U.K.: E & FN Spon.
44. Shchukin, E. D., Rybakova, L. M., Kuksenova, L. I., and E. A. Amelina. 1997. X-ray determination of residual internal stresses in cement. *Kolloidnyi Zh.* 59: 96–101.
45. Amelina, E. A., Kuksenova, L. I., Parfenova, A. M. et al. 1997. The mechanism of action of organic additives on the properties of cement curing structures. *Kolloidnyi Zh.* 59: 102–108, 729–735.
46. Rybakova, L. M., Amelina, E. A., Kuksenova, L. I., and E. D. Shchukin. 1999. Investigation of residual internal stresses of the I and II modes in cement-hardening structures. *Colloids Surf.* 160: 163–170.
47. Djordjevic, B. B., Rouch, L. L., Shchukin, E. D., Amelina, E. A., and I. V. Videnskiy. 2001. Nondestructive characterization of the hydration process in tricalcium silicate. In *Nondestructive Characterization of Materials*, R. Green (Ed.), pp. 279–285. Amsterdam, the Netherlands: Elsevier.
48. Shchukin, E. D. 2006. The influence of surface-active media on the mechanical properties of materials. *Adv. Colloid Interface Sci.* 123–126: 33–47.
49. Rehbinder, P. A. and N. V. Mihaylov. 1979. Physical- chemical mechanics—The scientific basis of optimal technology of concrete and reinforced concrete. In *Selected Works. Surface Phenomena in Disperse Systems: Physical-Chemical Mechanics*, E.D. Shchukin (Ed.), pp. 324–335. Moscow, Russia: Nauka.
50. Uriev, N. B. and A. A. Potanin. 1992. *Fluidity of Suspensions and Powders.* Moscow, Russia: Himiya.
51. Matijević, E. and R. S. Sapieszko. 2000. In *Surfactant Science Series*, T. Sugimoto (Ed.), Vol. 92. New York: Marcel Dekker.
52. Shchukin, E. D. 1992. *Advances in Colloid Chemistry and Physical-Chemical Mechanics.* Moscow, Russia: Nauka.
53. Huenger, K. J. and O. Henning. 1998. On the crystallization of gypsum from supersaturated solutions. *Cryst. Res. Technol.* 23: 1135–1143.
54. Chernov, A. A. and B. Y. Lyubov. 1965. Problems of the theory of crystal growth. In *Crystal Growth*, B. Veinshtein (Ed.), pp. 5–33. Moscow, Russia: Nauka.
55. Kontorovich, S. I., Lankin, Ya. I., Sporish, N. I., Amelina, E. A., and E. D. Shchukin. 1981. Investigation of the process of formation of phase contact in alumosilic sols. *Kolloidnyi Zh.* 43: 1076–1080.
56. Kontorovich, S. I., Lankin, Ya. I., Sokolova, L. N., and V. V. Matveev. 1984. On agglutination of polysilicic acid globules with different solid particles. *Kolloidnyi Zh.* 46: 352–355.

7 Interfacial Phenomena in Processes of Deformation and Failure of Solids

In this chapter we will address the role that physical-chemical interactions with the media play in the processes of the deformation and failure of solid materials. Among such interactions are various processes facilitating plastic flow and failure in solids caused by the reversible physical-chemical action of the media related to a decrease in the surface free energy of solids. The latter in turn lowers the work needed to form new surfaces in the course of deformation and fracturing. These phenomena were originally discovered by Rehbinder and are generally known as the *Rehbinder effect*. The peculiarities of the media-induced lowering in the strength of materials are associated with the fact that they manifest themselves under conditions when there is a combined action present: that of the active media and that of particular mechanical stresses. A reduction in the surface free energy does not by itself result in the growth of a new surface but makes it easier for that to occur under the influence of external forces. In Chapter 4, we argued that spontaneous dispersion is possible under conditions consisting of a very strong decline in the interfacial tension, down to the critical value $\sigma \sim \sigma_{cr} = \beta'kT/d^2$, where d is the particle size.

In the discussion of the reversibility of the Rehbinder effect, it was implied that there is a thermodynamically stable interface present between the mutually saturated solid phase and liquid medium and that the effect vanishes when the liquid medium is removed, for example, by evaporation. These two peculiarities make the Rehbinder effect principally different from the corrosion caused by the action of aggressive media. At the same time, one must realize that complete segregation is not possible: various processes can cover a fairly broad spectrum: from idealized cases involving purely mechanical failure to purely corrosive processes (or dissolution). The Rehbinder effect, which involves the adsorption-induced lowering of strength, stress-facilitated corrosion, and corrosive fatigue, often occupies intermediate positions in these series. In this type of phenomenon, the action of external forces and the action of chemically active media both contribute to the net result in certain proportions.

Depending on the nature of the solid and the medium and the conditions under which the interaction between the media and the solid takes place, the effects of the reversible physical-chemical action of the active medium can manifest themselves differently. Namely, they may be revealed as strength decrease, embrittlement, or the facilitation of plastic deformation (adsorption-induced plasticizing). The principal factors that influence the form and intensity of adsorption-induced processes may be subdivided into three main groups:

1. The chemical nature of the media and the solid, which defines the nature of intermolecular forces
2. The real structure of the solid, defined by the number and types of defects, including the size of the grains and the presence of microcracks
3. The conditions under which the deformation and breakdown of the solid take place, including the type, intensity, and temperature of the stressed state, the amount and phase state of the medium, and the duration of the contact between the medium and the solid

The role of all these factors is discussed in the first section of this chapter (Section 7.1), in which the principles, laws, and mechanisms of the surface phenomena that take place in solids in the course of deformation and destruction will be discussed. In Section 7.2, the focus will be on *the critical role of the surface*: its stability, defectiveness, and vulnerability to the initiation of degradation processes due to the influence of the active medium. The same section also describes a priority study on the defectiveness of the glass surface. In the final section of this chapter (Section 7.3), we will discuss problems associated with the adsorption-induced action of the active medium on the materials, as well as some useful applications of these effects. There we will also discuss in a greater detail studies on increasing the strength of catalysts and adsorbents.

7.1 INFLUENCE OF AN ACTIVE MEDIUM ON THE MECHANICAL PROPERTIES OF SOLIDS: THE REHBINDER EFFECT

This section includes two large essays on the Rehbinder effect, which involves a reversible physical-chemical (adsorption) action on the part of the medium on the mechanical behavior of solids, with an emphasis on the two main aspects associated with the effect. The first is the universal nature and selectiveness of the Rehbinder effect as a function of the chemical nature of the solid phase and the dispersion medium. The second aspect is related to the dependence of the extent and specific type of the effect on the experimental conditions and the real (defect) structure of the subject material. These two essays contain a summary of the studies conducted at the Russian Institute of Physical Chemistry and Moscow State University over the years.

Surface phenomena in solids are vividly revealed in the course of their deformation and fracture as they take place in surface-active media. This has been shown in numerous studies published since 1928—the year in which Rehbinder discovered the effect [1–9]. This essay mainly focuses on the reversible action of the medium, which is manifested in the partial compensation of the intermolecular forces at the newly formed surfaces. Such an action on the part of the medium may be referred to as "adsorption action," but one has to keep in mind that in addition to the adsorption layer, the formation of a layer of the *liquid* phase can also take place. The formation of such a layer would enhance the changes in the mechanical properties corresponding to very low values of interfacial free energy.

The combined influence of physical-chemical and mechanical factors on the processes of deformation and fracture is the subject of physical-chemical mechanics, which constitutes the foundation of the physical-chemical theory of the strength of solids. One can identify two distinctive areas of engineering that are closely associated with these processes:

1. The facilitation and improvement of the mechanical treatment of materials by pressure, cutting, or grinding, leading to improved productivity and quality of the formed surface, as well as controlling the processes of friction and wear using lubricants
2. The production of solid materials with a fine-disperse structure having an optimum balance of mechanical properties with other essential performance characteristics, such as enhanced durability under the particular conditions of use

The generation of new surfaces in the course of plastic flow and degradation constitutes the principal act taking place in the processes of the deformation, mechanical processing, and dispersion of solid materials. At the same time, all durable and strong materials must be fine disperse (fine grain or fine fibrous) with a uniform dense packing of the particles. Such materials include alloys, construction materials, materials made of filled polymers, finely crystallized glasses, and various composite materials. In all such materials, the internal interfaces are very well developed and the adhesive interactions at the interfaces play a critical role. Consequently, by controlling the interfacial phenomena in solid materials and disperse systems with the right combination of physical-chemical and mechanical actions, it is possible to achieve both of the aforesaid goals.

Over the years, numerous studies on the influence of surface-active media on the mechanical properties of solids were conducted and reported in the literature [10–30]. To acknowledge the role of significant contributors to the area, in addition to the group working at Moscow State University directed by Shchukin, we need to name Kishkin's and Potak's groups in Moscow, Karpenko's group in Ukraine, and Rostocker, Westbrook, and Westwood's groups in the United States. Some of the early traditional studies on the influence of media on materials were devoted to irreversible corrosion processes. More recently, studies on the behavior of materials in space have gained significant interest [24,28]. However, we believe that the adsorption-induced influence of media on the mechanical properties and durability of various materials is not covered in the literature to a significant extent. Here, we would like to partially fill this gap by addressing the principal ideas and directions, with some characteristic examples from published experimental works as an illustration.

7.1.1 Universal Nature and Selectivity of the Influence of the Medium on the Strength and Ductility of Solids with Different Types of Interatomic Interactions

The subject of primary interest here is the role played by foreign atoms, molecules, or ions in the processes of material fracture. These foreign atoms of the liquid medium penetrate into the pre-destruction zone of the solid, where they participate in interactions at the time when the chemical bonds are either broken or rearranged. The foreign atoms and molecules influence the interatomic interaction and participate in the compensation of newly formed bonds. A direct description of this process using the interaction potential of components is of a particular interest here.

The difficulty in estimating the strength, P_c, of a real solid, especially in the presence of an active medium, is related to the fact that structurally sensitive variable P_c is not a thermodynamic characteristic. At the same time, the specific surface free energy of the solid phase, lowered at the interface with a liquid phase, is a thermodynamic, structurally insensitive parameter that can be estimated for a given chemical composition of the phases. The specific surface free energy can either be estimated using the parameters of the interatomic interactions or be experimentally determined by direct measurements.

The theoretical strength of an ideal crystal in a vacuum, P_{id}, is a thermodynamic characteristic similar to the elasticity modulus, E, alongside with the σ/b ratio (b is a linear parameter comparable to the atomic dimensions) or the Q/V ratio (Q is the energy of sublimation, and V is the molar volume). All of these quantities bear the same physical meaning, namely, the limiting value of the energy density, and are thus of the same order of magnitude. However, the situation is different in the case when the mobile foreign atoms compensate the exposed bonds upon entering the fracture zone from the surrounding medium (or due to the adsorption from a solution on a given solid). In this case, the variables E and Q/V, both of which are volume-based characteristics, remain essentially the same, while the strength and surface free energy, σ, may change significantly. This indicates that these characteristics are related to the chemical bonds that are exposed on the forming surface, and not to the entire population of chemical bonds. At the same time, the use of σ does not interfere with the need to take into account the kinetics (i.e., nonequilibrium nature) of the bond-breaking process upon the penetration of foreign atoms [17].

Let us assume at this point that the interfacial free energy, σ, is the principal parameter responsible for the characterization of the interactions between the solid and the medium determined by their chemical composition. The following simplification allows one to obtain the relationship between the strength and the surface free energy of a solid body with a defect in the form of a microcrack.

Let us consider a solid body in the form of a plate of unit thickness to which an extending stress, p (N/m^2), is applied. In agreement with Hooke's law, the elastic deformation results in the accumulation of elastic energy with a density W_{el}:

$$W_{el} = \frac{p^2}{2E}$$

where E is Young's modulus.

In Section 3.1, we examined the mechanical properties of disperse systems that were capable of undergoing viscoplastic flow. In these cases, we considered the stressed state of shear with its characteristic parameters G, η, and τ^*. The strength of such systems could be characterized by the yield point. When we shift to describing the mechanical behavior of compact and primarily elastic–brittle solids, it is worth using the stressed state of a uniaxial extension in which we replace the shear stress, τ, with the extension stress, p; the shear modulus G with Young's modulus, E; and the resistance to tear, P_c, with the yield stress, τ^*, as the strength characteristic.

Let a crack (or incision) of length l form in the body. This results in a drop in the elastic deformation and in a decrease in the elastic energy density, W_{el}. Within a good degree of approximation, one may assume that this takes place within a narrow region of size l (Figure 7.1), so that a decrease of the stored elastic energy is proportional to the second power of the crack size. Namely,

$$\Delta F_{el} \sim \frac{-p^2 l^2}{2E}$$

At the same time, the crack opening is accompanied by an increase in the surface free energy due to the formation of a new surface with an area proportional to twice the length of the crack. Consequently, the dependence of the system free energy on the crack size can be expressed as

$$\Delta F \sim 2\sigma l - \frac{p^2 l^2}{2E}$$

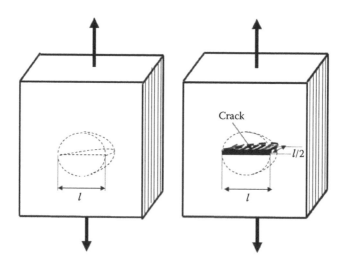

FIGURE 7.1   Schematic representation of how the critical crack size is calculated.

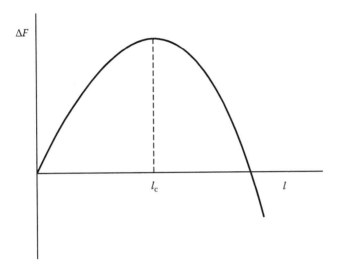

FIGURE 7.2   The free energy excess of the system as a function of the crack length.

According to Gibbs and Volmer, the free energy associated with the formation of a crack, ΔF, passes through a maximum. This maximum corresponds to the critical size of the crack (Figure 7.2):

$$l_{c} \sim \frac{\sigma E}{p^{2}}$$

Cracks of length greater than l_{c} are unstable and spontaneously enlarge, leading to the fracturing of the body. Cracks that are smaller than l_{c} should become smaller, that is, they should have a tendency to heal. However, in real solids, such crack healing may take place only in some exceptional cases due to the low rates of the diffusion processes, which are due to the adsorption of admixtures, and irreversible changes in the shape of the crack walls, which are due to plastic deformations. One known example of crack healing is the cleavage of mica in a high vacuum.

The earlier expression can be also written as

$$P_{c} \sim \left(\frac{\sigma E}{l}\right)^{1/2}$$

which was first obtained by Griffith from the calculations of the stressed state of the material in the vicinity of a crack of elliptical form [31]. According to the *Griffith equation*, the real strength, P_{c}, in a brittle–elastic solid body with a crack of size l is proportional to the square root of the surface energy and inversely proportional to the square root of the crack length. Taking into consideration the expression for the theoretical strength of an ideal solid, one can write that

$$P_{id} \sim \frac{\sigma}{b} \sim E \sim \left(\frac{\sigma E}{b}\right)^{1/2}$$

And then the Griffith equation can be rewritten as

$$\frac{P_{c}}{P_{id}} \sim \left(\frac{b}{l}\right)^{1/2}$$

This shows that the ratio between the ideal strength and the real strength of a solid is determined by the relationship between the size of the molecular dimensions, b, and the size of the defect, l.

The described scheme, illustrating the loss of crack stability due to the action of external extension stresses, is valid only in the case of an ideal brittle fracture of the solid. In more complex cases of fracturing involving plastic flow, the real work of the formation of new surfaces, the *effective surface free energy*, σ*, includes the work associated with plastic deformations, that is, the *energy of distortion, w,* in the vicinity of a crack:

$$\sigma^* = \sigma + w$$

The value of w may be higher than the true value of σ by several orders of magnitude. This is the so-called Orowan problem [32]. However, for a correct understanding of the aforementioned relationship, one needs to realize that the work w changes symbatically with the value of σ and consequently plummets with a significant drop in σ. The latter constitutes the *embrittlement* due to the action of a surface-active medium. The work w itself corresponds to the degree of distortion in the vicinity of the fracture surface that one needs to introduce in order to achieve a critical state. The Griffith equation can also be applied to the case of fractures of plastic bodies in which the microcrack nuclei are formed as a result of plastic strain.

The Griffith equation can also be used to compare the lowering of the surface free energy, Δσ, and the lowering of the strength, ΔP, of solids of different natures exposed to the action of an adsorption-active medium. Rehbinder and Shchukin pointed out that the largest decrease in the strength of a solid takes place when it enters into contact with a similar liquid medium that is similar to the deforming body with respect to the type of its existing interatomic interactions. Let us now turn to some typical examples illustrating the relationship between the lowering of the surface energy and the lowering of the strength in the presence of adsorption-active media [33–35].

7.1.1.1 Metals

In the case of solid metals, the liquid media with the most similar chemical nature are clearly melts of other, more fusible metals [1–4,10–19,27]. Numerous cases showing a drastic reduction in strength, embrittlement, and order-of-magnitude reduction of the durability of various metals and alloys on contact with liquid metals in the course of welding, soldering, overheating, and the melting of antifriction and anticorrosive coatings have been reported [1–4,10–19,27,33]. There are numerous specific examples. Various steels have displayed an appreciable decline in strength under the influence of solders and in the presence of cadmium, indium, and alkaline metals; the strength of copper was substantially reduced by the action of liquid bismuth, and a decrease in the strength of zinc and brass was caused by contact with mercury; and the strength of aluminum dropped sharply in the presence of gallium. We will describe the latter case in a greater detail. An experiment of the embrittlement of aluminum in the presence of a small droplet of liquid gallium can be used as an effective laboratory demonstration of this effect. In many observations, the decrease in strength was twofold or threefold, but in certain cases the reduction in the stress was 10 times or more.

If one is comparing similar types of fractures (i.e., brittle), the decrease in the strength by several times corresponds to a lowering of the surface free energy by about an order of magnitude. That is, if the starting value of σ for metals is about 10^3 erg/cm^2, contact with a melt will cause a decrease in σ to ~100 erg/cm^2. The effects of the strength lowering are especially drastic in cases when the metal in its initial state is highly plastic, where contact with a melt causes it to become very brittle. In cases when the starting metal is of low plasticity (and strength), the effect of the strength lowering is on the order of several times the initial strength. This is the case when the material is already brittle at low temperatures.

One characteristic example is that of monocrystals of zinc in the presence of a thin (several microns) liquid film of tin, mercury, or gallium (see Figures 7.3 and 7.4 and also Figure 7.18 later in the chapter [4,12]).

Theoretical calculations and quantum-mechanical approximations of the free surface energy at the interface between a solid metal and a melt have indicated that the surface free energy at the interface between relevant pairs, such as Cd–Ga, Zn–Hg, Zn–Sn, and Cu–Bi, can be lowered by an order of magnitude [36]. These theoretical calculations are mainly focused on illustrating the trends

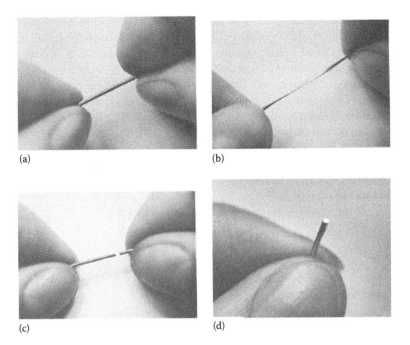

FIGURE 7.3 Stretching a single crystal of zinc (a and b) and stretching of a similar crystal covered with a film of mercury (c and d) (From Rehbinder, P.A. and Shchukin, E.D., Surface phenomena in solids during deformation and fracture processes, in: *Progress in Surface Science*, S.G. Davison (Ed.), Pergamon Press, Oxford, U.K., 1972, pp. 97–188; Shchukin, E.D. and Lichtman, V.I., *Doklady AN SSSR.*, 124, 307, 1959.)

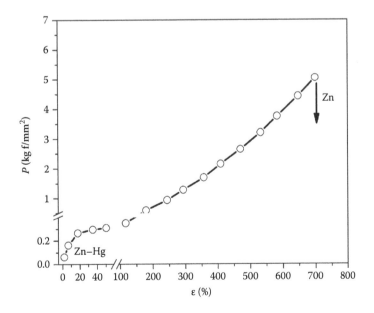

FIGURE 7.4 Stress, *P*; strain, ε; curves for a pure zinc single crystal and a zinc single crystal covered with mercury film. (From Rehbinder, P.A. and Shchukin, E.D., Surface phenomena in solids during deformation and fracture processes, in: *Progress in Surface Science*, S.G. Davison (Ed.), Pergamon Press, Oxford, U.K., 1972, pp. 97–188; Shchukin, E.D. and Lichtman, V.I., *Doklady AN SSSR.*, 124, 307, 1959.)

and providing crude estimates rather than yielding rigorous results. In this respect, one can also utilize the local-coordination (*quasi-chemical*) approximation. Although for metals this approach will only give an estimate good to the precision of an order of magnitude, it has the advantage of being universal, that is, applicable to a large number of systems.

The local-coordination approximation allows one to establish a simple relationship between the position of the liquidus, $x(T)$, and the solidus, $y(T)$, on the A–B phase diagram and the parameters describing the interatomic interactions between the components. The latter include the interaction energies between similar and dissimilar atoms, U_{AA}, U_{BB}, and U_{AB}, as well as the corresponding fusion temperatures and entropies of the components, T_A, T_B, q_A, and q_B. From this point on, it will usually be assumed that the melting point of component B is higher than the melting point of component A. The designations x and y are the mole fractions of the higher melting point component B in liquid and solid coexisting phases (Figure 7.5).

In this description, the *energy of mixing* plays a critical role, that is,

$$U_0 = z \left[U_{AB} - \frac{1}{2}(U_{AA} + U_{BB}) \right]$$

where
    z is the coordination number
    U_{AA}, U_{BB}, and U_{AB} the values that are related to the separate bonds
    U_0 is related to a single atom

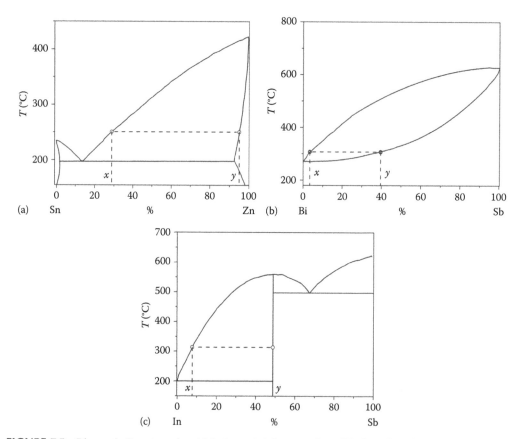

FIGURE 7.5 Binary A–B systems in which the embrittlement of a solid phase (y) takes place as the result of the action of a liquid phase (x): (a) system with a simple eutectic (Zn–Sn), (b) system with infinite mutual solubility of components (Sb–Bi), (c) and system with an intermetallic compound present (Sb–In). (From Yushenko, V.S. et al., *Fiz.-Him. Meh. Mater.*, 7, 85, 1971.)

The energy of mixing is essentially the only parameter that characterizes the nonideality of the system, that is, the difference in the interactions between similar and dissimilar atoms. Using the indices ' and " to designate the liquid and solid phases, respectively, one can write the following equations for the solidus and the liquidus:

$$U_0'(1-x)^2 - U_0''(1-y)^2 = -kT \ln\left(\frac{x}{y}\right) + q(T-T_B)$$

$$U_0' x^2 - U_0'' y^2 = -kT \ln\left[\frac{(1-x)}{(1-y)}\right] + q(T-T_A)$$

(7.1)

In the simplest case, it is assumed that both phases have the same structure, and hence $q_A = q_B = q$. This approach allows one to obtain approximate estimates for both the surface energy of the individual components and the interfacial free energy, σ:

$$\sigma = -1/2n'z_b' [U_{AA}'(1-x_1) + U_{BB}' x_1 + (2/z')U_0' x_1(1-x_1)]$$

$$-1/2n''z_b''[U_{AA}''(1-y_1) + U_{BB}'' y_1 + (2/z'')U_0'' y_1(1-y_1)]$$

$$+n^* z_b^* \{U_{AA}^*[1-1/2(x_1+y_1)] + U_{BB}^* 1/2(x_1+y_1) + (1/z^*)U_0^*[x_1(1-y_1) + y_1(1-x_1)]\}$$

$$+n' \sum_{i=1}^{\infty} \left\{ kT[x_i \ln(x_i/x) + (1-x_i)\ln(1-x_i)/(1-x)] - U_0'(x_i-x)^2 + (z_b'/z')U_0'(x_{i+1}-x_i)^2 \right\}$$

$$+n'' \sum_{i=1}^{\infty} \left\{ kT[y_i \ln(y_i/x) + (1-y_i)\ln(1-y_i)/(1-y)] - U_0''(y_i-y)^2 + (z_b''/z'')U_0''(y_{i+1}-y_i)^2 \right\}, \quad (7.2)$$

where

   n is the number of atoms per unit area in the surface layer parallel to the interface
   z_b is the number of neighbors of a given atom that are located in the adjacent layer
   x_i and y_i are the concentrations of the component B in the liquid and solid phases in the ith layer
      counting from the interface
   x and y are the mean values of the bulk concentration far away from the interface

The values relating to the interface itself are denoted by the * symbol. The values of x_i and y_i take into account the adsorption of components and the fuzziness of the interface.

In a very simple approximation, one can assume that an abrupt concentration change takes place across the interface, that is, $x_i = x$ and $y_i = y$. Furthermore, let us also assume that $U_{AA}^* = U_{AA}'$, $U_{BB}^* = U_{BB}'$, $U_0^* = U_0'$, and $n'z_b' = n_b'' = n^* z_b$. That is, there are no broken interatomic bonds left at the interface. With these assumptions in place, the expression for σ can be reduced to a form suitable for the estimation of the value of the surface free energy at the interface between the metal and the melt:

$$\sigma \approx \sigma_A \left(\frac{Q_{mA}}{Q_{vA}}\right)(1-y) + \sigma_B \left(\frac{Q_{mB}}{Q_{vB}}\right) y + (n''z_b''/z'')U_0''(y-x)^2$$

(7.3)

where

   Q_{mA} and Q_{mB} are the energy changes due to melting
   Q_{vA} and Q_{vB} are the energy changes due to evaporation
   σ_A and σ_B are the surface free energies of the pure components in a liquid state

The quadratic terms containing the energy of mixing as a prefactor characterize the deviation of the system from ideality. One can identify the following three characteristic cases:

1. The energy of mixing the components in both phases and the mean value of U_0^* at the interface are positive and high for a given temperature range, that is, $U_0', U_0'', U_0'' \gg kT$. In this case, the analysis shows that the value of σ is high and approaches in the limit the sum of the free surface energies of the corresponding phases. This result indicates that in the absence of an appreciable interaction between the components (consequently, their mutual solubility in the coexisting phases is very low), there is no significant lowering of the interfacial free energy upon the contact of a solid phase with a liquid. In such a system, there are no means that would allow for the *compensation* of the interatomic bonds that become unsaturated upon fracturing, and hence there should be no strength lowering. The phase diagram in this case shows the separation of the system into two immiscible liquid phases.

2. The energies of mixing are high and negative (strong negative deviation from ideality). This is the case corresponding to a strong chemical interaction and the formation of intermetallic compounds (or other compounds, if one does not restrict oneself to binary systems). Both Equations 7.2 and 7.3 yield negative values for the interfacial free energy, σ, which indicates that a two-phase system with an interface is not an equilibrium system. Chemical interactions may also include corrosion and dissolution. Such *irreversible* interactions with the liquid medium are beyond the scope of this book, but there are several peculiarities associated with these systems that we find noteworthy. First, the degradation (fracture) of the solid due to the influence of the liquid medium does not require the presence of mechanical stresses. To cause an appreciable change in the dimensions (cross section) of the solid object, a considerable amount of the liquid medium is necessary. Since this type of interaction process is volume based, it is characterized by a rather low rate of proliferation of the chemical reaction between the solid and the medium into the bulk of a solid phase. Second, the degrading action of the medium is amplified with an increase in temperature.

One may say that these trends are opposite to those that would be expected in the case of *reversible* strength lowering phenomena. However, one must use a certain degree of caution in making such a comparison, because it is related only to the idealized *limiting* cases. In real systems there is no such clear boundary.

A broad range of phenomena that involve a change in the mechanical properties of materials due to the action of the medium require mechanical activation for them to proceed with a noticeable rate. They include many chemical and electrochemical processes, for example, corrosion taking place under an applied voltage, corrosion-related fatigue, and corrosion-induced fracturing. It is noteworthy that in all of these cases, the act of adsorption takes place at the onset of the chemical interaction. There are also typical *adsorption-driven* phenomena in which the entire interaction with the media is localized at the interface and does not involve the bulk. This includes the process of the facilitated deformation of a fracture in metal in the presence of fatty acids (e.g., a solution of oleic acid in nonpolar oil). This process involves the act of the chemisorption of the oleic acid, which causes a substantial lowering of the interfacial free energy (as the work of the new surface formation), and consequently significant mechanical effects, in such an *irreversible* process.

In the case when an intermetallic compound is formed in a binary A–B system, the interactions between the intermetallic compound and the melt A are very important. A priori, various scenarios, ranging from drastic embrittlement to no influence at all, are possible here. Of particular interest are the diverse phenomena taking place when a third component C capable of active interaction with B is introduced into the liquid phase A. Depending on the specific conditions, a significant decrease in the strength as well as an attenuation in that effect can be observed. An example of the first case is the weakening of steel by liquid alkali metals in the presence of oxygen. The second situation offers one an opportunity to seek ways to prevent strength loss and to extend the durability of material. A good example of this is the Cu–Bi pair studies extensively carried out by Yushchenko et al. [36]. It has been shown that the introduction of tin into the liquid phase inhibits the strength reduction caused in copper by bismuth to a significant extent. The effect is visible at quite low concentrations of tin, which can form intermediates due to a chemical reaction with the copper. In contrast to this case, the introduction of lead, which is not capable of forming intermediates, starts to play a similar role only when present in melts at high concentrations.

3. The absolute values of the energies of mixing of components A and B are not high. This is the case when the liquid phase (enriched in A) is chemically and structurally similar to the solid phase. The principle difference of this case from the previous one is that here one is dealing with a thermodynamically stable two-phase system with a finite value of the interfacial free energy at the interface between the two phases. The interfacial free energy may be very low but needs to be positive. The most common example of such a system is the contact of a metal with its own melt. Calculations show that the interfacial free energy may experience a significant drop, of an order of magnitude or more, in comparison with the surface free energy of the individual condensed phases [37]. Such cases are of special interest for the analysis of nucleation during crystallization from melts. Such phenomena can be the reason for the appearance of visible and hidden cracks in the course of the welding of metal parts subjected to stress. Another common example is the embrittlement of ice that can take place upon the contact of ice with water. It was shown in a study by Naydich et al. that a significant decrease in the surface free energy and of the hardness of an organic crystal placed in contact with its own melt took place in the phenyl salicylate (salol) model system [38].

In the case when contact is established between a solid phase and a "foreign" melt, the deviation from nonideality is the least (the smallest energies of mixing) in systems with infinite mutual solubility in both the liquid and the solid phases. Such systems are characterized by cigar-shaped phase diagrams; see Figure 7.5b. A particular example of such a system characterized by the close natures of the solid and the melt is the Sb–Bi pair. This system reveals a substantial strength lowering on contact with the corresponding equilibrium liquid phase. In general, such systems are of limited interest, because the range of temperatures and compositions relevant to the strength decay are very narrow and are close to the melting point of the more refractory component. One interesting feature of these systems is that despite a significant reduction in σ, in the case when a brittle-to-plastic transition has already taken place at lower temperatures, the effects of strength lowering may not appear.

Of particular interest are binary A–B systems with a simple eutectic. In these systems, the solubility of the components in the solid phase is finite and rather low. These are the systems in which strong embrittlement of the hard metal B takes place upon contact with either melt A or with a melt having eutectic composition [39]. The energy of mixing of the components in a liquid phase, U'_0, is very close to the value of kT_e, where T_e is the eutectic temperature. The respective values for the solid phase, U''_0, are typically higher. The approximate estimates of the surface free energy obtained with Equations 7.2 and 7.3 show that a significant decrease in the surface free energy takes place at the interface. These data correlate well with the results of the experimental strength decrease measurements. Table 7.1 summarizes the values of the energies of mixing in the solid and liquid phases, U'_0, and U''_0, calculated from Equation 7.1, and the values of the interfacial free energy, σ, estimated from Equations 7.2 and 7.3, for some systems having a eutectic. It is implied that the concentration of component B is higher in the solid phase. The relationship between the strength lowering and the lowering of the interfacial free energy is the basis for selecting proper adsorption-active media that facilitate the mechanical treatment of hard materials. In agreement with this rule, a significant strength decrease is observed in Zn–Hg, Zn–Ga, Cd–Ga, Al–Ga, Cu–Bi, Fe–Zn, Ti–Cd, and Ge–Au systems.

TABLE 7.1

Comparison of the Energy of Mixing and the Interfacial Free Energy Values for a Number of Solid (B) and Liquid (A) Metal Pairs

B–A	U_0' 10^{-13} erg/atom	U_0''	σ erg/cm^2	B–A	U_0' 10^{-13} erg/atom	U_0''	σ erg/cm^2
Ag–Bi	1.59	1.87	130	Cu–Ag	2.08	4.56	180
Ag–Cu	1.68	2.90	90	Cu–Bi	2.70	N/A	220
Ag–Pb	1.46	3.89	190	Cu–Pb	4.08	5.81	350
Al–Sn	1.90	5.73	260	Ge–Sb	0.68	4.67	280
Al–Zn	2.54	3.51	60	Ge–Ag	2.24	N/A	210
Be–Al	4.78	N/A	360	Ni–Cr	1.36	1.59	150
Be–Si	4.81	N/A	130	Pd–Cr	0.84	1.10	110
Co–Bi	−0.14	N/A	110	Sn–Zn	0.89	3.31	40
Cz–Ni	0.67	2.44	130	Zn–Sn	1.22	4.81	150
Cz–Pd	2.94	8.29	220	Zn–Ga	0.60	1.84	110

Note: It is assumed that the concentration of component B is higher in the solid phase.

Systems with a simple eutectic diagram with *moderate* positive deviations from ideality reveal *symmetry*. The liquid phase, which causes a strong reduction in the interfacial free energy, can be in equilibrium with both solid phases: the one with a higher melting point (component B) and the other one with a lower melting point (component A). A relevant example here is the data on the ductility and strength lowering of tin in contact with liquid solutions of tin–zinc shown in Figure 7.6 [18].

One needs to take this into the account in the analysis of the reasons for the reduction in the high temperature strength of alloys that contain refractory components as admixtures. The decrease in the strength of both solid metals forming the eutectic can be clearly observed when they are brought into contact around the eutectic point, T_e (the so-called contact fusion).

The vapors formed by component A and by the compound A_nB_m can also belong to systems with eutectic. One example is the 40% strength lowering of a single crystal of indium antimonide upon contact with a drop of liquid indium.

There are not too many studies documenting direct measurements of the interfacial free energy at metal–melt interfaces. Bryuhanova et al. conducted measurements of the surface free energy of zinc and cadmium in the presence of small additives, such as gallium or tin [40,41]. The Tamman–Odin zero creep method (Chapter 1) was employed in these studies. The interfacial tension at the interface between cadmium and a film of gallium was ~150–200 erg/cm^2, which is several times lower than the surface free energy of cadmium. Westwood et al. [19] used the Obreimow–Gilman method to evaluate the surface energy of zinc in the presence of mercury. One can also find other experimental data obtained by indirect methods. Examples include measurements of the dihedral angle etched by a drop of a melt at the grain boundary that crops out at the surface (e.g., in the copper–bismuth pair) and studies on the dependence of the brittle-to-plastic transition temperature as a function of the grain size. In all these cases, a significant decrease in σ at the interface between a solid metal and a melt occurred.

Up to this point, we have been talking about the lowering of the interfacial energy, σ_{sl}, of a solid metal at the interface with a liquid phase. In the case when the amount of fusible component does not exceed its solubility in the solid phase, there is no film of liquid phase present at the surface. Under these conditions (thermodynamic equilibrium), only the adsorption layer may form at the solid surface. Such conditions favor an adsorption-induced strength lowering in the narrowest sense of this term. It was shown by Frumkin that for the case when there is a finite contact angle, there is

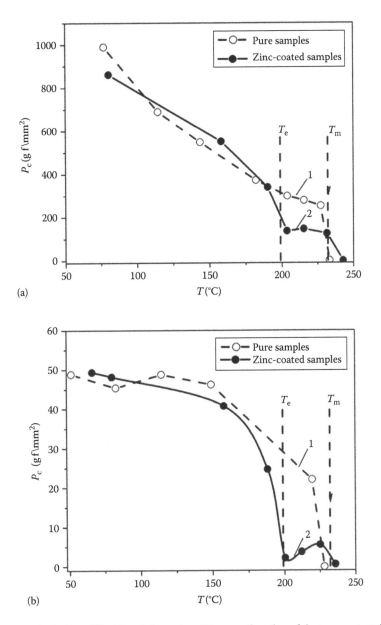

FIGURE 7.6 Strength, P, (a) and limiting deformation, (b) ε, as a function of the temperature for the stretching of (1) polycrystalline samples of pure tin and (2) samples of tin covered with zinc film. Tin melting point, $T_m = 232°C$. (Data from Shchukin, E.D., *Fiz.-Him. Meh. Mater.*, 12, 3, 1976.)

no continuous transition between the two states [29]. As the concentration of a fusible component is increased, the adsorption layer first reaches a particular limiting concentration. Then, in equilibrium with this adsorption layer, droplets of a liquid phase start to appear on the solid surface, forming a finite contact angle. It is essential for this effect that we are operating here with binary A–B systems in which both components have σ values of similar magnitude.

Within the local-coordination approximation, the system of equations describing a layer-wise distribution of the adsorbing component in the vicinity of the solid phase was developed by Pines and was numerically solved by Yushchenko [36,58,61,66]. The numerical simulations showed that when $\sigma_{A-v} \approx \sigma_{B-v}$, the adsorption-induced lowering of σ_{B-v} due to the introduction of small additives of component A was related to the positive deviation of the system from ideality.

The lowering of surface energy, $\Delta\sigma$, was on the order of $\sim 10^2$ erg/cm^2, which constituted 10%–20% of the initial value of σ. This result agrees well with the zero creep experimental data on the adsorption-induced lowering of σ in the presence of small additives introduced. It is worth pointing out that in the case of the adsorption of zinc on the surface of solid tin, a modest lowering of σ of that sort was indeed observed; zinc is a more refractory component with a higher value of the surface energy.

Strong lowering in the value of σ upon physical adsorption in binary metallic systems can be expected in cases when there is a large difference between the values of σ of the individual components. However, it is worth pointing out that strong lowering of σ can be also achieved in the case of chemisorption on the metal surface, for example, in the case of the adsorption of diphilic molecules of organic acids. A strong lowering in the surface free energy in this case is related to the fact that the energy of the chemical bond between the polar group and the metal surface is high, while the surface itself constitutes a layer of hydrocarbon chains with low surface tension. In the case of solid nonmetals, one example of the strong adsorption-related lowering of the interfacial energy is that caused by the adsorption of water on the surface of ionic crystals.

7.1.1.2 Solids with Covalent Bonds

Similar to solid metals, crystals with covalent bonds may experience a strong strength decrease when in contact with metal melts. The strength of germanium single crystals decreases upon contact with metallic gallium or molten copper. The pair germanium–gold can serve as a good model example here. This system is characterized by a simple phase diagram with a low-melting-point eutectic. When heated in air, single crystals of germanium acquire a noticeable ductility and hence become stronger. The experiments conducted by Pertsov et al. [3,5,18,41,42] showed that the strength of germanium crystals decreased when they were placed in contact with a drop of molten gold saturated with germanium at a given temperature. As a result, the crystal returned to the brittle state and the stress needed to rupture the crystal was greatly reduced (Figure 7.7).

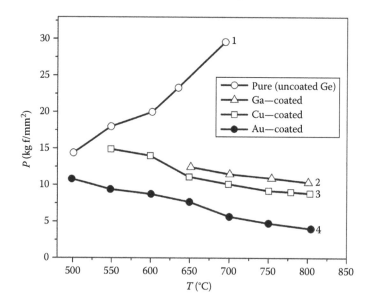

FIGURE 7.7 The strength of germanium single crystals as a function of temperature: (1) pure crystals and crystals coated with a thin film of (2) gallium, (3) copper, and (4) gold. (From Shchukin, E.D., *Kolloidnyi Zh.*, 25, 108, 1963; Shchukin, E.D., *Fiz.-Him. Meh. Mater.*, 12, 3, 1976; Skvortsova, Z.N. et al., Mechanics of fracture of cohesive boundaries with different concentrations of foreign inclusions, in: *The Successes of Colloid Chemistry and Physical-Chemical Mechanics*, E.D. Shchukin (Ed.), Nauka, Moscow, Russia, 1992, pp. 222–228.)

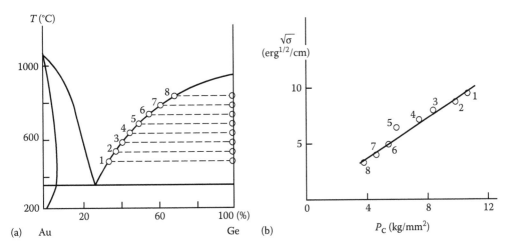

FIGURE 7.8 Phase diagram of Ge-Au system (a). A direct comparison between the experimentally determined values of the strength, P_c, and the estimated values of the interfacial free energy, σ, for germanium single crystals tested at various temperatures and in the presence of a drop of gold saturated with germanium (b). (From Shchukin, E.D., *Kolloidnyi Zh.*, 25, 108, 1963; Shchukin, E.D., *Fiz.-Him. Meh. Mater.*, 12, 3, 1976; Skvortsova, Z.N. et al., Mechanics of fracture of cohesive boundaries with different concentrations of foreign inclusions, in: *The Successes of Colloid Chemistry and Physical-Chemical Mechanics*, E.D. Shchukin (Ed.), Nauka, Moscow, Russia, 1992, pp. 222–228.)

For the series of experiments shown in Figure 7.7 the local-coordination approximation produced values of the energy at the interface between the solid and liquid phases [41]. A comparison of the results of the theoretical calculations with the results of the strength measurements produces a straight line passing through the origin in the $P_c - \sigma^{1/2}$ coordinate system (Figure 7.8). The latter confirms the validity of the Griffith equation and thus indicates that the fracture forming upon contact with a melt is of a brittle nature. Furthermore, one has confirmed one more time that it is useful to use σ as a single macroscopic parameter for the description of interatomic interactions.

We have mentioned earlier the case of the strength decrease in indium antimonide in the presence of liquid indium. These types of phenomena are of significance in the soldering of semiconductor crystals under conditions of applied mechanical stresses.

Based on the nature of the interatomic interactions, the effects of the medium on strength can be classified as belonging to several different types. We can illustrate this using the pool of data collected by Savenko et al. on the strength of graphite in contact with different liquid metals [43–46]. Namely, one can outline the following characteristic cases:

1. Tin, lead, gallium, and other metals that do not interact with carbon and do not wet graphite have no impact on the strength of graphite.
2. Iron, vanadium, molybdenum, tungsten, and other metals capable of forming carbides and dissolving carbon cause fractures even in the absence of stresses. The probability of fracturing increases with the decrease in cross-sectional area of the sample.
3. Zirconium and titanium react with carbon to form strong and stable carbides; the formation of carbide film on the surface of graphite takes place upon exposure of the latter to the molten zirconium and titanium. Such films reveal a protective action, and no strength lowering is observed.
4. There is the case of a *moderate interaction*, such as in contacts between liquid aluminum and carbon. In this case, the unstable carbides form in a limited temperature interval and cause a strong decrease in the strength of graphite in particular temperature intervals. Similar, but even stronger, effects are observed upon the contact of graphite with liquid alkali metals at elevated temperatures.

It is also possible that the lowering of the interfacial energy plays a determining role in the impact of iron on the mechanical properties of diamond. This is supported by the data on the eutectic melting of diamond in the presence of iron, by the data on the strength of diamond grains used in the mechanical treatment of steel, as well as by direct observations on the strength lowering of diamond grains in the presence of iron.

The family of solids that are impacted by the action of low-temperature metallic melts under conditions of intensive mechanical action and upon the exposure of the juvenile regions of the surface is substantially larger than one could possibly have imagined originally. The effectiveness of polishing hard materials, such as tungsten carbide and titanium carbide with cobalt binders using diamond grinding wheels, is increased in the presence of metallic melts. Under these conditions, both the strength of the carbide grains themselves and their adherence to the binder decrease and the wetting of the surface by the melt increases. Furthermore, under these conditions, the diamond polishing tool is not harmed by the action of the melt: fusible metals are not surface active toward the diamond.

According to studies by Bravinsky, fusible metals can significantly (~1.5 times) decrease the strength of vacuum-tight alumina ceramics. It is interesting that under normal conditions these metals do not wet alumina. Nevertheless, studies carried out with ceramic samples have revealed that the juvenile fracture surface is well wetted by the melt, which results in strength decrease.

7.1.1.3   Solids with Ionic Bonds

The media that are chemically similar to ionic crystals are salt melts and electrolyte solutions. Here, one assumes that the liquid medium is saturated with the component of the solid phase and that in the absence of mechanical stresses the system is a thermodynamically stable two-phase system with a distinct interface.

A characteristic example is that of the influence of aluminum chloride melt on the mechanical behavior of a sodium chloride single crystal. It is well known that under extension stress at room temperature, a NaCl single crystal easily undergoes brittle fracture along the cleavage plane. The same crystal becomes ductile at high temperatures. Prior to rupturing at around 400°C–500°C, the formation of a neck takes place and the strength increases. It has been shown in the studies by Pertsov et al. that in the presence of molten aluminum chloride at the same temperatures, sodium chloride crystals return to the brittle state with a corresponding decrease in strength [41–43] (Figure 7.9).

Throughout this book, we have described numerous observations on the lowering of the strength of ionic crystals in the presence of water and electrolyte solutions and upon introducing small additives of surface-active substances. The effect of adsorption-induced strength lowering was originally discovered in studies dealing with cleavage of calcite and rock salt. Of special interest are the studies carried out with mica. It was shown by Loginov that in the presence of water vapor or in liquid water, in addition to a decrease in the force needed for mica cleavage, there is also an elastic aftereffect that gradually increases with time (over several days of observation). The elastic aftereffect is of adsorption-induced origin and is completely reversible with respect to the deformation, which slowly but completely vanishes with the removal of the stress.

Similar phenomena take place in silicate glasses and in glass fiber. It is well known that glass cutting is facilitated by moisture: glass fractures more easily when water is applied to the scratch made with a diamond cutting tool. Water also influences the long-term stability of glass. Making the glass surface hydrophobic prevents the action of atmospheric humidity and hence benefits its durability.

Decreasing the strength of mineral rocks through the action of water and dissolved adsorbing substances (*hardness modifiers*) is essential in the facilitation of the drilling of oil wells in hard rocks and in increasing the durability of drilling tools and plays a significant role in hydraulic fracturing. The impact of the melts of various chlorides and silicates on rock strength, studied by Pertsov [47–50], allows one to understand the origin of some geological phenomena and lays the foundation for *physical-chemical geomechanics*.

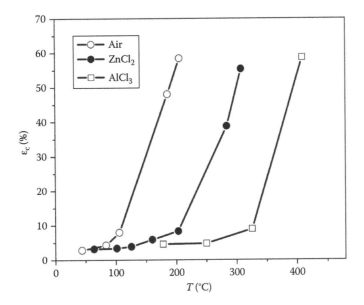

FIGURE 7.9 The limiting elongation of sodium chloride crystals as a function of temperature: in air and upon contact with zinc chloride and aluminum chloride. (From Skvortsova, Z.N. et al., Mechanics of fracture of cohesive boundaries with different concentrations of foreign inclusions, in: *The Successes of Colloid Chemistry and Physical-Chemical Mechanics*, E.D. Shchukin (Ed.), Nauka, Moscow, Russia, 1992, pp. 222–228; Traskin, V.Y. et al., *Doklady AN SSSR*, 191, 876, 1970; Traskin, V.Y. and Skvortsova, Z.N., Thermodynamic activity of water in electrolyte solutions and their impact on the strength of solids, in: *Surface Water Films in Dispersed Structures*, E.D. Shchukin (Ed.), Izd. MGU, Moscow, Russia, 1988, pp. 197–202.)

The importance of the chemical similarity between the medium and the solid becomes apparent when the nature of the medium is varied. Studies on the strength of polycrystalline samples of potassium chloride were carried out using a continuous series of mixtures with varying polarity—some of heptane with dioxane and others of dioxane with water [44,45]. As shown in Figure 7.10, the strength was the highest in nonpolar media, that is, in heptane. The trend in the strength closely follows the trend in the adsorption isotherms: it gradually decreases with an increase in medium polarity and reaches its lowest value in the most polar medium, that is, in water.

Similar trends were also observed in experiments with saturated alcohols. The data in Table 7.2 illustrate the effect that saturated alcohols have on the strength of polycrystalline sodium chloride (p, dipole moment of molecules; d [g/cm^3], density; M, molecular mass [g/mol] of a liquid).

Assuming that the brittle fracturing process follows the Griffith equation, that is, that $P_c/P_0 = (\sigma/\sigma_0)^{1/2}$, and combining it with the Gibbs equation in the low C limit,

$$\Gamma = -(RT)^{-1} \frac{d\sigma}{d\ln C}$$

we arrive at the expression relating the adsorption (mol/cm^2) to the strength

$$\Gamma = -2\left(\frac{\sigma_0}{P_0^2}\right)(RT)^{-1} P_c (dP_c/d\ln C)$$

This relationship allows one to estimate the maximum adsorption, Γ_{max}, from the dependence of the strength on the concentration of the active component. From the value of Γ_{max}, one can then estimate the limiting area per molecule, a_m, that is occupied by an adsorbing molecule on a newly formed surface in the deformed body, $a_m = 1/N_A\Gamma_{max}$. This calculation yields a reasonable estimate for the

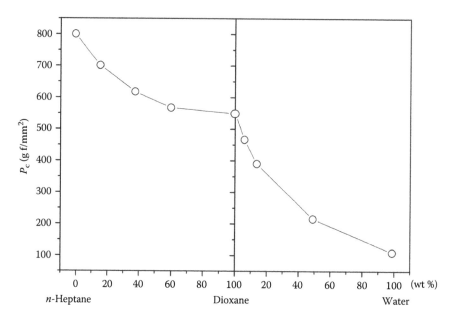

FIGURE 7.10   Strength of polycrystalline sodium chloride specimens as a function of the medium composition in *n*-heptane/dioxane and dioxane/water systems. (Redrawn from Kuchumova, V.M. et al., *Kolloidnyi Zh.*, 69, 357, 2007; Kuchumova, V.M. et al., *J. Mater Sci.*, 27, 5516, 1992.)

TABLE 7.2

Effect of Saturated Alcohols on the Strength of Polycrystalline Sodium Chloride

Alcohol	$(P_0 - P_A)/P_0$	$\dfrac{P_0 - P_A}{P_0}\dfrac{M}{pd}\times 10^{-18}$
Methanol	0.58	13.8
Ethanol	0.52	17.3
n-Propanol	0.25	12.0
n-Butanol	0.23	12.8
n-Hexanol	0.20	16.0

---

value of a_m, for instance, for the case of the adsorption of water from dioxane on the KCl surface. This, in turn, confirms the validity of explanations regarding the role of adsorption in the lowering of the strength upon the exposure of solids to a liquid medium.

The attempt to perform an approximate estimation of the influence of the polarity of the medium on the strength of ionic crystals at the atomic level has been made by Traskin [41–43]. The model system that was used in this study was the apex of a wedge-like crack into which a molecule with a dipole moment had penetrated. By analyzing the interaction of this molecule with the surrounding positive and negative ions, one can estimate the component of the force acting normally to the walls of the crack and determine the decrease in the strength. These numerical simulations yielded data that were in an agreement with experiments and confirmed the proportionality between the observed effect and the dipole moment (per unit volume).

7.1.1.4   Disperse Porous Materials

Disperse porous materials are of interest in the analysis of adsorption-induced strength lowering mechanisms, including porous bodies with ionic structure. Convenient model objects for studying porous bodies are the magnesium hydroxide crystallization structures obtained in the course of the hydration hardening of magnesium oxide. Such structures have high surface areas (tens of m²/g), which allows one to determine the value of the adsorption by the mass using a different method: the adsorption of a single molecular monolayer of water results in a mass increase of around 1%. A series of such samples were aged in desiccators over various sulfuric acid solutions, producing a given water vapor pressure, p. The $\Gamma = \Gamma(p)$ dependence can be determined by the change in mass, which in turn allows one to determine the extent of the surface energy lowering for magnesium hydroxide due to the adsorption of water vapor:

$$-\Delta\sigma = \sigma_0 - \sigma = RT \int_0^p \Gamma \, d\ln p$$

The $-\Delta\sigma$ values obtained can be compared with the results of independent measurements of the strength lowering in these moist samples, P_c. These values can be compared to the starting strength, P_0, determined in the absence of water vapor. Since such bodies undergo brittle fracturing, the Griffith relationship is valid and one can write that

$$\frac{P_0^2 - P_c^2}{P_0^2} = -\frac{\Delta\sigma}{\sigma_0}$$

Indeed, when plotted in the proper coordinate system, the experimental data fall on a straight line passing though the origin. For σ_0 one finds an estimate of ~300 erg/cm² (Figure 7.11). These results are of principal and priority importance for physical-chemical mechanics: they represent the first rigorous proof of the nature of the Rehbinder effect as the adsorption-induced strength lowering of solids.

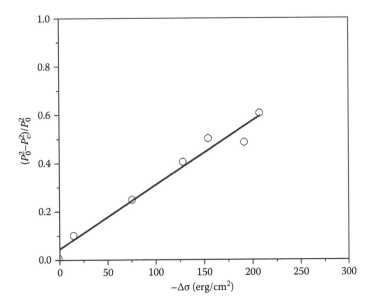

FIGURE 7.11 The correlation between independently measured values of the surface free energy lowering and the values of the strength decrease for magnesium hydroxide specimens in the presence of water vapor. (Data from Yushchenko, V.S. et al., *Zh. Fizicheskoy Himii,* 53, 1556, 1969; Shchukin, E.D. et al., *Doklady AN SSSR,* 175, 882, 1967; Shchukin, E.D. et al., *Doklady AN SSSR,* 182, 394, 1968; Kontorovich, S.I., Physical-chemical regularities of strengthening of the fine-disperse porous structures—Catalysts and adsorbents. Science thesis, Institute of Physics and Chemistry, Russian Academic Science, Moscow, Russia, 1990.)

Specialized experiments using nuclear magnetic resonance (NMR), which were conducted in collaboration with Chuvaev, indicated that the adsorbed water was indeed present in the adsorption layers and was not associated with the liquid phase. In these experiments, there was no localized dissolution of the stressed material taking place in the apex of the crack (i.e., in contacts between the crystals) [51].

The adsorption of a water monolayer results in a twofold decrease in the strength of a fine-disperse structure of magnesium hydroxide, which in turn corresponds to approximately a fourfold decrease in the surface energy. The surface energy decrease in this case is caused by adsorption, because the adsorbing component has a low surface tension in comparison with the surface energy, σ, of the adsorbent.

Similar results were also obtained in other systems, for example, in the case of the adsorption of ethanol on magnesium hydroxide. The adsorption of water is the reason for the reduced strength of all hydrophilic construction materials, such as cements and gypsum dehydrate.

The adsorption-induced strength lowering is even more pronounced in disperse structures with point-like *atomic* contacts that are formed upon the removal of a liquid phase from the adsorption layers present in the coagulation structures (see Chapters 2 and 3). This is the main cause for the poor water resistance in certain grounds, the formation of the so-called quick grounds.

The adsorption of water vapor, ethanol, and benzene causes a decrease in strength of porous adsorbents and zeolites granulated with clay-based binders. However, it was demonstrated that these porous disperse materials act as catalysts under *unfavorable* conditions. This is the case because the principle of heterogeneous catalysis requires a decrease in the surface energy of a catalyst during the catalyzed reaction. An unfortunate result is the decrease in the durability of the catalyst, as will be discussed in more detail in Section 7.3.

7.1.1.5 Organic Molecular Crystals

Molecular crystals of organic compounds, in particular those of hydrocarbons, are in some sense the opposites of ionic crystals. For materials based on molecular crystals, the media that are the most similar in structure to the crystals are polar and nonpolar organic liquids. The studies conducted by Pertsov et al. [17,52–54] involved anthracene, naphthalene, urotropin, urea, and many other organic solids. Some characteristic studies were conducted with single crystals of naphthalene. Exactly the same patterns as those observed with pure and amalgam-modified zinc crystals were reproduced with this system. Under normal conditions naphthalene single crystals are very ductile. If the initial crystal orientation is favorable, naphthalene crystals can be stretched into a thin ribbon that ruptures only after very high elongations, on the order of 1000%. A completely different picture is observed upon the exposure of naphthalene crystals to a drop of benzene or heptane saturated with naphthalene. After some ageing, the naphthalene crystals become brittle and undergo easy rupture along the cleavage plane under a rather small limiting strain (Figure 7.12).

Direct measurements of the surface energy of naphthalene were based on Obreimow's crystal cleavage method. In these measurements, the crystals were cleaved along the cleavage plain and the stress necessary to cause the split was measured. The main difficulty associated with the use of this method on molecular crystals is their softness. This issue was overcome by using a special way to mount the crystal samples, as illustrated in Figure 7.13. The sample was a thin plate (1) with a thickness of 1.6–2 mm, one side of which was glued to a much stiffer metallic plate (2), which served as an elastic element, and with the other side attached to a thick glass tile (3). The interference pattern visible through the tile allowed one to register the apex position of the crack. The surface free energy determined in this way is given by

$$\sigma = \frac{3F^2c^2}{Eabh^3}$$

where
   F is the maximum cleaving force, at which the crack undergoes a rapid growth
   E is the elastic modulus of the metallic plate (2)
   The meaning of dimensions a, b, c, and h is evident from Figure 7.13

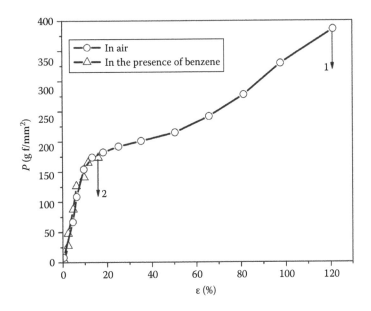

FIGURE 7.12 Stress–strain curves of naphthalene single crystals (1) in air and (2) in the presence of benzene. (From Pertsov, N.V. and Rehbinder, P.A., *Doklady AN SSSR*, 123, 1068, 1958; Skvortsov, A.G. et al., *Doklady AN SSSR*, 193, 76, 1970; Pertsov, N.V. et al., *Doklady AN SSSR*, 179, 633, 1968.)

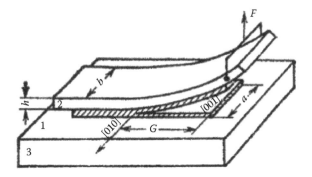

FIGURE 7.13 The determination of the surface energy of naphthalene crystals using Obreimow's method. (From Pertsov, N.V. and Rehbinder, P.A., *Doklady AN SSSR*, 123, 1068, 1958; Skvortsov, A.G. et al., *Doklady AN SSSR*, 193, 76, 1970; Pertsov, N.V. et al., *Doklady AN SSSR*, 179, 633, 1968.)

In inactive media, such as air and water, the free energy of the naphthalene was the same and maximal, $\sigma_0 = 60$ erg/cm^2. In benzene the surface energy was the lowest, $\sigma \sim 10$ erg/cm^2. Among all liquids tested, benzene had a structure closest to that of naphthalene (with respect to the type of intermolecular interactions). A lesser, but still significant, effect of surface energy lowering was observed in other liquids: in pentane, hexane, and heptane, the surface energy $\sigma = 15$ erg/cm^2; in butanol, $\sigma = 20$ erg/cm^2; and in more polar ethanol and methanol, σ was around 30 erg/cm^2. These data are in good agreement with the data from the mechanical strength testing. The experiments carried out with fine-grain polycrystalline samples of naphthalene in the presence of various organic liquids revealed strength decreases from 30% to 50%, and in some cases, even from 70% to a three times decrease was observed. The most effective organic liquids in causing strength lowering were benzene and butanol. Short-chain aliphatic alcohols also caused a significant reduction in strength. A weaker effect was observed with more polar liquids having relatively large functional groups. Water, which is an extremely polar liquid with the least degree of similarity to naphthalene

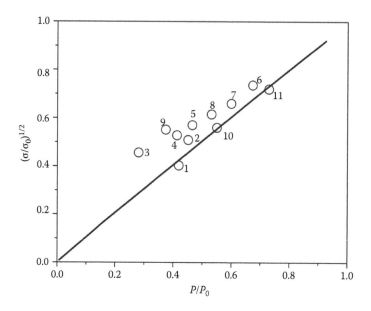

FIGURE 7.14 Correlation between a relative decrease in the surface free energy of naphthalene single crystals and a relative decrease in the strength of polycrystalline naphthalene specimens in (1) benzene, (2) heptane, (3) methylene chloride, (4) chloroform, (5) carbon tetrachloride, (6) methanol, (7) ethanol, (8) propanol, (9) *n*-butanol, (10) *t*-butanol, and (11) 0.2 M solution of *n*-butanol in water. (Data from Pertsov, N.V. and Rehbinder, P.A., *Doklady AN SSSR*, 123, 1068, 1958; Skvortsov, A.G. et al., *Doklady AN SSSR*, 193, 76, 1970; Pertsov, N.V. et al., *Doklady AN SSSR*, 179, 633, 1968.)

showed, as expected, no impact on naphthalene strength. Characteristic trends were also observed with a series of alcohols. The impact of the medium on the strength of naphthalene decreased with increasing polarity in the following series of alcohols: *n*-propanol, propylene glycol, and glycerin, respectively. This indicates that the impact on the strength and surface energy of nonpolar crystals produced by a medium with increasing polarity is opposite to that established for hydrophilic ionic crystals. Since sample fracture must be brittle, the validity of the Griffith equation could be verified. A straight line shown in Figure 7.14 corresponds to the Griffith equation, while the small deviations observed for actual experimental points may be related to differences in the behavior of single crystals and polycrystalline solids.

The experiments carried out with aqueous solutions of typical surfactants, such as fatty acids and alcohols, showed that upon the transition to each new member of the homologous series, the concentration of alcohol at which a given strength lowering could be achieved decreased by a factor of 3–3.5 (Figure 7.15). This is in good agreement with the Traube rule establishing the change in surface activity as the hydrocarbon chain length increases. One can also make the transition to an adsorption isotherm if the data on the decrease in strength as a function of surfactant concentration are available. Such a transition is possible because the Griffith equation is valid. The value of the maximum adsorption, Γ_{max}, allows one to estimate the lowest area per *active* molecule in a monolayer. The value of $a_m = 20\text{–}125$ Å^2 is in good agreement with the geometry of the adsorbent, that is, with the area corresponding to the elementary cell of naphthalene entering the basis plane [54].

7.1.1.3.4 Polymeric Materials

Significant studies on the fracture of polymeric materials upon exposure to surface-active media were conducted by Zuev et al. and Bartenev et al. [55]. It was experimentally shown that in the fracture of organic glass, the propagation of cracks was facilitated by acetone. It is also known that the lubricants

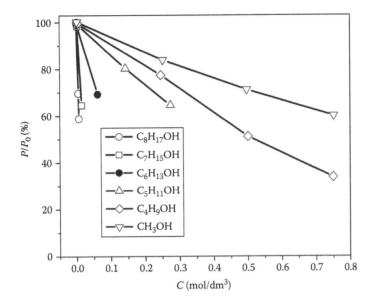

FIGURE 7.15 The relative strength decay of polycrystalline naphthalene specimens in aqueous solutions of saturated aliphatic alcohols as a function of the alcohol concentration. (Data from Pertsov, N.V. and Rehbinder, P.A., *Doklady AN SSSR*, 123, 1068, 1958; Skvortsov, A.G. et al., *Doklady AN SSSR*, 193, 76, 1970; Pertsov, N.V. et al., *Doklady AN SSSR*, 179, 633, 1968.)

used in hydraulic systems and various fuels can impact the strength and durability of carrier lines made of various polymers and plastics, including Teflon, which is the most inert material known.

Organic polymeric materials include both nonpolar hydrocarbons, such as polyethylene, and substances containing polar groups, which generate the mosaiclike surface, such as polymethylmethacrylate (PMMA). The fracture of these materials may involve the cleavage of the intermolecular bonds between polymer chains, as well as the polymer chains themselves. In media with strong surface activity, one may expect to see the prevailing role of weakened intermolecular interactions.

In studies that were conducted in collaboration with Tynniy et al. [56,57], polystyrene (PS) and PMMA were used as model substrates. For PS the effect of strength lowering increased upon the transition from polar liquids to nonpolar liquids with structure similar to PS. In experiments on the deformation by extension of PS samples immersed in ethanol, the strength decreased by 10%; in hexanol, it decreased by 40%; and in hydrocarbons, the tensile strength was lowered by 50%–60% (or 2–2.5 times) compared to that in air.

The results with PMMA were more complex and more versatile, due to the mosaiclike nature of the fracture surface of this polymer. The most interesting observations were made with chemical substances of the same type, such as homologous series of aliphatic alcohols. The efficiency of the strength lowering in the PMMA samples decreased upon a transition from methanol to octanol. This dependence was opposite to the one observed for PE (Figure 7.16).

The studies by Kozlov and Kargin established the existence of a plastifying effect caused by surface-active impurities that were insoluble in a given polymer (incompatible with it). In contrast to common plasticizers with molecular solubility, these impurities were adsorbed at the internal interfaces between the particles in the disperse microheterogeneous structure of the polymer. The adsorbed impurities increased the mutual mobility of the phase elements of the structure. Research in this direction was further developed in studies by Bakeev and Volynskiy. Among the most interesting results, one can cite a study on the conditions causing "crazing" upon the straining of polymers in active media [58].

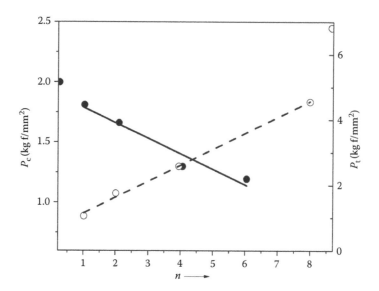

FIGURE 7.16 The effect of aliphatic alcohols ($C_nH_{2n+1}OH$) on the strength, P_c, and on the limiting stress, P_t, corresponding to the fracturing time, $t_f = 100$ s, for (1) polyethylene and (2) polymethylmethacrylate. (Redrawn from Shchukin, E.D. et al., *Fiz.-Him. Meh. Mater.*, 7, 33, 1971.)

7.1.2 Role of the Actual Structure of the Solid and Role of External Conditions in the Manifestation of the Rehbinder Effect and the Deformation of Solids

It has been illustrated that essentially for all types of solids there is a medium akin with the solid in structure and chemical composition that is capable of effectively compensating the broken chemical bonds at the new surface formed in the course of fracturing. This results in a strong lowering of the surface energy of the newly formed surface and hence can cause a strong decrease in the strength of a given solid, which explains both the generality and the selectivity of these phenomena. (It is worth recalling here that we are dealing with *moderately* reversible interactions under conditions of a stable two-phase solid-medium system with a clearly defined interface [2–5,8].)

At the same time, the physical-chemical conditions leading to the strong lowering in the surface energy establish only the *necessary*, but not the *sufficient*, conditions for observing strength lowering effects. The nature and extent of the influence of the medium on the mechanical properties of the solid may also depend on a large number of different factors, among which the structure of the defects present in the solid is very important.

So far we have introduced one structure-related factor that plays a determining role in the strength. This is the size of the *critical nucleus of fracture*, l, which is included in the Griffith equation. In a number of cases and in particular in those involving the fracturing of brittle objects (e.g., glass), such microcracks can exist within the solid even before stresses are applied. The presence of microcracks can be related to the presence of surface defects, such as scratches. In porous bodies the obvious defects are the pores. In the latter case, in agreement with the Griffith equation, the largest pores play the most detrimental role. However, even in the absence of porosity, there are microscopic defects present in any real body. These can include weakened grain boundaries, boundaries between the particles of the different phases (especially when brittle fillers are used), and various other micro- and macro-heterogeneities. The size of these heterogeneities determines the effective value of the parameter in the denominator of the Griffith equation.

It is worth pointing out that the adsorption-active medium does not, by itself, create defects in the solid body; it is only capable of facilitating their formation. For this reason, the ideal threadlike defect-free single crystal may be insensitive to the action of the medium. In bodies capable of plastic

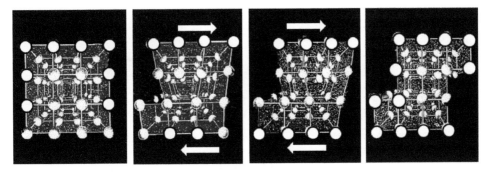

FIGURE 7.17   The shift in a crystal due to a shift caused by an edge dislocation. (From Shchukin, E.D., Rehbinder effect, in: *Science and Mankind*, 1970, pp. 336–367.)

flow (metals), the defects can form at the onset of plastic strain. For example, in single crystals of zinc, such defects form in the course of elongations of about 10%, which precedes the brittle fracture of samples coated with mercury.

Plastic strain in crystalline solids is related to the appearance and movement of the special linear defects of the structure referred to as *dislocations* [2–5,8,17]. In a sliding plane, the dislocation separates the parts of the crystal in which a shift in the position of atoms by one interatomic distance has already occurred and those in which such displacement has not yet taken place (Figure 7.17). The movement of dislocation throughout the entire crystal results in a shift of the sliding plane by one interatomic distance. A defect in which an extra half plane of atoms is introduced midway through the crystal, distorting nearby planes of atoms, is referred to as an *edge dislocation*.

The movement of dislocations may be retarded by various defects present in the crystal lattice. These include foreign atoms, inclusions, other dislocations, the boundaries of blocks of single crystals, twin boundaries, and grain boundaries in polycrystalline solids. The retardation of the dislocation movement may result in the accumulation of dislocations, that is, result in substantial strain nonuniformities. This in turn results in the accumulation of local stresses, which can further promote the formation of microcracks. The size of these microcracks and the strength of the entire body depend on the size of the grains. For this reason, the strength of polycrystalline materials is in many cases proportional to the square root of the grain size, d, that is, in this case $l_c \sim d$ in the Griffith equation. We thus arrive at the general conclusion that has been validated in numerous experiments:

As a rule, the higher the strength of a real heterogeneous material, the more finely disperse its structure and the smaller the probability of the presence of large nonuniformities.

The principles of making fine-disperse structures, which include modern nanotechnology, are the basis for nearly all practical means of increasing the strength and durability of numerous materials, including ceramics, construction materials, and tool-grade alloys. The known methods of alloying, hardening, dispersion hardening, and cold hardening all have the purpose of achieving a finer structure. An excellent example of a very strong finely dispersed composite material is live bone tissue.

This general principle for increasing the strength of a material may, however, yield to the opposite phenomenon of the weakening of a material due to the action of the active medium.

7.1.2.1 Influence of the Real (Defect) Structure on the Adsorption-Induced Strength Lowering

7.1.2.1.1 Specific Role of Grain Boundaries

The grain boundaries present in a polycrystalline solid provide one with the most characteristic example illustrating the effect of real defect structures on the extent of the adsorption-related lowering of strength. (In our discussion, we use the concept of *active medium* implying both the medium

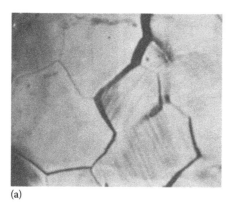

(a)   (b)

FIGURE 7.18   The fracture of zinc grains in the presence of gallium taking place along grain boundaries for specimens strained in extension by (a) 0.63% and (b) 0.83%. (From Zanozina, Z.M. and Shchukin, E.D., *Inzhenerno-Fizicheskiy Zh.*, 5, 86, 1962.)

that causes a significant decrease in the surface free energy upon contact with a similar solid and the medium from which adsorption of the active component can take place. The term "adsorption-induced strength lowering" also applies to both of these cases.

Since grain boundaries have an excess of free energy, σ_g, associated with them, they often serve as the main sources for the proliferation of cracks, even in the absence of an active medium. This is observed in the case of a brittle fracture process in which the value of σ_g is the principal contributor to the total energy balance. Consequently, when the work required to cause fracture is greatly reduced in the presence of an active medium, the role of the *energy source*, σ_g, becomes especially important. Namely, polycrystalline solids in most cases undergo the so-called intercrystalline fracturing, down to complete destruction along the grain boundaries, as was observed in the studies by Polukarova et al. [14,59] and is illustrated in Figure 7.18. It is noteworthy that the low-temperature brittle fracture of polycrystalline zinc itself is mainly transcrystalline, while the brittle fracture of the same poly-crystalline zinc caused by liquid gallium occurs exclusively along the grain boundaries.

At the same time, if a solid has strong initial plastic character (e.g., metal single crystal), the transition into a brittle state with reduced strength may not occur, even at low σ. For example, liquid bismuth does not cause an embrittlement of copper single crystals. The grain boundaries in poly-crystalline solids, while acting as a strengthening factor, can cause a reduction in plasticity due to the retardation of the dislocation movement. For this reason, for a polycrystalline solid, the reduc-tion in σ may be already sufficient for a transition to the brittle state to take place. Indeed, liquid bismuth is known to cause an embrittlement of *polycrystalline* copper.

General laws governing the influence of the structure of a solid on the intensity of the adsorption-induced action of the medium are related to the fact that there is an excess of free energy associated with the structure defects. In the polycrystal, this excess of energy may manifest itself as the energy of the grain boundaries. In the presence of an adsorption-active medium , this energy excess facili-tates the development of cracks along the defects. While under normal conditions, a polycrystalline solid undergoes fracture across the body of the grains, in the presence of active melts, it will frac-ture predominantly along the grain boundaries.

The limiting case of a facilitated propagation of cracks along the grain boundaries is dictated by the Gibbs-Smith condition (Chapter 1), which represents the conditions needed for the thermody-namically favorable formation of a liquid-filled gap along the grain boundary. If the Gibbs-Smith condition holds valid for a majority of grain boundaries, the liquid phase spontaneously propa-gates along the system of the grain boundaries without the need for any external forces. This was observed in such systems as polycrystalline zinc—liquid gallium and alkali metal chlorides—and the corresponding salt solutions. The data obtained by Skvortsova and Traskin [41,43] show that such an intercalation of the liquid can sometimes occur at substantial rates (~1 cm/day) and can result in the formation of a peculiar disperse system in which the individual grains are separated from each other by very thin (~10 nm) layers of the dispersion medium. These phenomena may commonly take place in the salt formations.

Due to the excess of the free energy associated with the grain boundaries, various surface-active impurities adsorb on them, both by diffusion and in the course of crystallization from solutions or melts. Depending on the particular set of conditions, different scenarios are possible. On the one hand, a fusible impurity with low surface tension may be strongly surface active and capable of adsorption, both at the outer surface and on the grain boundaries. In this case, the adsorption on the grain boundaries causes strong embrittlement because fracturing results in the formation of low-energy surfaces (e.g., phosphorous in iron). On the other hand, the impurity may adsorb at the grain boundaries but may not be very surface active at the free surface, such as a refractory impurity with high surface energy. Such are the effects of the presence of carbon and boron in iron. The alloying of steel with these elements may be beneficial, as they are capable of displacing detrimental impurities from grain boundaries. This allows one to optimize the casting process by using modifiers capable of promoting the formation of small grains by inhibiting the growth rate (the so-called modifiers of the first kind) instead of modifiers that increase the number of crystallization centers (the modifiers of the second kind).

7.1.2.1.2 Some Laws Governing the Formation and Development of Cracks

Defects in the structure of solids improve the hardness, strength, and durability of materials. At the same time, these defects result in a decrease of the thermodynamic stability of the material because of the accumulation of the excessive free energy associated with them. For this reason, a material that is strong and durable under normal conditions may undergo deterioration upon exposure to an active medium. Conversely, in cases when high strength is associated with the absence of defects and with proximity to the ideal state (e.g., in threadlike crystals), one may expect weaker effects from the medium. Indeed, a decrease in the effect of mercury on the strength of threadlike crystals of zinc as their diameter decreased was observed by Rozhanskiy [60]. Similar observations were also made by Westwood et al. [19] on large zinc single crystals that did not have noticeable surface defects. However, in the latter case, strength lowering was observed in bicrystals or crystals with a damaged surface.

The studies conducted by Shchukin with large, carefully grown and prepared zinc single crystals allowed one to observe a typical plastic-to-brittle transition in amalgamated samples by introducing a specialized *dosed* defect, that is, by puncturing the surface with a diamond indenter under a given pressure (Figure 7.19). Similar experiments were also conducted with glass [85].

The studies conducted after the puncture with the indenter were carried out on a round mount by pressing with a cylindrical punch, that is, under conditions involving comprehensive compression in the plane of a sample. In this way, there was no impact from the natural boundaries of samples that were defective a priori. This allowed one to gradually increase the amount of damage introduced and to identify the threshold level at which it started to play a role. This threshold in turn corresponds to the natural *damage* level of the sample surface. The next important step is the assessment of the influence of the active medium (which is humidity in this case) on a decrease in the strength of the sample with an unknown degree of surface damage.

One example of the quantitative analysis of the combined effects of the medium and defects introduced is the scheme developed by Shchukin in collaboration with Lichtman for describing the fracturing of single crystals with different starting orientations that underwent more or less significant deformation prior to fracture [2–5,12].

The experiments conducted with metallic, ionic, covalent, and molecular single crystals indicated that for a correct understanding of the mechanisms behind the medium-induced crystal fracture, one needs to analyze two stages of the process. The first stage involves gradual nucleation and the development of microscopic cracks due to the local concentration of stresses and strains related to the original structural defects and to micro-nonuniformities in the plastic deformation due to dislocations.

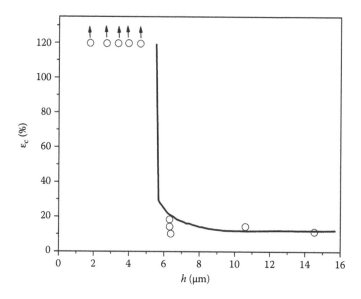

FIGURE 7.19 The influence of defects of various depths, h, introduced by a diamond indenter on the non-damaged virgin surface of zinc single crystals, on the embrittlement of the specimen: a decrease in the limiting strain, ε_c, in the presence of a thin film of mercury. (The experimental setup allowed one to achieve a maximum specimen elongation of only 120%.) (Data from Rehbinder, P.A. and Shchukin, E.D., *Uspehi Fiz. Nauk.*, 108, 1, 1972; Rehbinder, P.A. and Shchukin, E.D., Surface phenomena in solids during deformation and fracture processes, in: *Progress in Surface Science*, S.G. Davison (Ed.), Pergamon Press, Oxford, U.K., 1972, pp. 97–188; Zanozina, Z.M. and Shchukin, E.D., *Inzhenerno-Fizicheskiy Zh.*, 5, 86, 1962.)

The second stage involves the relatively fast propagation of cracks, which lose their equilibrium nature over the entire cross section of the crystal. In contrast to the Griffith microcracks with an elliptical cross section and a constant cross-sectional specific surface free energy, we have been always considering microcracks with sharp blinds and a sharp drop in the surface energy, σ, at these edges all the way down to zero from the highest (maximum) value in the wide portion of the crack.

The growth of that cracks that takes place in the first stage and is controlled by the plastic deformations and, consequently, by the shear stress τ can be described by the following approximate relationship:

$$c = \frac{\beta \tau^2 L^2}{G\sigma}$$

(7.4)

where

c is the maximum crack length under the given conditions

G is the shear modulus

L is the parameter of the structure corresponding to the maximum distance in the slipping plane within which the formation of dislocation micro-nonuniformities takes place

β is the dimensionless coefficient that characterizes a fraction of the elastic energy that is accumulated in the vicinity of the nonuniformity formed

This energy is converted into the work that is spent on the formation of the new surface during crack propagation. In the case of a brittle fracture or, to be precise, a small plastic deformation preceding the fracture, the numerical value of the coefficient β approaches 1. Upon transition from a brittle fracture to a plastic fracture, it may drop to $\sim 10^{-3}$ and less. The $c = c(\tau)$ dependence described does a good job of predicting the experimental results of the direct observation of crack sizes formed on the thin sections of amalgamated (mercury-modified) zinc single crystals (Figure 7.20) [1–4,45,61].

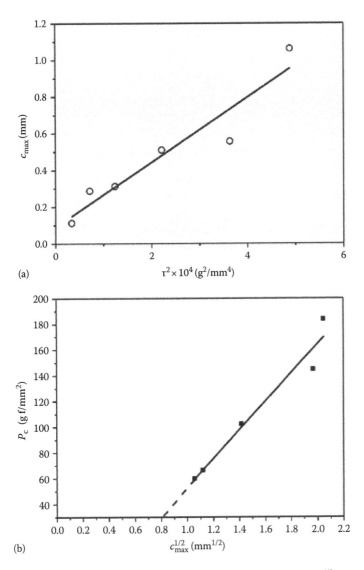

FIGURE 7.20 The value of (a) c_{max} as a function of τ^2 and (b) p_c as a function of $c_{max}^{-1/2}$ for amalgamated zinc single crystals. (From Kochanova, L.A. et al., *J. Mater Sci.*, 27, 5516, 1992.)

The onset of stage 2, that is, the fracture process itself, can be described as the conversion of the crack of length c into a nonequilibrium crack once the normal component of the extension stress, p, reaches its critical value:

$$p_c = \alpha \left(\frac{E\sigma}{c} \right)^{1/2} = \alpha' \left(\frac{G\sigma}{c} \right)^{1/2} \tag{7.5}$$

where α and α' are dimensionless coefficients close to 1. This indicates that the value of σ plays an essential role at both stages.

Direct comparison of the expressions for $c(\tau)$ and $p_c(c)$ reveals the fact that the product of the normal and shear components of the stress is a constant quantity:

$$p_c \tau_c = \text{const} = K^2, \quad K = \left(\frac{\alpha'}{\beta^{1/2}} \right)^{1/2} \left(\frac{G\sigma}{L} \right)^{1/2} \tag{7.6}$$

This expression establishes the condition of the brittle fractures in single crystals, which is invariant with respect to their orientation and takes into account both stress components, in contrast to well-known relationships with p_c = const (Sohncke's law) and τ_c = const (Schmid's law).

The $p_c\tau_c$ = const condition is referred to as the *Lichtman–Shchukin condition*. This condition manifests itself in a most explicit way in single crystals in which a single distinct sliding plane coincides with the cleavage plane. In this case, for the angle between the sliding plane and the longitudinal axis of the sample equal to χ, one can write that

$$p_c = K(\tan \chi)^{1/2}; \quad \tau_c = \text{const} = K(\cot \chi)^{1/2}$$

These expressions indeed hold true for the fracture of zinc single crystals having different orientations in the case of both brittleness caused by the presence of active liquid metal (mercury or gallium) and brittleness caused by a temperature change, that is, low-temperature brittleness (Figure 7.21). This opens the opportunity for a comparison of the surface energy values: in the first case, the value of K is 2–2.5 times smaller than in the second case, which, in turn, corresponds to approximately fivefold lowering of σ upon contact with a liquid metal.

These concepts describing the *equal* role of the shear and extensional stresses in the deformation processes can also be applied to polycrystalline materials.

7.1.2.2 Influence of Strain and Fracturing Conditions

7.1.2.2.1 Role of Temperature and Strain Rate

Temperature is the principal factor that, together with time, determines the kinetics of the fracture process. The effect of temperature can be manifested in various ways. Obviously, the described effects vanish upon the solidification of the medium. One can even observe some increase in the strength due to the formation of a solid surface film.

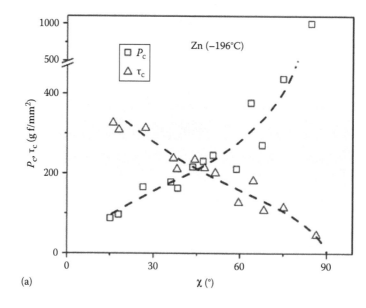

FIGURE 7.21 The dependence of the limiting normal, p, and the cleaving (tangential), τ, components of the stress tensor for the brittle rupture of single crystals of zinc with different orientations: (a) unmodified. (Data from Rehbinder, P.A. and Shchukin, E.D., Surface phenomena in solids during deformation and fracture processes, in: *Progress in Surface Science*, S.G. Davison (Ed.), Pergamon Press, Oxford, U.K., 1972, pp. 97–188; Shchukin, E.D. and Rehbinder, P.A., *Kolloidnyi Zh.*, 20, 645, 1958; Shchukin, E.D. et al., *Kolloidnyi Zh.*, 25, 108, 1963.)   (*Continued*)

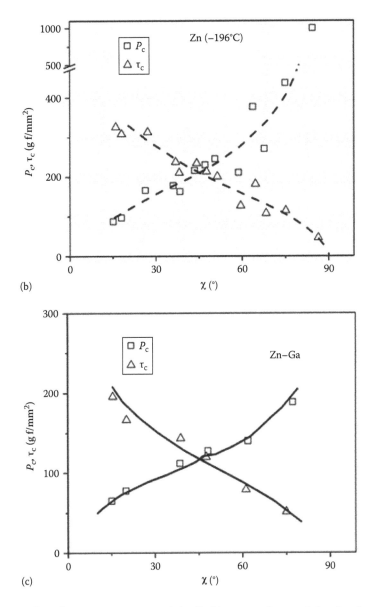

(b)

(c)

FIGURE 7.21 (Continued) The dependence of the limiting normal, p, and the cleaving (tangential), τ, components of the stress tensor for the brittle rupture of single crystals of zinc with different orientations: (a) unmodified, (b) coated with mercury, and (c) coated with gallium. (Data from Rehbinder, P.A. and Shchukin, E.D., Surface phenomena in solids during deformation and fracture processes, in: *Progress in Surface Science*, S.G. Davison (Ed.), Pergamon Press, Oxford, U.K., 1972, pp. 97–188; Shchukin, E.D. and Rehbinder, P.A., *Kolloidnyi Zh.*, 20, 645, 1958; Shchukin, E.D. et al., *Kolloidnyi Zh.*, 25, 108, 1963.)

At the same time, either the weakening or the complete disappearance of the strong physical-chemical influence of the surface-active medium on the strength of the solid is often observed when the temperature is increased up to certain critical values. We refer to this phenomenon as the *threshold of the forced cold brittleness*. This terminology is analogous to that established for the phenomenon of the *natural* cold brittleness observed at a sufficient lowering of the temperature in the absence of the active medium (Figure 7.22). Figure 7.23 shows the experimental data obtained

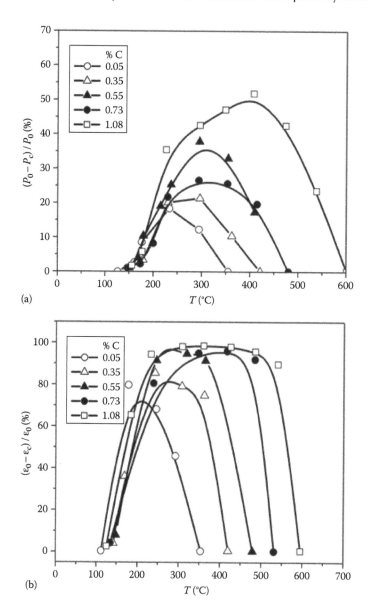

FIGURE 7.22 Relative strength lowering as a function of temperature (a) and relative plasticity lowering as a function of temperature (b) of carbon steels in the presence of tin. (From Lichtman, V.I. et al., *Physical-Chemical Mechanics of Metals*, Israel Program for Scientific Translations, Jerusalem, Israel, 1964.)

by Kochanova et al. [45] (full lines) for samples of various purities and orientations, as well as the theoretical $K(T)$ dependence (dash-dot lines); $t_1 = -39°C$, which is the freezing point for mercury.

The observed transition from brittleness to ductility evidences itself mainly in the facilitation of plastic deformations (movement of dislocations) and the overcoming of various types of defects that serve as obstacles in the sliding planes, that is, the dissolution of the deformation nonuniformities and the healing of the microcracks as temperature increases. As a result, at a sufficiently high temperature, even under conditions promoting a reduction in the surface energy, the ductility is restored. Consequently, there is a competition between the process of the infusion of dislocations into the growing crack and the process of the dissolution of accumulated dislocations (overcoming of the obstacles by dislocations). This approximate scheme allows one to come

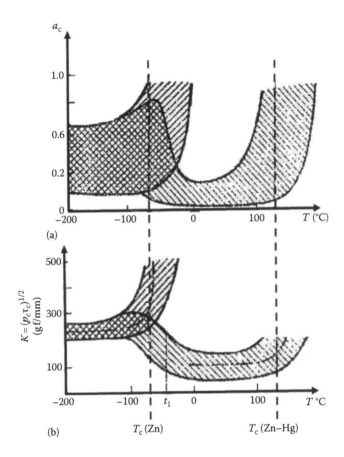

FIGURE 7.23   Temperature dependence of plasticity (crystallographic yield stress), a_c (a), and of the strength, $K = (p_c \tau_c)^{1/2}$ (b), for zinc single crystals—uncoated and coated with mercury. (From Rehbinder, P.A. and Shchukin, E.D., Surface phenomena in solids during deformation and fracture processes, in: *Progress in Surface Science*, S.G. Davison (Ed.), Pergamon Press, Oxford, U.K., 1972, pp. 97–188; Kuchumova, V.M. et al., *J. Mater Sci.*, 27, 5516, 1992.)

up with an expression for the strength of single crystals corresponding to the crossing of the cold brittleness thresholds—the *common* low-temperature one and the *forced one*:

$$(p_c \tau_c)^{1/2} = K = \left(\frac{\alpha'}{\beta(T)^{1/2}} \right)^{1/2} \left(\frac{G\sigma}{L} \right)^{1/2} = K_0 \left[\frac{\beta(T)}{\beta_0} \right]^{-1/4}$$

$$K_0 = \left(\frac{\alpha'}{\beta_0^{1/2}} \right)^{1/2} \left(\frac{G\sigma}{L} \right)^{1/2} ; \quad \beta(T) = \beta_0 \exp \left\{ -\exp \left[\frac{U(1/T_c - 1/T)}{k} \right] \right\}$$

where
   K_0 is a constant level of the strength at temperatures that are lower than the threshold of cold brittleness, that is, the minimum value of the strength
   T_c is the temperature corresponding to the brittle-to-plastic transition, that is,

$$T_c \sim U\left[k \ln(\nu\lambda b/(d\varepsilon/dt)hu)\right]^{-1}$$

where

λ is the number of points at which it is possible to have a "breakthrough" of a dislocation segment through a barrier (according to the Straw scheme)

ν is the frequency of atomic oscillations

b is the lattice parameter (the Burgers vector of the dislocations)

$d\varepsilon/dt$ is the strain rate

U is the activation energy describing the ability of the dislocations to overcome barriers

This activation energy is related to the plastic flow and the acting stresses that are accounted for and the possible drag of the dislocations by the atoms of the surface-active component resulting from the possible adsorption on the dislocations. The hu product reflects the nonuniformity of the deformation in both space and time and the localization of the latter in the slide lines located at distance h from one another; at any given moment, the number of active slide lines is smaller than their total number by the factor of u. In general, the hu factor can assume values in a rather broad range, for example, between 10^{-3} and 10^{-1} cm. For ordinary deformation rates, the ln term in the denominator is on the order of 30–35. In agreement with Equations 7.4 through 7.6, a sharp decrease in the value of β with an increase in the temperature within a specific temperature range means that only a small portion of the elastic energy of the deformation micro-nonuniformities is converted into the work of chemical bond cleavage and is used in the creation of a new surface in the crack. Most of the energy is dissipated by the thermal fluctuations. This dependence of β on T was directly observed in the experiments on measuring the length of the cracks on the sections of amalgamated zinc single crystals.

Along with metals, the threshold of forced cold brittleness is also observed in solids of all other kinds, that is, covalent crystals (e.g., in the system germanium–gold), ionic substances (e.g., sodium chloride in the melted aluminum chloride), and molecular crystals (e.g., naphthalene in liquid hydrocarbon). In the other words, there is only a limited interval of optimum temperatures in which the Rehbinder effect is observed. At temperatures that are too low, the effect is retarded by the excessive starting brittleness and the solidification of the medium, while at temperatures that are too high, it is retarded by the excessive plasticity of the solid. This temperature dependence is one of the principal features of the Rehbinder effect, which makes it very different from the chemical or corrosive action of the medium, both of which intensify as temperature increases.

The effect of temperature is not restricted to the earlier described forms. With an increase in temperature, the rate of the penetration of atoms of the media into the prefracturing zone at the apex of the growing crack also increases. This should result in an enhanced effect on the part of the medium. At the same time, with an increase in temperature, the rate of the diffusional *dissipation* of the atoms of the medium in the bulk of a solid is also higher, which weakens the medium-induced effects. Here, it is assumed that there is enough liquid medium present and that it has a noticeable solubility in the solid phase. On the one hand, the higher the temperature, the lower the interfacial tension at the interface between the solid and the melt, and hence the strength lowering can be higher. On the other hand, with the increase in temperature, the adsorption of the active component from a solution or from a vapor phase can become lower, which weakens the effectiveness of the medium action. Heat can significantly influence the initial structural (and hence mechanical) properties of solids, causing recrystallization and phase transitions, which in turn affects the degree of the manifestation of adsorption effects.

The decrease in the strain rate may have an effect similar to that caused by a temperature increase, because, over longer periods of time, thermal oscillations are capable of supporting longer duration of plastic deformations (Figures 7.24 and 7.25). Conversely, with a significant increase in the rate of crack loading, cracks start undergoing branching, the active melt is intensively consumed by the newly forming surfaces, and the propagation of the main crack may be slowed down. If at

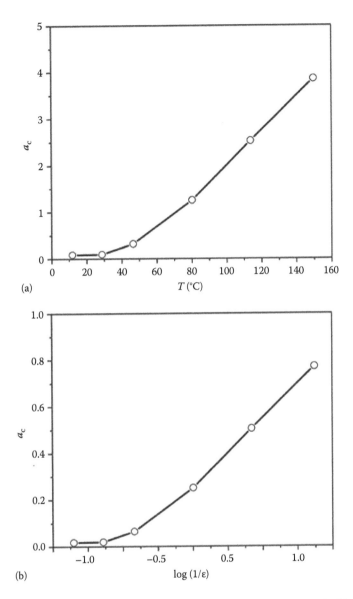

(a)

(b)

FIGURE 7.24 Temperature dependence of the ductility of zinc single crystals coated with a layer of gallium at a constant strain rate of 30% min⁻¹ (a) and the dependence of the ductility of zinc single crystals coated with a layer of gallium on the strain rate (% min⁻¹) at constant temperature $T = 50°C$ (b). (Data from Rehbinder, P.A. and Shchukin, E.D., Surface phenomena in solids during deformation and fracture processes, in: *Progress in Surface Science*, S.G. Davison (Ed.), Pergamon Press, Oxford, U.K., 1972, pp. 97–188; Shchukin, E.D. and Rehbinder, P.A., *Kolloidnyi Zh.*, 20, 645, 1958; Shchukin, E.D. et al., *Kolloidnyi Zh.*, 25, 108, 1963.)

large strain rates the cracks are forced to grow so fast that the active melt does not have enough time to penetrate into them, the effect disappears.

The joint action of the thermal fluctuations and mechanical stresses results in a gradual decrease in the time required for the sample to fracture with an increase in stress. This was established by Zhurkov et al. for various types of solids [62]. In the case of contact with the active medium, such as in the case of the contact of a metal with a melt, high stresses can cause a sharp decrease in durability, while low stresses will have practically no impact on it. The durability curve reveals a kink related to the same mechanism as the brittle-to-plastic transition upon active elongation with a given deformation rate (Figure 7.26).

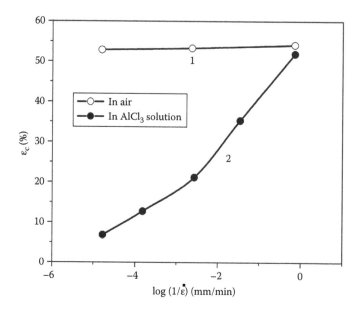

FIGURE 7.25 The dependence of ductility, expressed as a limiting deformation on the rate of strain for single crystals of silver chloride, (1) in air and (2) in a solution of aluminum chloride. (Data from Rehbinder, P.A. and Shchukin, E.D., Surface phenomena in solids during deformation and fracture processes, in: *Progress in Surface Science*, S.G. Davison (Ed.), Pergamon Press, Oxford, U.K., 1972, pp. 97–188; Shchukin, E.D. and Rehbinder, P.A., *Kolloidnyi Zh.*, 20, 645, 1958; Shchukin, E.D. et al., *Kolloidnyi Zh.*, 25, 108, 1963.)

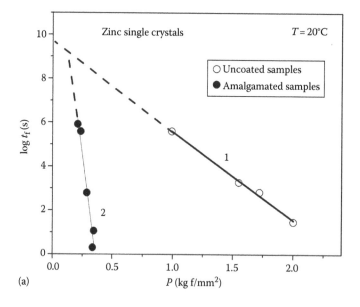

FIGURE 7.26 (a) The long durability, t_f, of zinc single crystals as a function of the stress P at 20°C (1) for uncoated specimens and (2) for amalgamated specimens. (Data from Rehbinder, P.A. and Shchukin, E.D., *Uspehi Fiz. Nauk.*, 108, 1, 1972; Rehbinder, P.A. and Shchukin, E.D., Surface phenomena in solids during deformation and fracture processes, in: *Progress in Surface Science*, S.G. Davison (Ed.), Pergamon Press, Oxford, U.K., 1972, pp. 97–188.) *(Continued)*

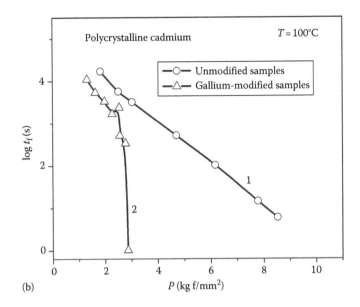

(b)

FIGURE 7.26 (*Continued*) (b) the durability, t_f, of polycrystalline cadmium as a function of the stress P at 100°C (1) for uncoated specimens and (2) for gallium-coated specimens. (Data from Rehbinder, P.A. and Shchukin, E.D., *Uspehi Fiz. Nauk.*, 108, 1, 1972; Rehbinder, P.A. and Shchukin, E.D., Surface phenomena in solids during deformation and fracture processes, in: *Progress in Surface Science*, S.G. Davison (Ed.), Pergamon Press, Oxford, U.K., 1972, pp. 97–188.)

7.1.2.2.2 Laws Governing the Propagation of the Active Medium

When the strength of a solid is decreased due to the action of a surface-active medium, the rate with which cracks propagate is limited by the rate at which atoms are supplied from this medium, that is, by the rate of the surface diffusion or by the rate of the liquid flow from the walls of the crack toward the apex. In a set of significant studies by Goryunov et al. [4,15], the rate of crack growth in bending metal plates upon the application of a drop of liquid metal was compared to the rate of propagation for the same metals on the free surface of similar plates free of oxide films. It was established that these two phenomena obey the same laws.

When the melt penetrates into the crack, its further propagation follows various mechanisms. For a large portion of the crack, with the exception of the parts that are adjacent to the crack entry and apex, the process of liquid propagation can be viewed as a quasi-stationary spreading of a thin phase layer of a liquid on the free surface driven by surface tension forces. The surface tension gradient is determined by the gain in the net surface free energy due to wetting of the solid surface by the melt. The essential feature of the liquid phase that is present in the crack is that it can be deformed without applying any substantial stress. Because of this, the gain in the force needed to separate one part of the solid from another may be realized. This gain in force is related to a lowering in the interfacial free energy at the interface in comparison with the surface free energy of the same solid in a vacuum. Near the apex of the crack, the 2D diffusion, that is, the surface migration of active atoms, is the process that most likely prevails. The quantitative description of the kinetics of crack propagation is based on the description of the competition between the propagation of the medium along the walls of the crack and the diffusion-driven absorption of the medium atoms by the bulk of the solid phase that is adjacent to the wall. Consequently, the main stage of the viscous spread of a liquid metal droplet of mass m, resulting in the formation of a thin phase film along a "groove" of size a on the free surface of the metal, can be described by the approximate relationship:

$$x(t) \approx \left[\left(\frac{3m}{2a} \right) \left(\frac{f}{\eta \delta \kappa} \right) \right]^{1/3} t^{1/3}$$

where

x is the distance that the front of the liquid has traveled up to the moment of time t

η and δ are the viscosity and the density of the liquid phase, respectively

$\kappa \sim 10$ is a dimensionless coefficient that takes into account the profile of the layer near its front

$f = K\sigma_{sv} - (\sigma_{lv} - K\sigma_{sl})$ is the decrease of the total energy of the system upon the wetting of 1 cm², that is, the driving force of the spreading process

K is the coefficient of roughness of the surface

The end distance X by which the front of the spreading liquid propagates and the time t_m by which the spreading has completed are, respectively,

$$X \approx 1.3 (C_0 D^{1/2})^{-2/5} \left(\frac{f}{\eta \delta \kappa} \right)^{1/5} \left(\frac{m}{2a} \right)^{3/5}$$

$$t_m \approx 0.7 (C_0 D^{1/2})^{-6/5} \left(\frac{f}{\eta \delta \kappa} \right)^{2/5} \left(\frac{m}{2a} \right)^{4/5}$$

where

C_0 is the highest concentration of the adsorption-active component in the surface layer of the solid phase

D is the bulk diffusion coefficient

The kinetics of the crack growth in a plate of thickness d upon the introduction of a droplet of melt of mass m and along final crack length L are given by

$$l(t) \approx \left[\left(\frac{3m}{2d} \right) \left(\frac{f}{\eta \delta \kappa} \right) \right]^{1/3} t^{1/3}$$

$$L \approx 2.56 (C_0 D^{1/2})^{-2/5} \left(\frac{f}{\eta \delta \kappa} \right)^{1/5} \left(\frac{m}{2d} \right)^{3/5}$$

These theoretical expressions are in good agreement with the experimental data, as shown in Figure 7.27 [15]. Similar types of relationships are also valid for another form of medium propagation, that is, for the processes of the surface diffusion of adsorption-active atoms along the free surface of a solid and along the crack walls.

Along with the kinetics of the propagation of active medium in the processes of adsorption-driven strength lowering, another very important factor is the establishment of a reliable primary contact between the liquid phase and a bare section of the solid surface. This means that there needs to be good wetting of the oxide-free solid surface by a liquid phase. The latter can be achieved by etching; even small amount of alkali is sufficient to facilitate the wetting of a zinc plate by mercury over an area on the order of several mm². The propagation of cracks is initiated on this area, and the newly formed surface absorbs the entire drop. At the same time, the local contact is often provided by either mechanically removing or damaging the surface film. The influence of the process of the formation of juvenile surfaces in the course of fracturing that favors wetting is also a question of interest. Under such peculiar nonequilibrium conditions *mechanical-chemical* phenomena may play a significant role.

Relatively fast penetration of noticeable amounts of the adsorption-active component into the bulk of the solid may also take place in the absence of external forces during the liquid's migration along the grain boundaries. It has been pointed out earlier that the limiting case corresponding to the penetration of liquid phase along the grain boundaries is possible if the Gibbs-Smith condition of a zero dihedral angle is obeyed, that is, $\sigma_g > 2\sigma_{sl}$. This type of spreading along the grain boundaries takes place at sufficiently high temperatures upon the contact of gallium with zinc and of bismuth with copper [4,15].

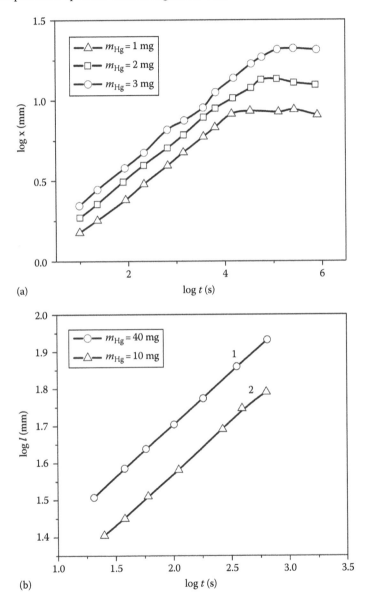

FIGURE 7.27   (a) The distance x (mm) of the propagation of mercury along a groove of size $a = 1$ mm on the zinc surface (mercury propagates only to one side from the applied drop) as a function of time, t (s), for a mass of the mercury drop, $m = 1, 2, 3$ mg; (b) the growth of the crack length, l (mm), in a zinc plate of thickness $d = 1.85$ mm as a function of time, t (s), at the main stage of crack development (the crack propagates only on the one side relative to the spot where it was initiated) for a drop mass (1) $m = 40$ mg and (2) $m = 10$ mg. (Data from Rehbinder, P.A. and Shchukin, E.D., Surface phenomena in solids during deformation and fracture processes, in: *Progress in Surface Science*, S.G. Davison (Ed.), Pergamon Press, Oxford, U.K., 1972, pp. 97–188; Shchukin, E.D. et al., *Kolloidnyi Zh.*, 25, 108, 1963.)   *(Continued)*

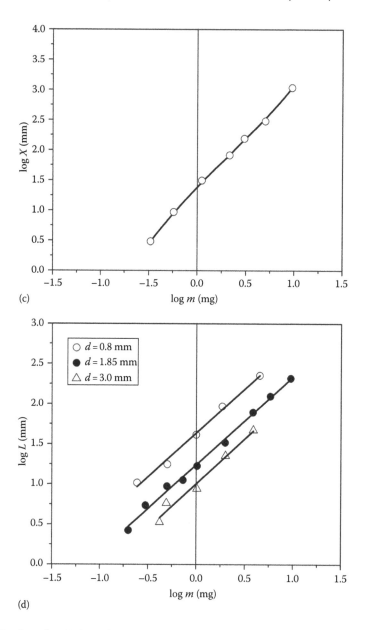

FIGURE 7.27 (Continued) (c) the ending distance X (mm) by which the mercury propagated along a groove of size $a = 1$ mm present on the zinc surface (into one direction from the initial drop position) as a function of drop mass, m (mg); (d) the final total crack length, L (mm), as a function of a drop mass, m (mg), in zinc plates of various thicknesses (1) $d = 0.8$ mm, (2) 1.85 mm, (3) and 3.0 mm. (Data from Rehbinder, P.A. and Shchukin, E.D., Surface phenomena in solids during deformation and fracture processes, in: *Progress in Surface Science*, S.G. Davison (Ed.), Pergamon Press, Oxford, U.K., 1972, pp. 97–188; Shchukin, E.D. et al., *Kolloidnyi Zh.*, 25, 108, 1963.)

As a result of the adsorption taking place on the grain boundaries in the process of such *impregnation* by a liquid, the solid is already weakened prior to testing. In the course of further testing, after contact with a given medium, the solid reveals lower strength as compared to the situation when the stress is applied immediately after wetting. A characteristic example of such an *impregnation* is shown in Figure 7.28, which illustrates the behavior of polycrystalline zinc after a rather long contact with either mercury or liquid gallium. Similar phenomena are also observed if polycrystalline naphthalene is placed in contact with benzene and with polycrystalline sodium chloride

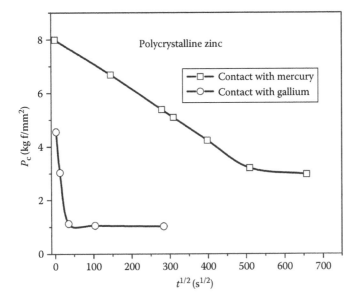

FIGURE 7.28   The strength of polycrystalline zinc as a function of the time of contact with (1) mercury and (2) gallium prior to the application of stress. (Redrawn from Shchukin, E.D., *Adv. Colloid Interface Sci.*, 123–126, 33, 2006.)

in a saturated aqueous sodium chloride solution. Nevertheless, even these specific examples do not give a clear reason that identifies the third, *diffusion*, mechanism, which describes the influence of a melt on the strength of a solid. The two basic mechanisms include various forms of corrosion and adsorption effects. Indeed, these forms of the transport of the active component represent the necessary condition for both corrosion and strength lowering. However, by itself such transport of foreign atoms can't be regarded as a direct cause of the chemical bond breaking in a solid that is facilitated.

Finally, it is worth emphasizing the role of the stressed state. The combined action of the medium and mechanical stresses forms a common group of the adsorption effects considered. One characteristic feature of this group is the influence of stressed states in which the predominant role is played by the extension component. If the stresses are of a compressive nature (especially within surface layers), the influence of the active medium may not necessarily be revealed. Consequently, various stress concentrators, such as recesses or threads, may have an especially negative impact on strength and durability.

The degree of the manifestation of adsorption-induced strength lowering depends on many factors and may cover a very broad range. This is a reason for the various apparent contradictions found in the literature regarding either the influence or noninfluence of the media in various cases.

Consequently, in cases when contact with an adsorption-active melt is inevitable, it is necessary, in order to prevent adsorption-related strength lowering, to ensure that at least one of the described conditions is not fulfilled. Conversely, in order to take an advantage of these effects, one needs to choose the most favorable combination of conditions. Namely, the polycrystalline metal needs to be in a complex stressed state with a high concentration of sheer and extension stresses produced by vibrations at high frequencies. However, the vibration frequencies need to be such that they can ensure that the newly open juvenile surfaces have enough time to be covered with thin layers of active medium. Such conditions are achieved in the course of the cutting and drilling of metallic and nonmetallic materials in the presence of a medium that is strongly surface active toward the given material (Figure 7.29).

7.1.2.3 Spontaneous Dispersion of Solids under Conditions of a Very Strong Reduction in the Free Interfacial Energy Facilitation of Mechanical Dispersion

The formation of thermodynamically stable colloidal systems by spontaneous dispersion is very common in nature, for example, in biological processes. However, in most cases, one encounters a

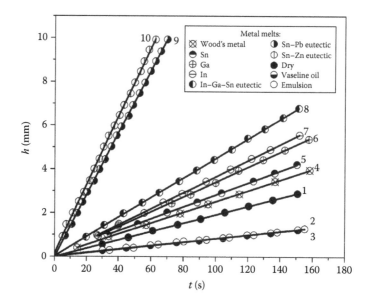

FIGURE 7.29 The immersion depth of a bit, h, as a function of time (1) in the course of the drilling stainless steel that is dry, (2) in Vaseline oil, (3) in emulsion, and in metallic melts: (4) Wood's alloy, (5) tin, (6) gallium, (7) indium, (8) In–Ga–Sn eutectic, (9) Sn–Pb eutectic, and (10) Sn–Zn eutectic. (Data from Shchukin, E.D. (Ed.), *The Successes of Colloid Chemistry and Physical-Chemical Mechanics*, Nauka, Moscow, Russia, 1992.)

situation when the lowering of the surface energy of a solid at the interface with a medium is strong, but not yet sufficient to cause spontaneous dispersion. It has been shown that under these conditions, the processes of mechanical dispersion are facilitated [63–66]. For example, there has not been much success in attempts to grind zinc granules in a ball mill. Zinc is a plastic metal that does not undergo efficient grinding and typically becomes only slightly crumpled if one attempts to grind it. At the same time, the presence of only 1% gallium leads to a 200-fold increase in the grinding rate (Figure 7.30). Gallium has a similar effect on the grinding of tin, cadmium, aluminum, and bismuth. The interaction of gallium with tin leads to deep structural changes even in the absence of external mechanical stresses. These changes are caused by internal stresses and the energy of structural defects, or they accompany the formation of a solid solution. A single crystal of tin covered with a film of liquid gallium undergoes gradual spontaneous transformation into a polycrystal, as evidenced by characteristic changes in the x-ray diffraction pattern.

A noticeable increase in the grinding rate is also observed in some organic media and in aqueous solutions of typical surfactants.

The peptization of lead upon the addition of oleic acid to a kerosene suspension of lead granules was observed by Gurvich. However, in this case, the process was driven by a chemical reaction—the formation of a soft film of lead oleate that could be easily removed by friction. This case establishes the basis for the chemical polishing of glass and metals.

In general, fine grinding cannot be a solely mechanical process. Since it is associated with the formation of a new surface with a large area, physical-chemical factors inevitably play an important role in controlling the phenomena taking place at the interfaces. In this case, the role played by adsorption-induced strength lowering is revealed not only in the facilitation of disintegration into small particles but also in the destruction of the coagulation contacts, which in turn prevents aggregation. The destruction of the developing structure is necessary in subsequent operations with the fine-disperse matter produced, as it facilitates the formation of a dense and uniform particle pack in the manufacturing of various materials.

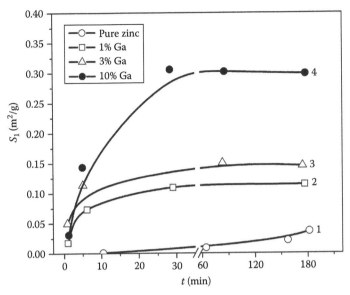

FIGURE 7.30 The specific surface area of a powder, S_1, as a function of the grinding time, t, (1) for pure zinc and for zinc in the presence of gallium: (2) 1%, (3) 3%, (4) and 10%. (Redrawn from Pertsov, A.V., The studies of dispersion processes under conditions of the strong decrease in the free interfacial energy, Canadian Science thesis, Izd. MGU, Moscow, Russia, 1967.)

7.1.2.4 Facilitation of Plastic Deformations in Solids under the Influence of an Adsorption-Active Medium

The manifestation of adsorption effects depends on a number of factors, such as the nature of the interatomic interactions, the real structure of the solid, and the deformation conditions. When the energy at the interface with the medium is lowered, but any one of the requirements for a sharp decrease in strength and embrittlement is not met, *plasticizing*, that is, the facilitation of plastic flow of crystal, may take place.

One characteristic example of the plasticizing action of the medium is the lowering of the yield point and the hardening coefficient of metal single crystals in solutions and vapors of organic surfactants. This is often accompanied by significant changes in the sliding patterns (Figure 7.31) [1–5,8,55,67].

We have come to the conclusion that the observed facilitation of plastic flow is the result of the lowering in the potential barrier, which needs to be overcome by the dislocations upon the migration of their entry points to the crystal surface [4,6,50,68]. This potential barrier is associated with the restructuring and rupture of interatomic bonds. The latter, in turn, result in the formation of new elementary cells of the surface in the course of the formation and development of sliding lines. The lowering of the surface free energy due to the adsorption of surface-active molecules corresponds to a lowering in the amount of work consumed in this process. Under the appropriate conditions, an increase in the mobility of the dislocation exit points leads to the facilitation of the movement and multiplication of dislocations within the metal layer close to the surface. The depth of this layer is comparable to the average length of the dislocation segments present in the crystal. In relatively thin single crystals of plastic metals of low dislocation density, the entire volume of the sample starts to "feel" the surface, especially around the yield point, where small changes in the resistance to strain produce a noticeable increase in the flow. By comparing the rate at which dislocations accumulate in the near-surface layer and the rate at which the dislocations overcome the potential barrier, one finds the dependence of the optimum temperature of plasticizing, T_{opt}, on the strain rate

$$T_{opt} \approx \frac{U_a}{k \ln[\nu b / (d\varepsilon / dt)hu]}$$

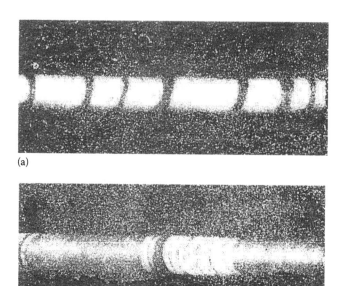

(a)

(b)

FIGURE 7.31 Traces of sliding formed on the surface of single crystals of tin upon stretching (a) in air and (b) in a solution of oleic acid in Vaseline oil. (From Lichtman, V.I. et al., *Physical-Chemical Mechanics of Metals*, Israel Program for Scientific Translations, Jerusalem, Israel, 1964.)

where the activation energy, U_a, is close to the work of formation of the elementary surface cell, that is, $U_a \approx b^2\sigma$. The parameters h and u take into account the nonuniformity of the deformation in the space and time domain, that is, the localization of the deformation within the sliding lines. The hu product may assume values between 10^{-3} and 10^{-1} cm. Assuming that $v \approx 10^{12}$ s^{-1} and that b is on the order of several Å, one finds that $\ln(vb/hu) \approx 12-17$.

The earlier relationship can be compared with the data on the influence of temperature and the deformation rate on the facilitation of the plastic flow of tin single crystals in solutions of surfactants in nonpolar Vaseline oil. According to these data, at room temperature, the effect is the highest at $d\varepsilon/dt \approx 10^{-3}$ s^{-1}, while at 100°C, the effect is the highest at $d\varepsilon/dt \approx 10^{-1}$ s^{-1}. Substituting these values into the this relationship, one gets the following estimates: $\ln(vb/hu) \approx 15$, and $U_a \approx 0.9\times10^{-12}$ erg, which is of the same order of magnitude as predicted by this scheme. Indeed, $b^2\sigma \approx 10^{-15}$ cm$^2 \times 10^3$ erg/cm$^2 \approx 10^{-12}$ erg.

The surface barrier overcome by the dislocations may also be changed by other physical-chemical means. For example, the surface energy can also be lowered by supplying an electric charge to the surface (in agreement with the Lippmann equation $d\sigma/d\phi = -e_s$). It has been shown that there is a complete analogy between plasticizing caused by surfactant adsorption and plasticizing caused by the electric charging of the surface (Figure 7.32) [2–4,9,68–70].

Adsorption-induced plasticizing is a universal effect that can take place in all types of solids. Various organic media, such as hydrocarbons, containing relatively large molecules accelerate the plastic flow of naphthalene single crystals if the masking effects of the lowering of the strength and plasticity are removed. Gypsum single crystals reveal a characteristic creep behavior when exposed to water vapor. Furthermore, even such elastic–brittle materials as inorganic glasses display an irreversible creep in the presence of water in the course of indentation when compressive stresses prevail.

The dissolving medium may cause an increase in the plastic deformations in brittle bodies for a reason other than a reduced resistance to plastic flow: an increase in the strength may take place due

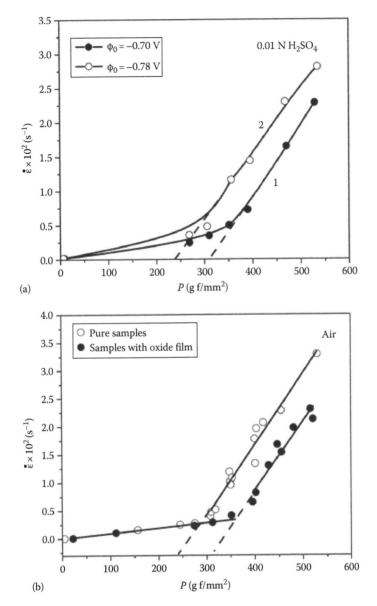

FIGURE 7.32 The rate of plastic deformation of lead single crystals, $d\varepsilon/dt$, at the onset of the creep region as a function of the applied voltage, P, (a) in 0.01 N solution of H_2SO_4 ([1] at a potential of zero charge $\phi_0 = -0.70$ V and [2] at a potential of zero charge $\phi_0 = -0.78$ V) and (b) in air ([1] for pure samples and [2] samples coated with 0.25 μm thick oxide film). (Data from Lichtman, V.I. et al., *Physical-Chemical Mechanics of Metals*, Israel Program for Scientific Translations, Jerusalem, Israel, 1964; Rehbinder, P.A. and Shchukin, E.D., *Uspehi Fiz. Nauk.*, 108, 1, 1972; Rehbinder, P.A. and Shchukin, E.D., Surface phenomena in solids during deformation and fracture processes, in: *Progress in Surface Science*, S.G. Davison (Ed.), Pergamon Press, Oxford, U.K., 1972, pp. 97–188; Rehbinder, P.A., *Selected Works. Surface Phenomena in Disperse Systems: Physical-Chemical Mechanics*, Nauka, Moscow, Russia, 1979; Shchukin, E.D. et al., Electric surface effects in solids plasticity and strength, in: *Modern Aspects of Electrochemistry*, R.E. White, B.E. Conway, and J. O'M. Bockris (Eds.), 1993, pp. 245–298; Shchukin, E.D. et al., *Colloids Surf.*, 142, 175, 1998; Shchukin, E.D. et al., *Fiz.-Him. Meh. Mater.* 3, 49, 1998.)

to the dissolution of the surface layer containing defects, that is, cracks. This is known as the *Ioffe effect*. In this case, the effect of the medium is manifested to the extent that it is not saturated with the active substance.

The plasticizing action of surface-active substances has an application in the treatment of metals by pressure. The surface-active components of lubricating greases play an important role at the points of friction, helping one to automatically control the conditions of wear.

In the case of intensive repetitive actions, the facilitation of plastic deformation in the surface layer may at some point result in the opposite effect, namely, an additional strength increase due to the accelerated accumulation of distortions in the metal structure. Direct observations by electron microscopy, conducted by Kostetskiy et al., indicated a significant increase in the dislocation density in the surface layer. Under the appropriate conditions (temperature, stress, velocity, etc.), such a peculiar *sample training* may be used in the improvement of the structure and the mechanical properties of the surface layer. However, this already corresponds to the adsorption-induced fatigue region, studied in detail by Karpenko et al. These studies showed that at a certain level of stress the adsorption-caused acceleration of defect accumulation within the surface layer may lead to the premature development of cracks and partial failure after a certain number of cycles (*cyclic fatigue*).

The previous examples do not in any way limit the variety of the manifestations of the action of surface-active media on the properties of solids. The current level of understanding of the nature and mechanisms involved in these phenomena is far from being complete. It is essential to investigate in greater detail such *intermediate* phenomena as the influence of chemisorption on mechanical characteristics, the surface chemical reactions taking place in the course of mechanical treatment, and the electrochemical reactions. It is important to further study the influence of an active medium on the strength of the interfacial cohesive contacts in various materials, that is, the processes of adsorption-induced adhesive strength lowering.

The next section of this chapter describes the results of significant studies by Yushchenko et al. [71–74]. These studies provided insights into the molecular nature of the Rehbinder effect and constituted the first steps toward a numerical simulation of the elementary acts of deformation and bond rupture in the lattice of a solid. The studies also dealt with the influence of adsorption on this process (Sections 1.1 and 4.2). For detailed discussions of these molecular dynamic experiments, the reader is referred to the original works and reviews [71–83]. The quantum mechanical studies of Rehbinder effect were also done by Ab-initio calculations (c–c bond cleavage) [84].

7.1.2.5 Numerical Modeling of the Rehbinder Effect

A dynamic numerical experiment was used to study the molecular mechanism of the Rehbinder effect using molecular dynamic simulation [71–74]. A numerical experiment allows one to observe fine details of the process being modeled. However, the experiment is limited in terms of the size of the system (up to $\sim 10^3$ particles) and the observation time (for argon this time is $\sim 10^{-10}$ s). Consequently, processes involving a large number of particles, or processes that take place over periods significantly exceeding the maximum time of dynamic-type numerical experiments, can't be modeled.

It was shown previously that dynamic numerical simulation can be used successfully to analyze adsorption and crystallization [80,81]. The method was further developed to cover the processes of the deformation and fracture of a crystal [71,72]. Here, we will present the results of a study on the deformation and fracture of a crystal in the presence of the foreign atoms of the *adsorption-active medium*. A reasonably simple system was used in an attempt to cover a broad range of variations of the experimental conditions so that various cases of the medium influence could be observed.

Similar to the case described in [71], a 2D crystal with a cavity was used as a model system (Figures 7.33 and 7.34). The stresses were concentrated in the cavity. The atoms of the second component in the initial state were placed at the cavity walls (Figure 7.35). The chosen physical-chemical

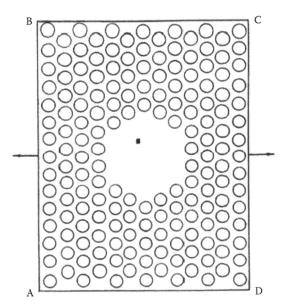

FIGURE 7.33 Initial position of the atoms. Black point indicates the precision of the calculations. (Redrawn from Yushchenko, V.S. et al., *Doklady AN SSSR,* 215, 148, 1974.)

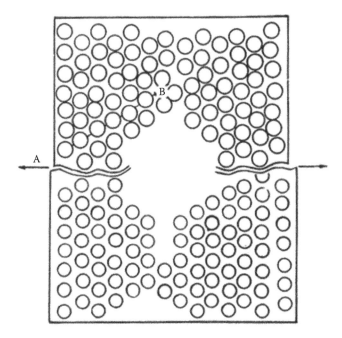

FIGURE 7.34 Plastic sheer at elevated temperatures (upper part) and brittle crack at low temperatures (lower part). The first observation of the formation of dislocation AB by MD simulations. (Redrawn from Yushchenko, V.S. et al., *Doklady AN SSSR,* 215, 148, 1974.)

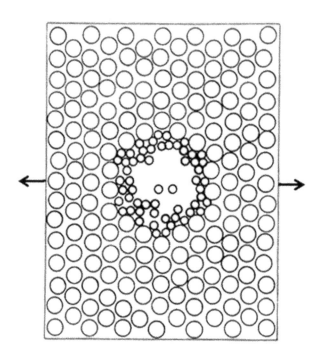

FIGURE 7.35 The adsorption of the surface-active component. (Redrawn from Yushchenko, V.S. et al., *Doklady AN SSSR,* 219, 162, 1974; Shchukin, E.D. and Yushchenko, V.S., *J. Mater Sci.,* 16, 313, 1981.)

conditions were optimal for the adsorption-induced lowering in the strength to be revealed. In such a system, it is quite difficult to draw a precise boundary between the formation of the adsorption layer of foreign atoms and the appearance of the liquid medium. For this reason, the following conditions fitting both of these cases were chosen.

1. The molecules of the admixture (A) and the main component (B) are of the same nature: they are single-atom molecules, the interaction between which can be described by the Lennard–Jones potential.
2. The A–A interactions are significantly (two to six times) weaker than the B–B interactions. This means that the medium A is more fusible and has a lower surface tension than the main component, B. The parameters of the interaction potential of the B–B atoms and the atomic mass of B are those of argon. The parameters of the A atoms are varied over a broad range and do not correspond to any particular substance.
3. The energy of mixing $u_0 = 1/2(\varepsilon_{AA} + \varepsilon_{BB}) - \varepsilon_{AB}$ is chosen in a range from 0 to $2kT$ (ε is the depth of the potential well corresponding to a given pairwise interaction). These types of systems conform to the majority of the cases of the adsorption-induced strength lowering upon contact with a liquid phase [83].
4. The temperature range where the lattice B is stable and the atoms of the medium maintain their mobility is chosen. Under these conditions, the atoms of the medium may desorb from the cavity surface.

A series of numerical simulations were also conducted, in which, in addition to the masses and the number of atoms of A, the A–A and A–B interaction parameters also differed in temperature, deformation rate, and initial conditions,. As in the experiments with a single-component system [71], the deformation first emerged in the elastic region, followed by the formation and accumulation of defects. The subsequent restructuring of the lattice was accompanied by stress relaxation.

No influence of the medium was observed in the elastic region, where the lattice exhibited an ideal behavior. The influence of the medium was mainly observed at stages when the accumulation of defects and lattice restructuring took place. The observed strength lowering was on the order of 10%–15%. (The sample strength was estimated as the maximum force acting on the box walls while they were moved apart at constant velocity). The observed reasonably small extent of the strength lowering is likely due to the absence of defects in the lattice in the initial state. Also, the lattice needed to be sufficiently distorted for the manifestation of the embrittlement action of the medium. In our case, the magnitude of stresses corresponding to the formation of the first defect is close to the strength of sample without the medium.

While in a single-component system at selected temperatures and deformation rates, the observed sheer is plastic (Figure 7.34), in the presence of foreign atoms (*adsorption-active medium*), the formation of a brittle crack takes place, that is, one directly observes the Rehbinder effect (Figure 7.36). Each given experiment is characterized by its individual deformation, but in general falls into one of the three main categories:

1. In the case when the physical-chemical conditions favor high migration mobility of the medium atoms, particles of A penetrate into the lattice of the principal component loosened by deformation. A crack filled with atoms of A is formed (Figure 7.36, lower part). This crack develops as the foreign atoms migrate toward its apex. At the same time, the migration is activated by the mechanical stresses. (Under the conditions of the numerical experiment, the diffusion of the particles of A into the lattice in the absence of stresses was not observed over the time of the simulation). It is noteworthy that the interaction between the atoms of the admixture with the particles of B results in the separation of the latter from the matrix. A two-component mixture is formed as a result. Since the mobility of the particles is high, this two-component mixture can be viewed as a liquid. Upon stress relaxation, the atoms of B either remain in solution or deposit on the walls, resulting in a lattice build-up. No such dissolution is observed in the absence of strain.

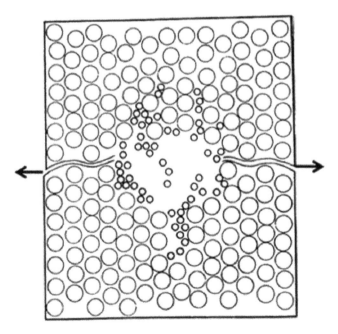

FIGURE 7.36   The development of cracks upon the penetration of the active component (at low and high temperatures). (Redrawn from Yushchenko, V.S. et al., *Doklady AN SSSR,* 219, 162, 1974; Shchukin, E.D. and Yushchenko, V.S., *J. Mater Sci.,* 16, 313, 1981.)

2. In contrast to the case described earlier, when the medium facilitates the entire crack propagation process, the influence of the second component is revealed only at the stage of crack initiation. This takes place under conditions of limited migration mobility of the atoms or at a high deformation rate. As a result, a crack that is not filled with atoms of the admixture may be formed. The admixture atoms, which originally favored crack nucleation, may either remain present in the mouth of the crack, forming a kind of a "jumper," or may spread into nearby regions.

Special simulations carried out with the fixed atoms of a lattice indicated that the admixture atoms tend to migrate between the particles of B and to pull them apart. This force undergoes high fluctuations, is quite substantial, and is, in general, comparable to the strength of the interatomic bond. One can likely regard this force as being analogous to the 2D pressure of the adsorption layer.

3. A numerical experiment allows one to observe various possible paths of the process. The same system may undergo different paths for different sets of initial conditions. Besides the clear examples of the embrittlement action of the medium, various intermediate cases, and even the absence of an embrittlement, were observed. For example, a brittle crack was formed in one half of the crystal, while in the other half of the crystal a plastic deformation took place. When comparing the results of the simulations with a given macroscopic system, one has to remember that numerous elementary acts of deformation and fracture take place in a real sample, and the processes may follow many different paths. For this reason, the crack, once formed, will undergo further propagation. The probability of plastic sheer in the numerical simulation may be compared to the plastic deformation accompanying quasi-brittle fractures in a macroscopic sample. It is worth pointing out that the numerical simulation experiment allows one to determine only the *sufficient conditions* needed for the manifestation of the Rehbinder effect, and the absence of a brittle fracture in the numerical simulation experiment does not necessarily imply the absence of the adsorption-induced strength lowering in a macroscopic system.

The influence of the medium makes it possible for the energy-consuming process of the interatomic bond restructuring of the principal component to take place. Figure 7.37 shows the change in the course of the deformation of the total energy of interaction between B atoms (1), the total system energy corresponding to the work performed by an external force in pulling apart the box walls (2), the kinetic energy of the system (3), and the net energy of the interaction between atoms of different types (4).

The presence of foreign atoms allows one to accumulate higher amount of energy than what can be introduced by the work of an external force. This energy is accumulated in the form of the elastic stress energy and the energy of the broken interatomic bonds. As the deformation progresses and the atoms of A penetrate into the lattice, the number of the A–B bonds increases, and the released energy facilitates the rearrangement of the B–B bonds.

The A–B interaction energy goes through a minimum (i.e., the interaction is the strongest between the atoms of different types) at the moment when the system overcomes the potential barrier in the course of crack nucleation. Stress relaxation results in a partial desorption of the A atoms, and as a result, the number of the A–B bonds decreases. The difference between the initial and final levels of the interaction energy between the B–B and A–B atoms can be compared to the work of new surface formation and to the lowering of the surface energy upon contact with an active medium, respectively. The maximum and the minimum in these curves correspond, respectively, to the potential barrier of crack formation and the lowering of this barrier due to the action of the medium.

The numerical simulations described made it possible to observe a microscopic picture of the deformation and fracture in the presence of foreign atoms. These experiments showed that the fast local processes taking place in the presence of the adsorption-active atoms, such as the migration into a stressed lattice and the 2D pressure, may result in the transition from a plastic deformation to a brittle fracture.

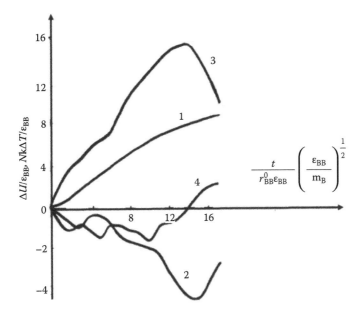

FIGURE 7.37 The change in the energy-related parameters of the system in the course of the deformation. (Redrawn from Yushchenko, V.S. et al., *Doklady AN SSSR,* 219, 162, 1974; Shchukin, E.D. and Yushchenko, V.S., *J. Mater Sci.,* 16, 313, 1981.)

7.2 INFLUENCE OF A SURFACE-ACTIVE MEDIUM ON THE MECHANICAL STABILITY OF THE SURFACE OF A SOLID SURFACE DAMAGEABILITY

In this section, the determining role of primary surface damage as the necessary condition for causing fracture will be addressed. Surface damage is encountered in a large variety of processes taking place in nature and technology. Some obvious examples include wear, the smoothing of sand in deserts and of gravel at the sea coast, the formation of harmful dusts in mines, explosions, the wear on bearings and all sorts of mechanisms and tools, and the wear on the surface of teeth due to excessive grinding. Consequently, numerous methods for protection against surface damage are utilized in friction nodes, in the processes of rolling, drawing, etc. These involve the use of various liquid and solid (e.g., graphite) lubricants and coverings, alongside with plasticizing the material surface. Certain lubricants are created by nature, such as the perfect lubricants in our joints, saliva, and tears. At the same time, in the processes involving treatment or processing of various materials (e.g., in metalworking), surface damage constitutes the necessary initial stage of any treatment process.

In nearly all cases fracture is initiated by surface damage. A rather obvious, and yet very general and determining, factor is the fact that nucleation and the development of surface damage serve as a precondition for the initiation of local fracture processes leading to the loss of stability in solids. This, in turn, means that a solid surface is most vulnerable to the action of the ambient medium. In other words, the surface is the *source* of both: the free energy excess due to the presence of uncompensated chemical bonds and the exposure to the active components of the medium [10].

Over the years, numerous versatile studies in the area of surface damageability have been carried out and significant progress has been achieved in the elucidation of the mechanisms and trends involved in fracture and stability loss in solid bodies of any nature and structure. Critical and priority studies were carried out by Shchukin in collaboration with Savenko, Kochanova, and Kuchumova [44–46,59,68,70].

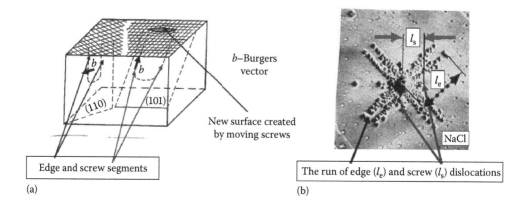

(a)   (b)

FIGURE 7.38 Edge and screw dislocations in a cubic crystal (a). Etching of the points where dislocations emerge in the course of the indentation of the NaCl surface (100) (b). (From Shchukin, E., *Physical-chemical mechanics of solid surfaces*, in: *Encyclopedia of Surface and Colloid Science*, 2nd edn., Taylor & Francis, New York, 2012, pp. 1–23; Kuchumova, V.M. et al., *J. Mater Sci.*, 27, 5516, 1992; Savenko, V.I. and Shchukin, E.D., *Kolloidnyi Zh.*, 69, 834, 2007.)

New surface cells are created by the screw components of the dislocation half-loops when they "pierce" the surface (Figure 7.38). This is associated with performing the work $b^2\sigma$ per cell. In contrast to this, the movement of edge components that are perpendicular to the surface is not associated with forming a new surface, and hence no work is consumed. When an edge component encounters a "foreign" surface, such as an oxide film, more work is required to pierce the surface, and the magnitude of this extra work can be much more severe for the screw component. Conversely, in the case when the surface is exposed to the active medium, less work is required because of surface energy lowering. There are no such effects for the perpendicular movement of the edge components, that is, the mobility of the edge dislocations is not significantly influenced by the active medium.

Let us present our rather simplified scheme of the application of a thermodynamic approach, following the Gibbs–Volmer–Griffith model, to the Hertz problem, described in Section 5.2. Let an indenter leave a hemispherical pit hole with a diameter equal to $2a$ and a depth equal to h in the plane of the sample surface (Figure 7.39).

The order of magnitude of the deformations can be estimated as $\varepsilon_{aver} \sim h/a$, which is the only dimensionless combination of the parameters involved. Similarly, the stresses by order of magnitude are $\sigma_{aver} \sim F/\pi a^2$. Within the framework of Hooke's law, we have $\sigma_{aver} \sim F/\pi a^2 \sim E\varepsilon_{aver} \sim h/a \sim Eh/a$. In order to find the final solution, we need one more equation establishing the relationship between the geometric parameters. This relationship is obtained using the Pythagorean theorem, namely, $a^2 = h(2R - h)$, or for $h \ll 2R$, $a^2 \approx 2Rh$, that is, $h \approx a^2/2R$. Substituting this expression into Hooke's law, $F/\pi a^2 \sim Eh/a$, one finds that $a \sim (FR/E)^{1/3}$, $\sigma \sim (E^2F/R^2)^{1/3}$, and $h \sim (F^2/E^2R)^{1/3}$. The work of elastic deformation is given by

$$W_{el} = \int F\,dh \sim Fh \sim \left(\frac{F^5}{E^2R}\right)^{1/3} \sim \frac{a^5E}{R^2}$$

By performing a cumbersome integration or by employing dimension analysis, we can come up with the same mean values of the work of elastic indentation as a function of the measured radius, a, of a hole. This free energy excess is proportional to a^5. The precision of the estimated mean values is within an order of magnitude, or sometimes even within the precision of dimensionless

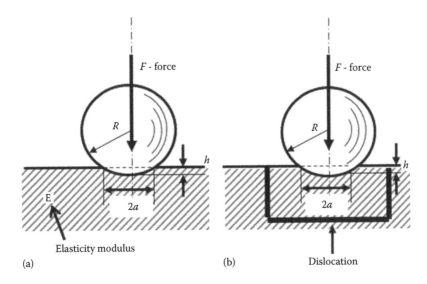

FIGURE 7.39 A simplified approach to the Hertz problem. (a) The indentation with a sphere results in an elastic deformation of the substrate; (b) a portion of the elastic energy can be used to create a dislocation.

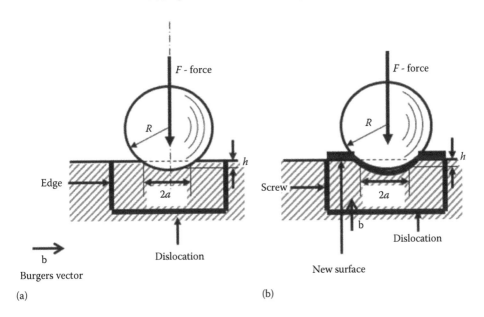

FIGURE 7.40 Formation of dislocations in the course of indentation of a surface with a sphere. (a) Edge components enter the surface layer without the formation of a new surface, and (b) screw components cause a formation of a new surface both in the vicinity and underneath the indenter.

coefficients, which are not very different from 1. Let a dislocation half-loop with the Burgers vector b parallel to the surface (i.e., with edge components) be created as a result of this work (Figure 7.40). This half-loop has a linear tension $\ae = kb^2E$ ($k < 1$) and a length $l_e = na$ ($n > 1$). The change in the free energy of the system $\Delta F \sim W_{disl} - W_{el} \sim kb^2Ena - a^5E/R^2$.

The condition $d\Delta F/da = knb^2E - 5a^4E/R^2 = 0$ yields the following critical values: $a_c \sim (Rb)^{1/2}$, $h_c \sim b$, $\varepsilon_c \sim h_c/a_c \sim (b/R)^{1/2}$, $P_c \sim \varepsilon_c E \sim (b/R)^{1/2}$, and $F_c \sim a_c^2P_c \sim b^{3/2} R^{1/2}E$. The height of the potential barrier is $W_c \sim F_ch_c \sim b^{5/2}R^{1/2}E$. For $R = 25$ µm, $E = 0.8 \cdot 10^{12}$ Pa, and $b = 3$Å: $a_c \sim 0.1$ µm, $e_c \sim 0.004$, $P_c \sim 3 \cdot 10^8$ Pa; $F_c \sim 0.3$ dyn $= 0.3 \times 10^{-5}$ N; $W_c \sim 10^{-15}$ J.

If the Burgers vector crosses the surface (screw components), a new unit surface step with the length $l_s \sim ia$ ($n > 1$) is formed. This act requires that the work $W_s \sim bma\sigma$ is spent. In this case, one can write that $W_c \sim W_{disl} + W_s$ and F_c increases. The latter means that the screws are less mobile.

The adsorption from the active medium decreases the surface energy, σ, which results in a lowering of the force F_c and hence in an increase in the mobility of screws (Figure 7.41) [46,68]. It is typical that such an obvious difference in the histograms of the propagations (runs) of the leading edge and screw dislocations is evident at small loads, that is, within a thin near-surface layer.

Within the present scheme, the work needed to create a dislocation half-loop is proportional to a^1. Such work does not depend on σ for the edge component, while for the screw component it does depend on σ. Competition between the positive term, which has the argument raised to the lower power, and the negative term, which has the argument raised to the higher power, determines the critical size of the hole and the critical load on the indenter. Both of these values are in good agreement with the experimental values.

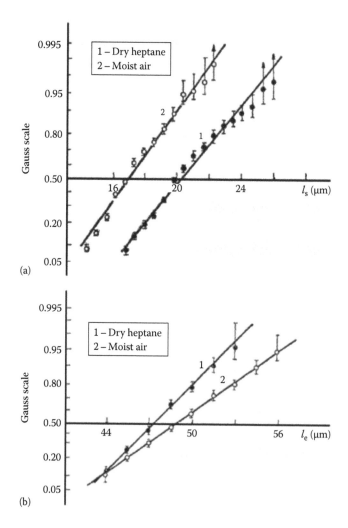

FIGURE 7.41 Histograms of the distances covered (a) by the leading screw dislocations, l_s, and (b) by the edge dislocations, l_e, in the indentation rosettes on the surface of NaCl single crystals in dry heptane and in moist air. The indenter load was 0.3 g force. (Redrawn from Shchukin, E., Physical-chemical mechanics of solid surfaces, in: *Encyclopedia of Surface and Colloid Science*, 2nd edn., Taylor & Francis, New York, 2012, pp. 1–23.)

The traditional microindentation of the surface of ionic and covalent crystals allows one to study the effect of adsorption on the movement of the screw components of the dislocation half-loops formed, but only outside the contact zone. The capabilities are broadened with the use of the micro-sclerometric and ultramicrosclerometric (scratching) methods developed by Savenko and coworkers [46,68,70]. A step-by-step increase in the load applied to the indenter allows one to observe a transition from the reversible elastic contact to the appearance of the very first damage, that is, near-surface dislocations, and further to the development of plastic deformations, and then to microcrack nucleation (Figure 7.42). The adsorption taking place from the active medium can both facilitate damageability and retard it.

The extensive facilitation of the damageability of LiF crystals under the influence of water in comparison with the behavior of the same sample immersed in a hydrocarbon is shown in Figure 7.43, which shows the probability of the appearance of plastic deformation zones (microcracks). The same picture also shows the protective effect of a monolayer of octadecylamine. This effect is particularly significant at low loads on the indenter, that is, in the near-surface layer. The protective role of the monolayer coating of octadecylamine is also shown in Figure 7.44, illustrating the probability of plastic deformation of Mo single crystals as a function of the applied load.

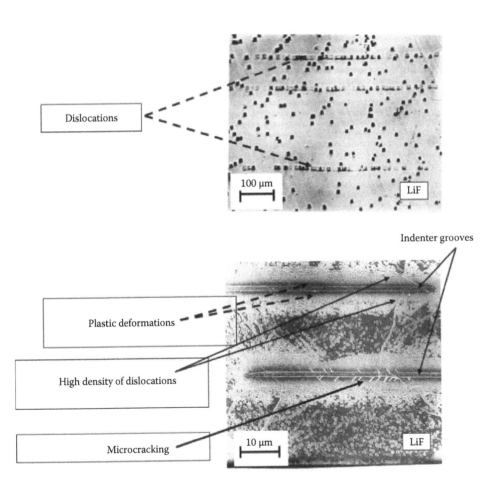

FIGURE 7.42 The very first signs of damage at a very small indenter load: the appearance of dislocations corresponds to a nanoscale-level transition from elastic to plastic contact. (From Kuchumova, V.M. et al., *J. Mater Sci.*, 27, 5516, 1992.)

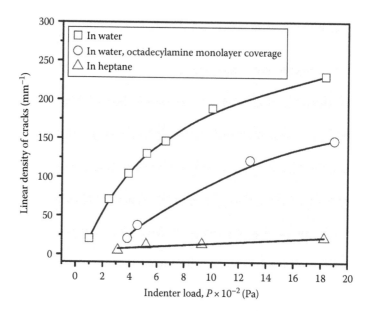

FIGURE 7.43 The linear density of cracks, n (mm), along the indenter path on the LiF (100) surface in water, in water in the presence of a layer of octadecylamine, and in heptane as a function of indenter load. (Redrawn from Shchukin, E., Physical-chemical mechanics of solid surfaces, in: *Encyclopedia of Surface and Colloid Science*, 2nd edn., Taylor & Francis, New York, 2012, pp. 1–23.)

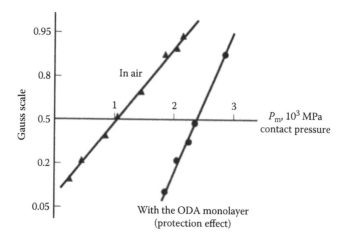

FIGURE 7.44 The probability of plastic deformation of Mo single crystals in air—for pure crystals and crystals coated with a monolayer of octadecylamine. (Redrawn from Shchukin, E., Physical-chemical mechanics of solid surfaces, in: *Encyclopedia of Surface and Colloid Science*, 2nd edn., Taylor & Francis, New York, 2012, pp. 1–23.)

Experiments in which the small compressive forces were measured in the approach and separation of *different* crystalline and amorphous samples showed the retraction of the adsorption layers as in the case of nonspecific adsorption, as well as the formation of rather strong monolayers that prevented both damage to and cohesion of the sample.

When the adsorption-active ingredients of the medium are present in the friction units, they reinforce both an initial break-in of the surfaces and a protection against wear at later stages. In processes involving pressure treatment, the same components plasticize the near-surface zone of the material.

Specific cases of fracture onset are observed when metals and high-strength ceramics are microscratched while in contact with metal melts. Of special interest is the investigation of the surface damage occurring under a combination of the medium effects and electric polarization. In this case, there is a possibility of using electrical-chemical-mechanical treatment, namely, liquid metal embrittlement in the absence of a liquid or a metal phase. The latter is achieved by the reduction of the active ions on the treated surface.

A comparison of these methods and the corresponding results enables one to examine the mechanisms of the Rehbinder effect. On one hand, this implies the return to the problem of generalization and the limitations associated with the thermodynamic approach when applied to the description of the conditions of surface stability loss according to Gibbs and Griffith. On the other hand, it is possible to apply Zhurkov's kinetic scheme to the description of the surface vulnerability. One of the problems here is the comparative analysis of the kinetic patterns of the influence of the adsorption effects on either enhancing the vulnerability of the surface or, conversely, on the protection of the surface along with its plastification. At the same time, in the limiting cases of liquid metal embrittlement, the reduction in the interfacial energy must be considered together with the manifestation of the high mobility of the active atoms.

Let us now focus on significant experimental work originally conducted by Kochanova, Polukarova et al., in which the residual damage in the surface of glass was examined [85].

7.2.1 LOWERING OF THE STRENGTH OF GLASS DUE TO MICROSCOPIC DEFECTS THAT WERE INTRODUCED TO ITS SURFACE

An understanding of the process of the fracturing of noncrystalline solids, and of glasses in particular, is of scientific and practical importance. This process has been clearly observed experimentally as a propagation of microscopic and macroscopic cracks. The well-known Griffith scheme [31] and the subsequent works of Frenkel [86], Elliott [87], and Barenblatt [88] on the analysis of the behavior of cracks in elastic–brittle bodies allow one to estimate the critical sizes of cracks that become "dangerous" in a given stressed state. In the end, the fracturing stress, P, is related to the crack length, l, via the relationship $(E\sigma/l)^{1/2}$, where E is the elasticity modulus and σ is the surface free energy. This shows that the fracture of glass under the common stresses (5–10 dyn/mm^2) is associated with the formation of cracks on the order of 10 μm. At the same time, experiments show that there are no such cracks present in the unloaded glass.

Similar contradictions in the theory of the fracture of crystalline bodies have been resolved by expanding our understanding of the role of structural defects in crystals, and in particular dislocations.

Any real crystal contains a variety of ultramicroscopic defects, such as dislocations or other structure distortions that can act as dislocation sources. In the process of even a very small deformation under some applied stress, the propagation, movement, and interaction of these structural defects, and primarily dislocations and dislocation groups, result in a sharp microheterogeneity of plastic deformation. The latter causes local stress concentrations and results in areas with high local free energy density. Such areas are the source of the formation of microcracks, which then reach "Griffith sizes" by gradual growth.

Most likely, similar processes take place in the fracture of glasses. It is well known that the structure of glasses formed from a supercooled liquid has an abundance of a variety of defects. The latter include nonuniformity in density and chemical composition, various near-order disruptions, and, conversely, the presence of microscopic regions in which a partial far-order one has been formed in the process of glass formation. It is possible that certain types of defects in the glass structure, such as pseudo-dislocations [89–92], have a certain degree of mobility and may group together like the arrays of dislocations in crystals. It is likely that under the action of mechanical stresses, some defects have an opportunity to develop into the local stress concentrators stipulating a primary breaking of the interatomic bonds and gradual formation of ultramicrocracks. In this respect, it is worth emphasizing the significance of the results obtained by Aleksandrov and Zhurkov [93], who showed that the long-term trends in the strength of glasses are the same as in metals.

One peculiar feature associated with the fracture of glasses is that it always originates on the surface. This means that the near-surface layers have the most favorable conditions for the formation and development of defects resulting in microcracks. Numerous studies indicated that the removal of the near-surface layers leads to a significant increase in the strength of glass.

The primary objective of that study was the quantitative description of the role played by defects in the glass structure. This was achieved by causing microscopic defects in the surface layer of glass. The decrease in strength caused by these defects was meant to imitate the lowering of strength caused by real defects. It was of interest to produce defects with a characteristic local concentration of residual deformations and a local concentration of stresses. One could then study the impact of these defects on the strength of the sample. Defects introduced with the diamond indentation pyramid using the microhardness meter PMT-3 were studied [85]. Glass samples were prepared by cutting glass microscope slides into 25×25 mm pieces 1.3–1.4 mm thick. The samples were thoroughly cleaned and were stored and tested in air under normal humidity. The puncture with the indenter was introduced in the center of the glass sample and the point of indentation was marked with a marker on the opposite side of the slide. The mechanical testing was conducted with a hydraulic press using a 15 mm steel ring support and a 3 mm punch. The glass was positioned between two filter paper sheets so that the indentation ink mark was in the center of the punch. The glass strength, P_c, represents the maximal stresses acting on the lower side of the sample at the point of fracture. The glass strength was estimated using the known equation, which for a given set of parameters with a precision of 1% can be reduced to the simple expression $P_c = F/d^2$, where F is the punch load and d is the glass thickness. Mean values of P_c were estimated by averaging the results of $n = 10$–20 measurements. The spread of the data was estimated as the mean square error of a single measurement, $\sigma = \pm [\Sigma_i(P_c - P_{ci})^2/(n - 1)]^{1/2}$, as shown by the error bars in Figure 7.45, or as the probability error of the mean, $r_A = \pm 0.6745[\Sigma_i(P_c - P_{ci})^2/n(n - 1)]^{1/2}$, as shown in Tables 7.3 and 7.4.

Figure 7.45a shows the values of the strength of the glass samples indented with the diamond pyramid under various loads, p, (h is the corresponding depth of the indent images). The shaded area in the vicinity of P_{c0} corresponds to the range of the strength of the unpunctured glass. Figure 7.45b illustrates similar data for the samples treated for 10 min in 15% hydrofluoric acid solution prior to indentation. When a substantially high load was placed on the indenter, a sharp decrease in the sample strength was observed. At the same time, at low values of p and h, the values of P_c were statistically indistinguishable from the strength of unindented glass. Punctured glass had typically fractured along the radii, forming four to six sectors, while the unpunctured glass was crushed into many small pieces along radial and concentric cracks. One may therefore conclude that the real defects existing in the glass were equivalent to the indentation with a pyramid under a particular range of loads (or range of image depths). For untreated glass this range was approximately 1–3 μm, while for glass etched with hydrofluoric acid it was about 0.5–1 μm.

The data in Figure 7.46 show how one can use the indenter-caused defects to estimate the distribution of the stresses in a sample, and for the assessment of the risk, they pose to the stability of the glass. In this case the indentation was not in the center but was imposed at various distances, r, from the center. The data show that the sample strength increased with an increase in r. However, even in the case of indentations close to the sample edge, the strength was still lowered considerably, which corresponds to a particular level of stresses applied at a given point.

The distortions originated from the indentation are of particular interest. Observations under an interference microscope clearly showed the result of the residual deformations caused by the indentation. Images taken in polarized light revealed the presence of a rosette of residual stresses (Figure 7.47).

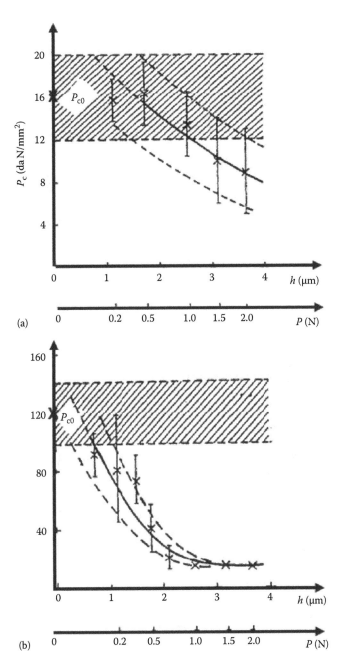

FIGURE 7.45 The strength of the glass, P_c, as a function of the indentation depth, h, using a diamond pyramid (or as a function of the indenter load, p): (a) untreated glass and (b) glass strengthened by HF treatment. P_{c0} is the strength of unindented glass. (Data from Shchukin, E.D. et al., *Fiz.-Him. Meh. Mater.*, 2, 128, 1965.)

It is difficult, however, to study the impact of this localized concentration of deformations and stresses on the strength, because the images of the diamond pyramid reveal diagonal cracks even at small loads (Figure 7.47a). The measurement of the length of these cracks as a function of the load allows one to determine the so-called glass microstrength. The challenge is to separate the effect of these cracks (as well as the impact of the image itself) from the impact of the residual deformations and stresses. Two different approaches were used to achieve this. One approach was to heat-treat the

TABLE 7.3

Effect of Indentation Using a Diamond Pyramid, Followed by Heat Treatment for 3 h at 350°C, on the Strength of Glass

Indenter Load, p (N)	$P_{c0} \pm r_A$ (dyn/mm^2)	$P_c \pm r_A$ (dyn/mm^2)	P_{c0}/P_c
	No puncture/no heat	With puncture/no heat	
0.2	102 $\pm$ 4	56.5 $\pm$ 2	1.8
1.0	102 $\pm$ 4	10 $\pm$ 0.1	10
	No puncture/with heat	With heat/with puncture	
0.2	70 $\pm$ 2.5	69 $\pm$ 3	1.0
1.0	70 $\pm$ 2.5	15 $\pm$ 0.2	4.7
	No puncture/with heat	With puncture/with heat	
0.2	70 $\pm$ 2.5	29 $\pm$ 2.5	2.4
1.0	70 $\pm$ 2.5	9.5 $\pm$ 0.1	7.4

TABLE 7.4

Effect of Indentation with a Diamond Cone or Corundum Needle on the Strength of Glass with an Indenter Load of 0.2 N and Heat Treatment for 3 h at 350°C

Indenter Load	$P_{c0} \pm r_A$ (dyn/mm^2)	$P_c \pm r_A$ (dyn/mm^2)	P_{c0}/P_c
	No puncture/no heat	With puncture/no heat	
Diamond cone	89 $\pm$ 6	61.5 $\pm$ 3	1.5
Corundum needle	87 $\pm$ 7	50 $\pm$ 5	1.5
	No puncture/with heat	With heat/with puncture	
Diamond cone	65 $\pm$ 2	64 $\pm$ 2	1.0
Corundum needle	72 $\pm$ 4	71 $\pm$ 3	1.0
	No puncture/with heat	With puncture/with heat	
Diamond cone	65 $\pm$ 2	40 $\pm$ 1	1.6

indented samples to the point when the residual stress rosette disappears (as visualized in polarized light), while the cracks and the indenter impression have not yet started to heal. This was achieved by heating for 3 h at 350°C. The healing of cracks could be monitored by measuring the crack length, l, after a brief treatment with 1.5% hydrofluoric acid. Some of these results are summarized in Table 7.3.

In order to standardize the experiments, the samples underwent the following treatment prior to testing: first, they were exposed to 15% hydrofluoric acid for 10 min, followed by ageing in air for 1 day; second, they were heat-treated at 350°C for 12 h and then left to cool in the furnace after it was turned off; third, the samples were exposed to a second 10 min treatment with 15% hydrofluoric acid. The samples were then split into five groups for the strength testing. Group A contained the control samples with no further treatment. Group B was heat-treated for 3 h at 350°C. Group C underwent indentation with a diamond pyramid without heat treatment. Group D experienced indentation and subsequent heat treatment at 350°C for 3 h. Group E was heat-treated at 350°C for 3 h with subsequent indentation. The results are summarized in Table 7.3. These data indicate that heat-treating the samples without subsequent indentation resulted in a decrease in strength (i.e., the surface became damaged), while the strength of indented samples was restored. For the indentation with a load of 1 N, partial restoration of the strength was observed, while for small loads, the strength was restored up to a level corresponding to the same heat treatment without prior indentation. One can thus draw a conclusion regarding the influence of the local residual

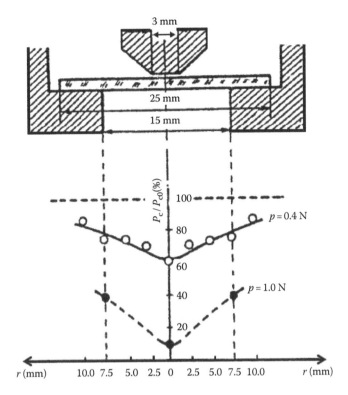

FIGURE 7.46 The strength of glass samples indented using a diamond pyramid in the center and at various distances from the center, r. The strength is shown as % of P_{c0}. Data are shown for two indenter loads: $p = 0.4$ and 1 N. (Data from Shchukin, E.D. et al., *Fiz.-Him. Meh. Mater.*, 2, 128, 1965.)

FIGURE 7.47 Microphotographs of the indentation images on glass for the diamond pyramid (upper) and the diamond cone (lower) at indenter loads $p = 1, 0.7, 0.4$, and 0.2 N (a). Near all of the impressions, with the exception of the cone at $p = 0.2$ N, one can see small cracks (b). The same microphotographs taken in polarized light with a 70× magnification. (Data from Shchukin, E.D. et al., *Fiz.-Him. Meh. Mater.*, 2, 128, 1965.)

deformations and stresses on the strength of glass. This conclusion is also supported by a comparison of the strength of the heat-treated samples after indentation (group D) and prior to indentation (group E). The additional strength lowering observed in the latter case would not have been possible if it had been solely related to the geometry of the impression and that of the resulting cracks. The geometry was not impacted by the heat treatment.

A second way of establishing the influence of local residual stresses and deformations is based on the selection of the impressions that did not reveal any cracks at all. This was the case for indentations at low loads using both a diamond cone and a corundum needle. Figure 7.48 shows the maximal crack length (developed by a light exposure to HF solutions) as a function of the load applied

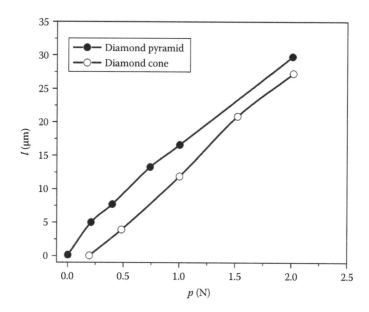

FIGURE 7.48   The maximum crack length, l, as a function of the indenter load, p, (1) for the diamond pyramid and (2) for the diamond cone. (Data from Shchukin, E.D. et al., *Fiz.-Him. Meh. Mater.*, 2, 128, 1965.)

to the diamond pyramid and the cone, and Figure 7.47 shows the corresponding microphotographs. Low loads applied to the cone produced oval impressions. These data show that there were no cracks that could significantly influence the sample strength. At the same time, as seen in Table 7.4, these defects also caused a substantial strength decrease in the samples. However, the strength was completely restored after subsequent heat treatment. The experiments summarized in Table 7.4 were conducted in the same way as those listed in Table 7.3 using an indenter load of 0.2 N. In this case, there was no direct influence on the imprint geometry at all, and the indentation-induced strength lowering was entirely caused by the residual deformations and stresses.

We have thus demonstrated that a simple method for introducing artificial defects using an indenter allows one to evaluate the role of real structural defects and investigate the role that local concentrations of deformations and stresses play in glass fracture. These results provide a basis for the comparison of fracture mechanisms in crystalline and noncrystalline bodies.

Most likely, the observed analogies in the fracture of crystalline and noncrystalline bodies cover a broad range of objects. Indeed, all modern methods of increasing the strength of metals involve the creation of a fine-disperse structure with a high density of equally distributed ultramicroscopic defects (such as grain boundaries, block boundaries, and phase inclusions). One can also use the same approach to analyze the mechanical characteristics of glass ceramics with fine-disperse microcrystalline frame structure. Such a structure is responsible for the high thermal strength and other important properties of these materials. Especially beneficial is the thermodynamically equilibrium structure of an *ideal* glass, which in the process of cooling completely maintains the structure of a homogeneous liquid without any kind of nonuniformities or ordering. This type of structure, which is saturated with defects on the order of atomic dimensions and does not contain any larger defects, can have very high strength.

The analogies between glasses and crystalline solids can also be observed in the processes of adsorption-induced strength lowering and facilitated deformation in the presence of surfactants that lower the surface free energy. It is well known that humidity and other surface-active media can significantly lower the strength of glass. The works by Aslanova and Rehbinder [94] showed that due to adsorption from active media, glass may reveal a significant elastic aftereffect once the load has been removed. The adsorption-active medium may facilitate the development of residual microdeformations in the surface layer of glass. One may consider conducting further studies using the method described on the influence of defects on the strength of glass exposed to various media. As in the studies on the adsorption-induced effects in metals, these studies will allow one to achieve a better understanding of the adsorption-induced strength lowering in noncrystalline solids.

7.3   REHBINDER EFFECT IN NATURE AND TECHNOLOGY

The ability of adsorption-active media to facilitate the fracture of solids has been utilized in grinding applications, for example, in the milling of ores and cements, and in various processes involving dispersion. As was pointed out in Section 7.1, fine grinding can't be achieved solely by mechanical means: the development of an enormous surface area requires the involvement of physical-chemical factors. The role of adsorption-induced strength lowering is essential not only in the facilitation of solid fracture but also in the prevention of aggregation and the destruction of coagulation contacts formed between the particles.

At the same time, adsorption-active components are commonly present in the formulations of the cutting fluids used in various metalworking applications, such as drilling, cutting, grinding, and polishing. All of these processes involve the dispersion of the treated material. The surface phenomena that take place in dispersion processes are not limited to a lowering of the surface energy and strength due to the physical adsorption of molecules. The use of cutting fluids is associated with *chemisorption* and *mechanical-chemical* phenomena. These are related to the degradation of organic molecules due to the combined action of mechanical stresses, high temperatures in the cutting zone, and the interaction of the molecules with the newly formed (juvenile) solid surface, which have an elevated chemical activity.

The pressure treatment of metals, such as drawing, stamping, and rolling, is based on the *plasticizing action* of the medium and commonly involves the use of surfactants. The presence of surfactants in lubricating greases results in the plasticizing of the surface layers of the treated metal. This lowers the energy necessary to perform pressure treatment, improves the quality of surfaces, and lowers the extent of cold hardening in the near-surface layer of metal. The adsorption layers formed prevent cohesion (prevention of phase contacts) between the tool surface and the treated material, because they are not displaced out of the contact zone.

Lubricating greases are essential for the performance of friction units in various machines and mechanisms. In these cases, surfactants play a dual role: at early stages of their use, they enhance wear and thus facilitate a breaking-in of the friction surfaces, while at later stages, they form protective adsorption layers that help to retard wear in the parts. The role of surface-active media in the processes of friction and wear is the subject of *tribology*, which constitutes a separate area of physical-chemical mechanics [95–106].

The prediction of a catastrophic failure in constructions due to the melting of the antifriction melts and the anticorrosion coatings in the processes of soldering and welding requires a detailed understanding of the mechanisms of liquid metal embrittlement and of strength lowering in hard metals in the presence of melts. In certain cases, it may be possible to protect metallic constructions that come into contact with liquid metals from the selective action of the melt at grain boundaries. This is achieved by alloying the solid metal with a component that is capable of concentrating at the grain boundaries and preventing the penetration of liquid metal, but is itself not a source of adsorption-induced strength lowering.

At the same time, the lowering of metal strength by melts can be used to facilitate the mechanical treatment of very hard, intractable materials (Figure 7.49).

One example of the utilization of adsorption-induced strength lowering in these processes is the use of small amounts of fusible surface-active metals in the treatment of hardened steels and solid alloys. The powder of a fusible metal is introduced into the polymeric binder of grinding wheels along with the diamond powder (Figure 7.50). Due to the increase in temperature in the course of the grinding, the microamounts of active metal are melted and lower the strength of hard metals and

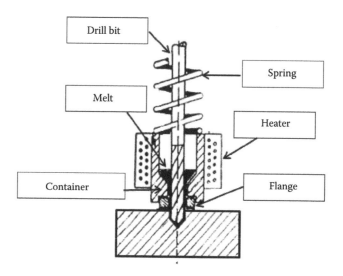

FIGURE 7.49   Schematic illustration of the application of liquid metal embrittlement in the drilling of very hard materials.

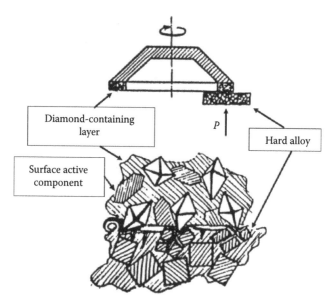

FIGURE 7.50   The introduction of an active component into the diamond grinding wheel. This additive facilitates the treatment of hard alloys. (Redrawn from Shchukin, E.D. (Ed.), *The Successes of Colloid Chemistry and Physical-Chemical Mechanics*, Nauka, Moscow, Russia, 1992.)

alloys (such as tungsten carbide). A sharp decrease in the strength of the materials treated allows one to increase the treatment rates with a simultaneous increase in the longevity of grinding wheels.

The presence of a liquid metal phase makes it difficult for one to study and utilize the effect of liquid metal embrittlement because of the need to employ high temperatures, the difficulties in ensuring good wetting conditions, and especially the need to remove the residual films and traces of the active component. A principally new approach to overcoming these issues was the use of liquid metal embrittlement in the *absence of a liquid metal phase*. This is achieved in an electrochemical cell by the cathode reduction of ions of the selected surface-active metal directly at the sample surface. The amount of metal formed can be controlled and fine-tuned down to the formation of monolayers. It is also possible for one to conduct these studies at room temperature [59,107,108] (Figures 7.51 and 7.52).

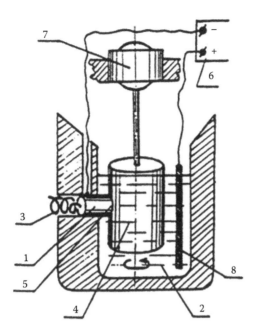

FIGURE 7.51 The experimental scheme of the electro-chemo-mechanical treatment of the extremely hard metal materials: (1) sample, (2) electrolyte, (3) loading, (4) grinding wheel, (5) treated surface, (6) power supply, (7) motor, and (8) anode. (Redrawn from Videnskiy, I.V. et al., *Colloids Surf.,* 156, 349, 1999; Shchukin, E.D., *Colloids Surf.,* 49, 529, 1999; Videnskiy, I.V. et al., *J. Mater. Res.,* 8, 2224, 1993.)

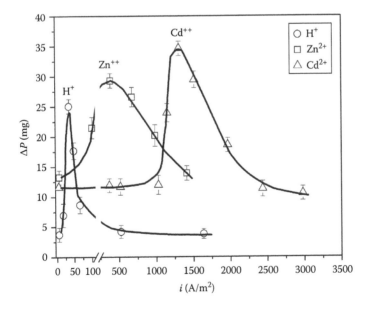

FIGURE 7.52 The mass of metal removed from the surface of hardened steel in the electro–chemo–mechanical treatment process as a function of current density, i. The required current density is several orders of magnitude lower than in the common processes of electrochemical treatment. (Data from Videnskiy, I.V. et al., *Colloids Surf.,* 156, 349, 1999; Shchukin, E.D., *Colloids Surf.,* 49, 529, 1999; Videnskiy, I.V. et al., *J. Mater. Res.,* 8, 2224, 1993.)

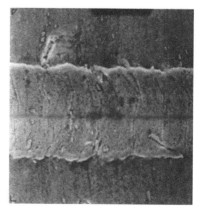

FIGURE 7.53 Electron microphotographs (500×) of microscratches on the surface of aluminum metal. (From Shchukin, E.D. et al., *Colloids Surf.,* 142, 175, 1998.)

Both the thermodynamic and the kinetic criteria for the Rehbinder effect are met in the electro-chemo-mechanical treatment (ECMT) process (Figure 7.51). The thermodynamic condition constitutes a contact between baring surface and the surface-active component, which results in a substantial lowering of the surface free energy of the treated metal. The kinetic factor constitutes the high mobility of the active atoms, which ensures their penetration into the fracture zone. The cathodic reduction of cadmium and zinc was used in place of the corresponding melts in order to enhance the effectiveness of hard steel polishing.

The principle of ECMT has led to the effective development of a method to study the selectivity and conformities of liquid metal embrittlement. A specialized experimental setup developed by Savenko and Vidensky allowed one to observed the stability and damageability of a solid metallic surface using a combination of microscratching and electrochemical production of the active component (the so-called electrochemical microsclerometry) [69,70,107–109]. The results of some studies using this method are shown in Figure 7.53. This figure shows the behavior of the surface of ultrapure (99.999%) aluminum in the solution of gallium salt.

The width of the groove obtained using cathodic polarization of the sample with a current density of 6.8 mA/cm^2 (right) is significantly larger than the width of a groove formed in the absence of the electric current (left). In this case, the initial ductility of the tested metal is very high, and the stressed state within a contact zone is rather mild. Under these conditions, the lowering of the mechanical strength of material results in *plastification* rather than in embrittlement. This effect is manifested as an increase in the groove width by nearly 20% and becomes more pronounced with the decreasing indenter load, that is, with the localization of deformations in a thin near-surface layer. Similar behavior was observed with a gold substrate treated with cathodically applied mercury. However, in the case of a material of high hardness (titanium aluminide), the active component (cadmium) causes the appearance of brittle cracks. It is possible that the ability to differentiate between these two manifestations of the adsorption-induced media's influence on the mechanical properties of metals can lead to potentially novel ways to optimize friction, wear, and lubrication. This method also allows one to study the influence of the reduction of nonmetallic cations, namely, hydrogen brittleness, the cracking of brass in the presence of ammonia, and anodic processes (voltage-induced corrosion).

Adsorption-induced strength lowering plays an important role in the use of various adsorbents and catalysts [7,110–112]. The adsorption or chemisorption on solid surfaces, which leads to a lowering of both the surface free energy and the strength, is the essential feature necessary for the proper functioning of these materials. Contact with a solid phase facilitates the breaking up and rearrangement of the interatomic bonds in the adsorbing molecules, which constitutes the effect of the mutual influence of the solid surface and media molecules on each other. The processes of the adsorption and rearrangement of adsorbate molecules lead, in turn, to the weakening of the chemical bonds in the surface layers of catalysts. Under these conditions, the excessive wear of the catalysts may occur due simply to the action of the internal stresses that originated in the process of making catalyst granules. The pressure exerted by the upper layer of the granules can accelerate the degradation process, which can be especially strong in the intensive regime of a boiling layer. The formation of an optimum condensation structure with phase contacts established between the particles making up the catalyst granule is an effective way to prevent accelerated catalyst degradation. The specific results of studies on the influence of active media on the strength and durability of catalysts is described in the following section, which concludes this chapter. These very significant studies include the direct experimental investigation of adsorption-induced strength lowering in porous disperse structures. These studies allowed one to apply the Griffith equation to the independently measured strength lowering and surface free energy of the sample. Furthermore, the studies on catalysts led to the development of instrumentation and experimental methods for studying the strength of materials with a porous structure.

7.3.1 Physical-Chemical Mechanics of Catalysts: The Strength and Durability of Fine-Porous Materials in Active Media

The mechanical degradation of granules, which is enhanced by exposure to the active medium, is the main reason for the loss of catalysts employed in numerous heterogeneous catalysis processes. The author and associates have developed a number of experimental techniques and devices for studying the mechanical characteristics of catalyst granules and various sorbents. The techniques developed made it possible to investigate the properties and stability of catalysts used in numerous industrial processes (synthesis of ammonia, crude oil cracking, making monomers for synthetic rubber, etc.) under various loading conditions and various work temperatures. The results of experiments carried out with MgO, Co–Mn, Ca–Ni, Ca–Ni–P, Al–Cr–K, and other systems clearly indicated that the strength and durability of catalysts degrade significantly during the catalytic process in comparison with the same properties measured in an inert medium using the same methods. This observation can be explained by the mutual influence of the solid phase and the medium, which takes place throughout the catalytic process. Indeed, the formation of new chemical bonds between the catalyst and the medium may result in the weakening and rupture of the chemical bonds between both the adsorbing molecules and the molecules of the solid surface. This is a manifestation of the Rehbinder effect. In a thermodynamic sense, this implies that the work of the elementary catalytic act favors the mechanical rupture of the interatomic bonds on the solid surface or can even be sufficient for rupture to take place. One can say that the failure of catalysts is a direct result of their use. However, the resistance of catalyst granules to wear can be significantly improved by implementing changes at every stage of the technology. In particular, this can be achieved by an optimum selection of the composition of the particles making up the granules and by enhancing the strength of the contacts between the particles. This can be achieved via the use of fine-disperse inert fillers, by the hydration hardening of mineral binders or by lowering the residual stresses. The rupture of chemical bonds in the course of catalysis may be accompanied by the formation of new surfaces and atoms. The latter in turn results in the acceleration of surface self-diffusion and particle sintering, leading to an increase in strength, as was shown for metals (Fe, Ni) and porous ceramic materials (aluminum oxide, zirconium oxide, or yttrium oxide).

The chemical and petrochemical industries consume millions of tons of catalysts utilized in various heterogeneous processes. Mechanical degradation and wear, rather than poisoning, are the main factors leading to the loss of these valuable materials. The idea that adsorption and chemisorption are responsible for enhancing the degradation of catalysts has been around for quite some time [34,113]. Numerous detailed studies exploring the thermodynamic nature of the Rehbinder effect and studies combining a thermodynamic approach with dislocation models of fracture have led to an understanding of the origin of numerous cases of the catastrophic failure of materials.

Thanks to the universal nature of the Rehbinder effect, similar studies and approaches were applied to the investigation of durability and wear in industrial catalysts and adsorbents [114–118]. In these processes, the required thermodynamic and kinetic factors of the Rehbinder effect are present. In particular, the lowering of the surface energy of the solid due to intensive processes of adsorption and chemisorption (involving the processes involving intermediates) takes place along with the presence of mechanical stresses. They include the stresses caused by the weight of the catalysts in the column, thermoelastic stresses caused by cyclic changes in temperature in catalysis, residual internal stresses remaining after the preparation of the granules, as well as collision and friction between the granules and the moving layer. The kinetics of adsorption-induced strength lowering may be especially fast in porous disperse structures with open porosity. The pores and contacts between the particles in the granules are easily accessible to the dispersion medium, the penetration rate of which into the granules is not a limiting kinetic factor in contrast with the propagation of a microcrack in a continuous material.

The investigation of adsorption-induced strength lowering in adsorbents and catalysts demonstrated that there was a demand for a new physical-chemical theory describing the strength of porous structures. Within the scope of the alternative theory, the mechanical characteristics of globular disperse structures were described in terms of the number and distribution of the contacts between particles and the strength of these contacts [34,48,113,119]. This approach had predetermined two principal directions of experimental and theoretical study. One direction was related to the development of structural models that allowed one to estimate the number of contacts per unit area, χ, as a function of porosity and particle size. A number of different models were offered, such as ordered, chaotic, and other models (Chapter 6) [34,48,113,119–121]. The other direction involved the development of methods for the direct measurement of the strength of the individual contacts between particles of different natures in a broad range of physical-chemical conditions [122–125]. As part of these efforts, a number of instruments and experimental methods were developed, specifically for the comprehensive investigation of the mechanical properties of catalysts, carriers, and adsorbents under conditions of static and dynamic loading and abrasion. The studies, which were conducted under realistic conditions, in real media and at relevant reaction temperatures, were especially important [48,126,127]. Some characteristic features of the mechanical behavior of catalysts are presented in Figures 7.54 through 7.57. Figure 7.54 illustrates typical strength histograms for catalyst granules under static loading conditions (at a constant rate of compression) [126–128]. These distributions are mainly very broad and asymmetric (non-Gaussian) and include a large fraction of weak granules, which can be regarded as a deficiency in the industrial manufacturing process.

The universal method for the dynamic testing of catalysts involves the crushing of granules with a vertical drop hammer [126,127,129]. These tests reveal another deficiency in the disperse porous structures, namely, their extreme brittleness and very low resistance to impact, which is orders of magnitude lower than the resistance to impact of compact solids. The data also reveal a large spread of values. Figure 7.55 illustrates the results of dynamic studies carried out with horizontally oriented cylindrical 4 × 4 mm samples. These data are presented in terms of the fraction of tablets, q, that *were not crushed* in an impact of kinetic energy $W = mv^2/2 = mgh$ as a function of the energy, W. The results obtained for loads of different mass m, with the corresponding values of the drop velocity, v, and the drop height, h, coincide with each other. The critical value, W_c, at $q = 50\%$, normalized by the sample cross section, can be viewed as a measure of the mean impact strength. For the majority of catalysts tested, the mean impact strength falls into the range between 10^2 and 10^3 g cm/cm². For the unfilled spherical aluminosilicate sample, this value is on the order of 10^3 g cm/cm². The spread of the data is reflected by the interval $2\Delta W$.

Figure 7.56 shows a schematic of the construction of the specialized mill used to study the abrasion of samples. This device consists of two drums: an outer cylinder (1) rotating at a rate between 30 and 200 rpm, equipped with short blades (2) for granule mixing, and an inner cylinder (3) rotating at a much higher rate of 1,000–10,000 rpm. The second cylinder has a rough surface but no blades [26,126,127,130]. This construction allows one to replicate the real conditions of the use of granules in the moving layers, that is, the abrasion of the granule surface without comminution.

The kinetics of the wear in a spherical aluminosilicate catalyst is shown in Figure 7.57a. This figure shows the relative percentage of mass removed as a function of time, $m(t) = [M_0 - M(t)]/M_0$, where M_0 is the initial material mass and $M(t)$ is the residual mass after milling time, t. The three curves labeled 1, 2, and 3 represent the abrasion process as a reaction of the zeroth, first, and second order, respectively. Curve 2, representing the exponential dependence $m(t) = 1 - \exp(-t/\tau)$, yields the best fit of the experimental data. The time constant, τ, may serve as a measure of the catalyst's resistance to abrasion.

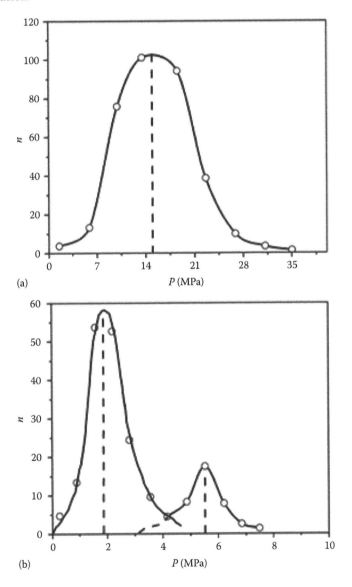

FIGURE 7.54 Histograms of the strength of four different catalysts under static loading; n is the number of specimens crushed in a given range of stresses, P. All four catalysts consisted of cylindrical granules. The granules were oriented either (a) horizontally or (b–d) vertically. The case shown in (b) corresponds to a mixture of two different batches of catalysts. (Data from Slepneva, A.T. et al., *Kolloidnyi Zh.*, 31, 281, 1969; Slepneva, A.T. et al., *Kolloidnyi Zh.*, 32, 251, 1970; Suzdaltseva, S.F. et al., *Doklady AN SSSR*, 201, 415, 1971.)

(*Continued*)

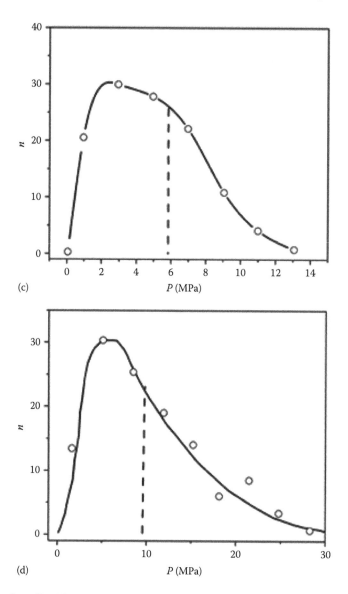

(c)

(d)

FIGURE 7.54 (*Continued*) Histograms of the strength of four different catalysts under static loading; n is the number of specimens crushed in a given range of stresses, P. All four catalysts consisted of cylindrical granules. The granules were oriented either (a) horizontally or (b–d) vertically. The case shown in (b) corresponds to a mixture of two different batches of catalysts. (Data from Slepneva, A.T. et al., *Kolloidnyi Zh.,* 31, 281, 1969; Slepneva, A.T. et al., *Kolloidnyi Zh.,* 32, 251, 1970; Suzdaltseva, S.F. et al., *Doklady AN SSSR,* 201, 415, 1971.)

The case of the more complex two-stage abrasion kinetics of a cylindrical catalyst is shown in Figure 7.57b. These data are for the catalyst used in ethylbenzene dehydration. The first (rapid) stage leads to the removal of about 30% of material in the process of the conversion of cylindrical samples into spherical ones. The second stage corresponds to the much slower wear process of the residual spheres. In this case, one needs two or three parameters to describe the resistance of the catalyst to wear. The flow reactor with the furnace was used to conduct measurements in the course of the catalytic reaction [130,131].

The methods described were used to investigate many different kinds of adsorbents and catalysts. As a result of these studies, the universal nature of chemisorption-induced strength lowering was established. The durability of the catalyst was decreased due to the combined action of the

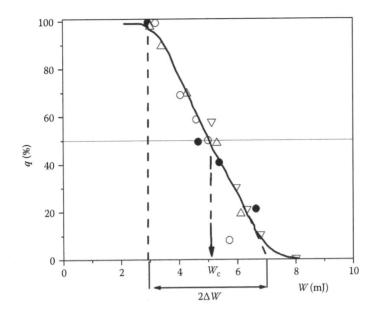

FIGURE 7.55 The estimate of the impact strength of catalysts in terms of the kinetic energy of the peen, $W = W_c$, corresponding to the fraction of uncrushed specimen, $q = 50\%$. (Data from Slepneva, A.T. et al., *Kolloidnyi Zh.*, 31, 281, 1969; Slepneva, A.T. et al., *Kolloidnyi Zh.*, 32, 251, 1970; Zhurkov, S.N. and Sanfirov, T.P., *Doklady AN SSSR*, 101, 237, 1955.)

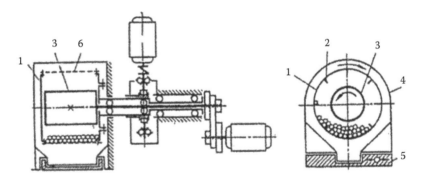

FIGURE 7.56 Schematic illustration of the mill for measuring catalyst wear and abrasion. (From Kutz, M. (Ed.), *Handbook of Environmental Degradation of Materials*, William Andrew, Norwich, NY, 2005; Slepneva, A.T. et al., *Kolloidnyi Zh.*, 31, 281, 1969; Slepneva, A.T. et al., *Kolloidnyi Zh.*, 32, 251, 1970; Grigoryan, V.A. et al., *Zh. Fizicheskoy Himii*, 40, 1144, 1966.)

active catalytic medium and the mechanical stresses, which included both residual internal stresses and structural defects [6,118,126,127,132,133]. Some typical data obtained for industrial and model samples are summarized below in Section 7.3.1.1.

7.3.1.1   Influence of Adsorption on the Mechanical Properties of Solids with Fine Porosity

The studies on the influence of the vapors of various liquids on the strength of porous magnesium hydroxide were the first examples illustrating the reversible physical-chemical influence of adsorption on the mechanical strength of a porous structure. In these studies, it was established that covering the surface of magnesium hydroxide with one to four monolayers of various adsorbates, such as water, ethanol, benzene, and cyclohexane, resulted in a significant decrease in strength, as shown in Figure 7.58 [118].

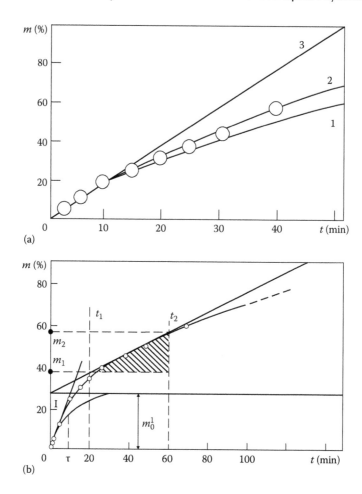

FIGURE 7.57 (a) The abrasion kinetics of a spherical catalyst and (b) the two-stage abrasion kinetics of a cylindrical catalyst. (From Slepneva, A.T. et al., *Kolloidnyi Zh.*, 31, 281, 1969; Slepneva, A.T. et al., *Kolloidnyi Zh.*, 32, 251, 1970; Grigoryan, V.A. et al., *Zh. Fizicheskoy Himii*, 40, 1144, 1966.)

The data in Figure 7.58 show that a decrease in strength by a factor of 1.5–2 was observed when the moisture content was increased by 1%–1.5%. This corresponds to the formation of saturated monolayers of water molecules on the inner and outer surfaces of the highly porous sample. It was discovered that the drop in the sample strength was not due to the condensation of the adsorbate molecules in contacts between the solid particles. Indeed, analysis by NMR spectroscopy revealed that water was not present as a separate phase in the magnesium hydroxide sample. The cause of the strength decrease was the lowering of the surface free energy of the solid due to the adsorption, which resulted in weakened bonds between the surface atoms. A further decrease in mechanical strength, observed at 3%–4% moisture content, was caused by further lowering of the surface free energy due to the multilayer adsorption. The increase in mechanical strength, observed at higher moisture content, and its subsequent drop at high water content, is associated with the peculiar structure of the porous body. Indeed, in contrast to a continuous solid, in the porous structure, the strength lowering can be partially compensated by the capillary attractive forces originating from the condensation of the vapors in the pores (Chapter 1). This is further illustrated in Figure 7.58b, which shows a much smaller increase in the sample strength in the case of ethanol. In the latter case, the capillary attractive forces are much lower than in the case of water due to the lower surface tension of ethanol. The adsorption-related nature of the strength lowering in the magnesium hydroxide sample in the presence of water vapor has

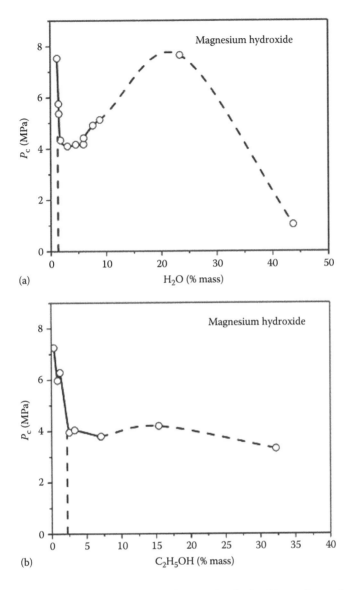

FIGURE 7.58 Change in the compressive strength of magnesium hydroxide specimens due to the adsorption of (a) water and (b) ethanol. (From Shchukin, E.D., Some colloid-chemical aspects of the small particles contact interactions, in: *Fine Particles Science and Technology*, E. Pelizzetti (Ed.), Kluwer Academic Publishers, Amsterdam, the Netherlands, 1996, pp. 239–253.)

been rigorously proven by a direct correlation between the relative decrease in the strength of the porous structure and the lowering of the surface free energy. This case was discussed in detail earlier (Section 7.1.1, Figure 7.11).

Similar trends were also observed in the studies on the mechanical strength lowering in pressed calcium carbonate, zinc oxide, sodium chloride, carbon black, and silica gel in the presence of the vapors of water, benzene, carbon tetrachloride, carbon dioxide, and nitrogen [134]. The more similar the polarity of the adsorbent to that of the medium, the larger the reduction in strength.

In these systems, the lowering of the mechanical strength was not related to dissolution or chemical interactions, that is, the strength lowering was caused by physical adsorption. Conversely, in catalysis the chemisorption of the reagents and the reaction products plays a predominant role in the strength lowering.

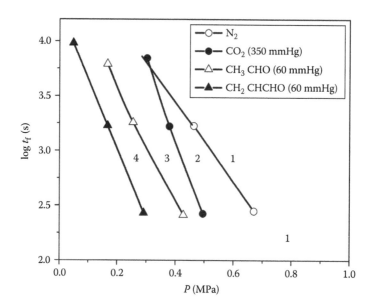

FIGURE 7.59 The durability of a cobalt–molybdenum catalyst as a function of the time to fracture, t_f, under the action of a constant compressive load, P_c, at 200°C in various media: (1) nitrogen, (2) carbon dioxide (at partial pressure 350 mmHg), (3) acetaldehyde (60 mmHg), and (4) acrolein (60 mmHg). (Data from Shchukin, E.D. et al., *Colloid and Surface Chemistry,* Elsevier, Amsterdam, the Netherlands, 2001; Lyubovskiy, M.P. and Barelko, V.V., *Kinetika i Kataliz,* 35, 376, 1994.)

The effect of chemisorption on both the short- and the long-term mechanical strength of dispersed porous structures was investigated in detail using a cobalt–molybdenum catalyst for propylene oxidation as the model system. These results are presented in Figure 7.59, showing the durability as a function of the stress applied. The durability is defined as the logarithm of the time to fracture under a constant applied load.

The universal relationship describing the time to fracture t_f as a function of the applied stress, P, and temperature, T, for a broad range of organic and inorganic, crystalline and amorphous materials, was derived by Zhurkov and is known as the thermofluctuation theory of durability (Chapter 5). This expression can be written as

$$t_f = t_0 \exp\left[-\frac{\left(U_0 - \gamma P\right)}{kT}\right]$$

where
   U_0 is the activation energy
   t_0 and γ are two other constant parameters

The adsorption lowers the surface free energy and thus lowers the activation barrier, U_0, necessary to rupture a chemical bond. It can also increase the *structural factor*, that is, the activation volume, γ [118,119]. Indeed, as seen in Figure 7.59, the data on the influence of chemisorption on the durability of catalyst granules follow Zhurkov's equation: ln t_f – P, the data fall on a straight line. The chemisorption causes a decrease in the strength by 30%–40% and an enormous decrease in the durability, up to several orders of magnitude. The extent of the disintegration and degradation of a catalyst depends on the surface coverage of the chemisorbed molecules and on the strength of the chemisorption bonds, that is, on the lowering of the surface free energy due to chemisorption.

7.3.1.2   The Effect of Catalysis on Mechanical Strength and Wear of Catalysts

The idea that a decrease in the interfacial energy in porous structures is the main reason for the lowering of the strength and durability of catalysts was, for the first time, expressed in the studies described in references [113–115]. A similar idea was also stated in studies on the interfacial phenomena in catalytic processes that involved the use of irreversible thermodynamics [136]. The formation of intermediates with a higher surface activity than the starting and ending reaction products results in the adsorption of these intermediates at the interfaces. This in turn results in the lowering of the surface free energy, the rupture of the chemical bonds, and the development of new surfaces. The experimentally observed decrease in the strength of amalgamated zinc and brass, which are used as catalysts for decomposing aqueous hydrogen peroxide solutions, was explained by the decrease in the surface free energy that takes place in the course of the catalysis [137]. The mechanical strength of a catalyst used in the oxidation of sulfur dioxide inside a reactor is significantly different from the strength of the same catalyst outside of the reactor [138].

The influence of catalysis on mechanical strength and durability has been systematically studied in the catalytic reaction of isopropanol dehydrogenation using a porous magnesium oxide catalyst [6,48,112,116]. The compressive strength of the catalyst pills was lowered by 1.5–2 times when the reaction was in progress for 20–30 s, even prior to reaching the steady state. In these experiments, the measurements were conducted after the samples had been pulled out of the reactor and dried.

Very significant data were obtained by conducting measurements *in situ*, that is, in the course of a catalytic reaction. For such experiments a specialized experimental setup was built. This device allows one to determine the time to fracture of the sample kept under a constant load at a given temperature and given gas flow rates. This setup is shown in Figure 7.60 [48,126,127]. The experimental results are presented in Figure 7.61, which shows a substantial decrease in the strength of the magnesium oxide catalyst in the course of isopropanol dehydrogenation. The same figure also shows a substantial decrease in the time to fracture, which corresponds to an acceleration in the wear and degradation of the catalyst.

Of special interest is the investigation of the influence of catalysis on the abrasive wear of catalyst granules, that is, the abrasion taking place in a real chemical reaction. It is not possible for one to use the traditional airlift because it is not possible to replace a strong stream of air by a stream of a reactive gas. A special mill described in [126,127] was used in these studies. A study on the abrasive wear of an aluminum–chromium–potassium oxide catalyst can be referenced as a characteristic example [130,131]. The studies on the wear of this catalyst in the course of the dehydrogenation of n-butane and butylene were conducted at elevated temperatures. Several different stages of catalyst wear were observed at various temperature intervals. At reasonably low temperatures, some strengthening of the sample took place due to the removal of moisture. With a gradual increase in temperature, which resulted in the start of the reaction, one observed significant wear. At even higher temperatures a new tendency toward strengthening was seen. The latter is related to the partial reduction of the chromium. Finally, at a very high temperature intensive wear of the catalyst took place. All these observations were made in reference to similar investigations conducted at the same temperatures but in an inert medium.

The influence of surface-active and catalytic media on the properties of the solid phase is not limited solely to the lowering of the surface free energy. One of the first observations in this area was that of the so-called catalytic corrosion, that is, the reconstruction of crystallographic edges on the catalyst surface [139]. On numerous occasions changes in the chemical composition of the near-surface layers, the migration of crystalline clusters, the disappearance of the small particles, and other phenomena were reported [79,140–148].

Various theoretical and experimental approaches to the investigation of the molecular-level mechanisms involved in the various interactions between solids and the medium have been described over the years [9,110,149–154]. The studies on the molecular mechanisms of adsorption, chemisorption, and heterogeneous catalysis contain data on the changes in the bond energies in the solid

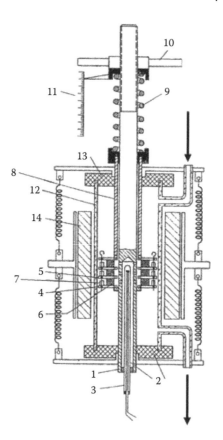

FIGURE 7.60 Schematic representation of a reactor for measuring the long-term durability of a catalyst in the course of a chemical reaction: (1) steel rod, (2) thermocouple insert, (3) thermocouple, (4, 5) flat perforated steel plates (plate 4 is immobilized, while plate 5 can move freely along the rod), (6) catalyst specimen, (7) pins that prevent the skewing of plate 5 in the event of the nonsimultaneous degradation of the specimen (6, 8) steel tube used to transmit an external load, (9) a spring that provides an external load, (10) a nut used to change the spring deformation, (11) measuring scale, (12) quartz tube to introduce reagents, (13) Teflon stoppers, and (14) tube furnace. (Redrawn from Shchukin, E.D. et al., *Physical-Chemical Mechanics of Natural Disperse Systems,* Izd. MGU., Moscow, Russia, 1985; Slepneva, A.T. et al., *Kolloidnyi Zh.,* 31, 281, 1969; Slepneva, A.T. et al., *Kolloidnyi Zh.,* 32, 251, 1970.)

phase. One example of such studies is the quantum-mechanical calculations showing a significant lowering of the bond energy in aluminum oxide due to the physical adsorption and chemisorption of alcohol at various temperatures [155].

In the works by Yushchenko et al. [79], an attempt was made to model the elementary act of the catalytic acid-phase cracking of hydrocarbons by using quantum mechanics. Ab initio calculations of the energy and force parameters of the single and double C–C bonds in the process of their mechanical rupture were conducted. The stress–strain curves were generated, and the influence of the proton on the hydrocarbon cracking on the surface of an oxide catalyst was established. These calculations showed that the addition of the proton can significantly lower the strength of the C–C bond (Figure 7.62).

Under some special conditions, the catalytic process may increase the strength of the interparticle contacts in a catalyst grain [151–153]. However, the lowering of the mechanical strength and the enhanced wear may be viewed as the primary and the most general reasons responsible for the loss of catalysts among the numerous and various effects related to the interactions between the solid phase and the medium in the processes of adsorption, chemisorption, and catalysis. One can say that the catalyst is a "thermodynamic victim" of its intended use. For this reason, the adsorption-induced strength lowering needs to be accounted for in the development of technologies aimed at increasing

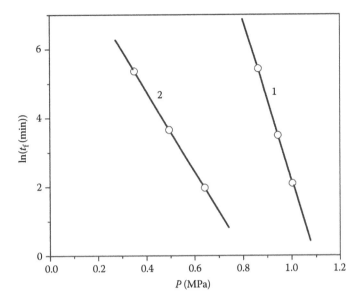

FIGURE 7.61 The durability of a MgO catalyst as a function of the time to fracture, ln t_f, under constant compression, P_c, at 395°C: (1) in the course of the catalytic reaction of isopropanol dehydrogenation and (2) in an inert medium. (Redrawn from Shchukin, E.D. et al., *Physical-Chemical Mechanics of Natural Disperse Systems,* Izd. MGU., Moscow, Russia, 1985; Slepneva, A.T. et al., *Kolloidnyi Zh.,* 31, 281, 1969; Slepneva, A.T. et al., *Kolloidnyi Zh.,* 32, 251, 1970.)

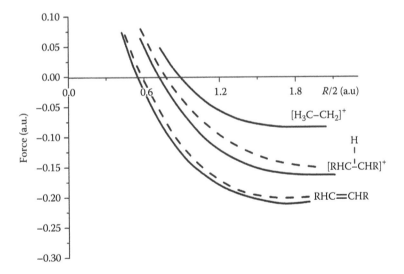

FIGURE 7.62 The influence of the proton on the dependence of the force originating in the deformation of the C=C bond on the distance between the carbon atoms: (–) ethylene molecule and $C_2H_5^+$ cation in symmetric and asymmetric forms (R=H); (---) molecule of trans-butylene-2 and the symmetric cation $C_4H_9^+$ (R=CH$_3$).

the mechanical strength, durability, and reliability of catalysts. Several different approaches can be used to achieve this [6,118].

On the one hand, the structure of the granules can be improved by achieving the highest possible degree of dispersion of the components leading to the maximum number of contacts. On the other hand, the strength of the contacts can be improved by ensuring the transition from coagulation contacts to phase contacts in the corresponding processes of sintering, the hardening of mineral

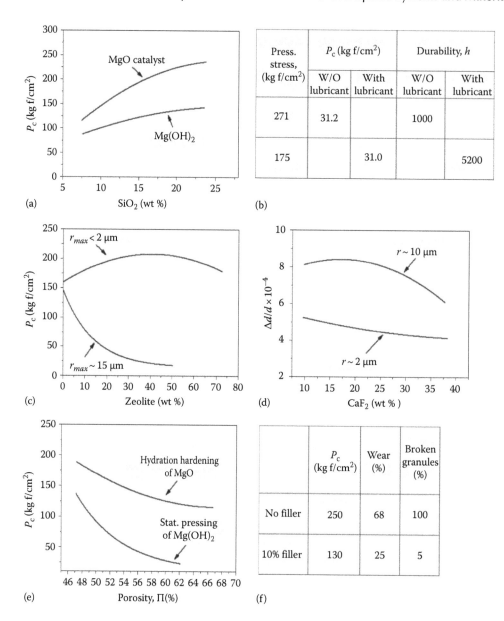

FIGURE 7.63 Some ways to enhance the strength and durability of industrial catalysts: magnesium oxide (a), calcium-nickel-phosphate (b), zeolite-filled aluminum silicate (c,d), magnesium oxide by hydration hardening (e), and sorbeats (f).

binders, and sol–gel transitions. One possible way of decreasing the sensitivity of the contacts to the influence of the medium is to choose proper composition ingredients. Some specific examples for improving the strength and durability of catalysts are described in Figure 7.63.

The suspended layer of an industrial microspherical aluminum–chromium catalyst for the dehydrogenation reaction in the synthesis of monomers for making rubber contains a coarse fraction (tens of microns) of aluminum oxide and finely dispersed clay minerals [145,156]. The clay mineral particles play the role of a binder, and the transition in the dispersion in this fraction from particles around 1 μms in size to particles around 0.1 μm favors a decrease in catalyst wear due to an increase in the number of contacts between the particles. Such a change in dispersion can be controlled by an optimization of the clay swelling process. At the same time, milling the aluminum oxide from 48 µm down to 15 µm also contributes to some increase in strength under static loading conditions due to the *thinning of defects.*

The role played by increasing the number of contacts can be illustrated by the example of enhancing the strength of the magnesium oxide catalyst by introducing small additives of finely dispersed silica (Figure 7.63a) [146,148,157]. In this case, a reduction in the sensitivity of the contacts to the presence of the medium also took place.

A study on the industrial calcium–nickel–phosphate catalyst used in the second stage of dehydrogenation illustrates the manifestation and elimination of the residual internal stresses [147]. Catalyst pellets that were originally intended to last for a year cracked after only 4 months of use due to the high level of residual stresses after the *overpressing* of the pellets. These stresses, in combination with the influence of the active catalytic medium, were the reason for the dramatic strength loss observed. A decrease in the pelletizing pressure and the use of some lubricating additives made it possible to extend the life of this catalyst up to 2 years (Figure 7.63b).

Another characteristic case illustrates one of the first attempts to fill the aluminum silicate cracking catalyst with zeolite. A large pilot batch of this material cracked after drying. The cause of failure was the presence of high extension stresses in the amorphous matrix that originated during gel setting due to the "uncompliant" structure of the crystalline particles. The problem was solved by milling down the zeolite and liquefying the dispersion by stirring in the presence of surfactants (Figure 7.63c and d) [146,148,157]. Effective liquefaction of the disperse system, thereby preventing the accumulation of residual stresses, may be achieved by vibration [158,159].

Another possible way to substantially modify the nature of contacts is to use hydration hardening of the magnesium oxide followed by dehydration instead of mechanical pellet pressing (Figure 7.63e) [120,149,160]. This allowed one to establish stronger phase contacts while simultaneously removing the residual internal stresses.

Studies in this area led to the development of concepts regarding the mutual character of the interactions between a solid phase and a medium in heterogeneous catalysis and to the discovery of the phenomenon of catalytic *acceleration of sintering.* The essence of such mutual influence can be illustrated by the example of ammonia synthesis on an iron catalyst (Figure 7.64) [107,111,150]. In this process, the iron takes part in the dissociative adsorption of hydrogen molecules, leading to the cleavage of the H–H bonds. At the same time, the new bonds, formed between hydrogen and iron, decrease the bond strength in the iron itself. This is the cause of the well-known phenomenon of the hydrogen brittleness of steel.

A similar phenomenon is also known for other catalysts, such as using a nickel catalyst for benzene hydrogenation. Calculations show that the energies of the elementary acts of such exothermic and endothermic catalytic reactions may be sufficiently high to cause the cleavage of the bonds in a solid phase and to free an atom from the crystal lattice. Such an atom plays the role of the free *adatom* taking part in the surface self-diffusion [155]. Examples of catalyst surface restructuring were also observed previously and were regarded as a type of *catalytic corrosion* [48]. The proposed

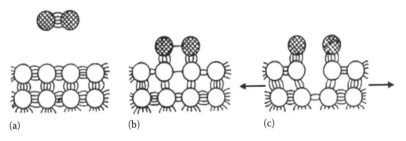

(a)   (b)   (c)

FIGURE 7.64 Schematic illustration of the mutual influence of the solid phase and the medium in heterogeneous catalysis: initially strong bonds (a); weakening of bonds in the course of chemisorption (b); facilitation of bond cleavage by stresses (c).

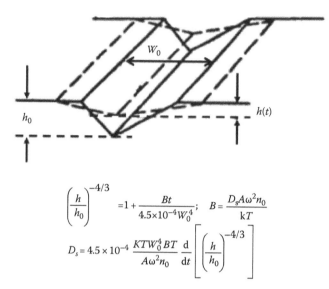

$$\left(\frac{h}{h_0}\right)^{-4/3} = 1 + \frac{Bt}{4.5\times10^{-4}W_0^4}; \quad B = \frac{D_s A\omega^2 n_0}{kT}$$

$$D_s = 4.5 \times 10^{-4} \frac{KTW_0^4 BT}{A\omega^2 n_0} \frac{d}{dt}\left[\left(\frac{h}{h_0}\right)^{-4/3}\right]$$

FIGURE 7.65 The measurement scheme for the surface self-diffusion coefficient, D_s, using the "groove healing" method, which plots the decrease in the depth of the groove with an initial width, W_0, as a function of time: $D_s = \text{Const } W_0 d/dt \left[(h/h_0)^{-4/3}\right]$.

explanation of this phenomenon has been confirmed by direct measurements of the surface self-diffusion coefficient of iron atoms using the *microcrack heal* method. Similar measurements were also conducted for the nickel catalyst used in benzene hydrogenation (Figures 7.65 and 7.66). The results showed that the coefficient of surface self-diffusion in the course of catalysis increased by four to five orders of magnitude. At these reasonably low temperatures, the equilibrium concentration of adatoms in the absence of catalysis is very low.

At the same time, this effect leads to an accelerated sintering, which in turn results in the growth of contacts between the particles and in structure strength enhancement (Figure 7.67) [110]. For powder technology, this means that in order for one to lower the sintering temperature of porous

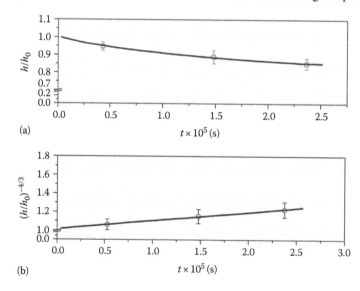

FIGURE 7.66 Self-healing of a groove, $h(t)$, on the surface of the nickel in a catalytic reaction of benzene hydrogenation: at 100°C the surface self-diffusion accelerates by 10^4–10^5 times. (Data from Shchukin, E.D. et al., *Fiz.-Him. Meh. Mater.*, 39, 28, 2003.)

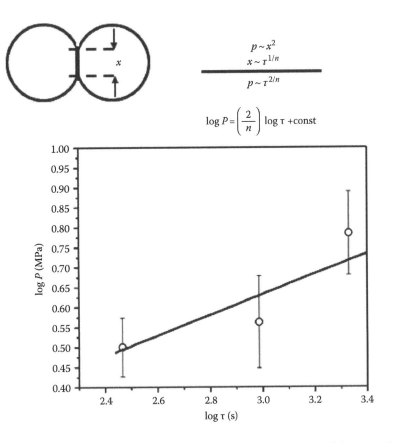

$$p \sim x^2$$
$$x \sim \tau^{1/n}$$
$$p \sim \tau^{2/n}$$

$$\log P = \left(\frac{2}{n}\right) \log \tau + \text{const}$$

FIGURE 7.67 The kinetics of the strength increase in the contacts as a function of time, revealing the role of surface diffusion. (Data from Shchukin, E.D. et al., *J. Mater. Sci.*, 28, 1937, 1993.)

materials while maintaining the same level of development and strength of contacts, *catalysis-enhanced sintering* can be used. This has been experimentally confirmed in studies by Shchukin et al., initially for metals, such as Fe and Ni, in the process of CO oxidation.

A substantial, six times or higher, increase in the strength of the pressed powdered sample of an iron catalyst in the course of CO oxidation is directly related to the development of interparticle contacts due to the mass transfer process. This in turn gives one an opportunity to significantly lower the sintering temperature, while preserving the overall strength of the contacts and the strength of the material. In the present case, it is possible to lower the temperature by about 300°C. Similar results were obtained for an analogous reaction utilizing powdered nickel samples (Tables 7.5 and 7.6). These studies were further extended to include the heat-proof oxides, such as MgO, Al_2O_3, ZrO_2, and Yt_2O_3 (Table 7.7) [154]. One may therefore assume that the effect of catalytic acceleration of sintering can also be used to improve the stability of catalysts at the relevant stage of the formation of the structure.

This concludes the discussion on the interactions between the catalyst surface and the medium. We have demonstrated that such interactions are indeed mutual, as they cause changes in both the adsorbed molecules as well as the catalysts. Such action of the medium on the mechanical properties of catalysts and adsorbents should be regarded as one of the manifestations of the Rehbinder effect, that is, the lowering of the mechanical strength of a solid due to the adsorption or chemisorption of the active components of the medium, causing the lowering of the surface free energy and the work of fracture. The critical factor in adsorption-induced strength lowering phenomena is the presence of mechanical stresses introduced by the weight of the catalyst column, collisions, and temperature gradients or the presence of residual internal stresses originating in the granulation process.

TABLE 7.5

Strength P (MPa) of the Iron Catalyst after Catalysis (CO Oxidation) and after Annealing in Various Media for 30 min

Composition of Reactants	P (MPa)	
	573 K	623 K
8% CO + 4% O$_2$ + 88% He (catalysis)	6.5	7.4
Air	2.4	2.9
Argon	1.4	1.5
8% CO + 92% He	0.3	0.4
H$_2$	N/A	0.9

Sources: Rehbinder, P.A. and Shchukin, E.D., Surface phenomena in solids during deformation and fracture processes, in: *Progress in Surface Science*, S.G. Davison (Ed.), Pergamon Press, Oxford, U.K., 1972, pp. 97–188; Berestetskaya, V. and Butyagin, P.Y., *Doklady AN SSSR*, 260, 361, 1981.

TABLE 7.6

Sintering Temperature under Conditions of Catalysis (T_c) and in an Inert Medium (T') Ensuring the Formation of Similar Contacts between Particles

Catalyst	U (kJ/mol)	T_c (K)	T' (K)	$\Delta T = T' - T_c$ (K)	ΔT_{exp} (K)
Ni	198.6	623	888	265	300–350
Fe	176.0	623	941	318	300–350

TABLE 7.7

Strength P (kg/cm^2) of ZrO$_2$ + Y$_2$O$_3$ Samples

Reaction	Temperature (°C)	P (kg/cm^2)		η (%)
		a	b	
No reaction	200	13.3	N/A	N/A
Methanol oxidation	700	29.8	42.5	43
Ethanol oxidation	530	21.9	29.6	35
Ethanol oxidation	700	34.8	38.5	11
Methanol dehydration	700	27.5	41.3	50
Methanol dehydration	800	43.0	76.8	79
Ethanol dehydration	700	27.5	45.3	65
Ethanol dehydration	800	430	81.2	88

Sources: Rozovskiy, A.Y., *The Catalyst and the Reaction Medium*, Nauka, Moscow, *Russia, 1988*; Johannesen, O., *Selected Topics in High Temperature Chemistry: Defect Chemistry of Solids* (Studies in Inorganic Chemistry), Elsevier Science, Amsterdam, the Netherlands, 1989.
Notes: a, Prior to; b, after catalytic conversion of alcohols; η (%) is the strengthening [142,143].

Detailed analysis of these phenomena from the standpoint of physical-chemical mechanics allows one to seek ways to improve the mechanical stability and durability of catalyst granules in the active medium. This can be achieved by optimizing the entire process of manufacturing disperse porous bodies. This includes the optimization of the composition by using coarse and fine fractions of particles, the careful selection of suitable pressing conditions by using hydration hardening of mineral binders, as well as modifying the contacts between the particles and lowering their sensitivity to the medium. The use of vibration and surfactants are an effective means for achieving control over the rheology of prepared materials. All these steps make it possible for one to enhance the life cycle of catalysts.

It was shown that the weakening of the interatomic bonds at the catalyst surface resulted in a sharp increase in the concentration of mobile adatoms and the acceleration of surface self-diffusion. This intensifies the sintering process and allows one to significantly decrease the sintering temperature.

The approach presented provides one with the means for the better selection of catalysts and the optimization of the manufacturing technology, enhancing the mechanical strength and durability of catalysts and adsorbents, improving the conditions of ceramics sintering, and minimizing the negative impact from the adsorption-induced strength lowering.

REFERENCES

1. Shchukin, E. D. 2006. The influence of surface-active media on the mechanical properties of materials. *Adv. Colloid Interface Sci.* 123–126: 33–47.
2. Lichtman, V. I., Shchukin, E. D., and P. A. Rehbinder. 1964. *Physical-Chemical Mechanics of Metals.* Jerusalem, Israel: Israel Program for Scientific Translations.
3. Rehbinder, P. A. and E. D. Shchukin. 1972. The surface phenomena in solids in the process of their deformation and fracture. *Uspehi Fiz. Nauk.* 108: 1–42.
4. Rehbinder, P. A. and E. D. Shchukin. 1972. Surface phenomena in solids during deformation and fracture processes. In *Progress in Surface Science*, S. G. Davison (Ed.), pp. 97–188. Oxford, U.K.: Pergamon Press.
5. Shchukin, E. D. 1977. Environmentally induced lowering of surface energy and the mechanical behavior of solids. In *Surface Effects in Crystal Plasticity*, R. M. Latanision, and J. F. Fourie (Eds.), pp. 701–736. Leiden, the Netherlands: Noordhof.
6. Shchukin, E. D., Pertsov, A. V., Amelina, E. A., and A. S. Zelenev. 2001. *Colloid and Surface Chemistry.* Amsterdam, the Netherlands: Elsevier.
7. Shchukin, E. D., Bessonov, A. I., Kontorovich, S. I. et al. 2006. Effects of adsorption-active media on the mechanical properties of catalysts and adsorbents. *Colloids Surf.* 282–283: 287–297.
8. Shchukin, E. D. and P. A. Rehbinder. 1958. Formation of new surfaces during deformation and fracture of a solid matter in a surface active medium. *Kolloidnyi Zh.* 20: 645–654.
9. Rehbinder, P. A. 1979. *Selected Works. Surface Phenomena in Disperse Systems: Physical-Chemical Mechanics.* Moscow, Russia: Nauka.
10. Shchukin, E. 2012. Physical-chemical mechanics of solid surfaces. In *Encyclopedia of Surface and Colloid Science*, 2nd edn., pp. 1–23. New York: Taylor & Francis.
11. Potak, Y. M. 1955. *Brittle Fractures of Steel and Steel Parts.* Moscow, Russia: Oborongiz.
12. Shchukin, E. D. and V. I. Lichtman. 1959. On brittle fracture of zinc single crystals. *Doklady AN SSSR.* 124: 307–310.
13. Rostoker, W., McCaughey, J. M., and H. Markus. 1960. *Embrittlement by Liquid Metals.* P. Somasundaran (Ed.), New York: Reinhold Publishing.
14. Zanozina, Z. M. and E. D. Shchukin. 1962. Destruction of metals along the grain and grain boundaries in the presence of strong adsorption-active metal melts. *Inzhenerno-Fizicheskiy Zh.* 5: 86–90.
15. Shchukin, E. D., Goryunov, Y. V., Denshchikova, G. I., Pertsov, N. V., and B. D. Summ. 1963. On the propagation of liquid metals on the surface of solid metals in connection with the adsorption effect of the decreased strength. *Kolloidnyi Zh.* 25: 108–114, 253–259.
16. Westwood, A. R. C. and N. S. Stoloff (Eds.). 1966. *Environment-Sensitive Mechanical Behavior.* New York: Gordon & Breach.
17. Shchukin, E. D. 1970. Rehbinder effect. In *Science and Mankind*, Moscow: Politechnic Museum Publishing, pp. 336–367.
18. Shchukin, E. D. 1976. Lowering of the surface energy and change of mechanical properties of solids under the influence of the environment. *Fiz.-Him. Meh. Mater.* 12: 3–20.
19. Westwood, A. R. C., Ahern, J. S., and J. J. Mills. 1981. Development in the theory and applications of chemomechanical effects. *Colloids Surf.* 2: 1–35.
20. Springer, G. S. 1981. *Environmental Effects on Composite Materials.* Boca Raton, FL: CRC Press.
21. Latanision, R. M. and J. R. Pickens. (Eds.). 1983. *Atomistic of Fracture.* New York: Plenum Press.

22. Kamdar, M. H. (Ed.). 1984. *Embrittlement by Liquid and Solid Metals.* Warrendale, PA: AIME.

23. Jones, R. H. (Ed.). 2001. *Environmental Effects on Engineered Materials: Corrosion Technology.* New York: Marcel Dekker.

24. Akishin, A. I. (Ed.). 2001. *Effects of Space Conditions on Materials.* Huntington, New York: Nova Science Publ.

25. Brimblecombe, P. (Ed.). 2003. *The Effects of Air Pollution on the Built Environment.* London, U.K.: Imperial College Press.

26. Kutz, M. (Ed.). 2005. *Handbook of Environmental Degradation of Materials.* Norwich, NY: William Andrew.

27. Malkin, A. I., Polukarova, Z. M., Zanozin, V. M. et al. 2006. Embrittling effect of metal melts on the deformation of highly ductile metals. *Deformatsiya i Razrushenie Material.* 7: 36–40.

28. Kleiman, J. I. (Ed.). 2006. *Protection of Materials and Structures from the Space Environment.* Berlin, Germany: Springer.

29. Frumkin, A. N. 1938. About the phenomena of wetting and adhesion of bubbles. *Zh. Fizicheskoi Khimii.* 12: 337.

30. Watt, J., Tidblad, J., Kucera, V., and R. Hamilton (Eds.). 2009. *The Effects of Air Pollution on Cultural Heritage.* Berlin, Germany: Springer.

31. Griffith, A. A. 1920. The phenomenon of rupture and flow in solids. *Philos. Trans. R. Soc. Lond. A.* 221: 163–198.

32. Orowan, E. 1944. The fatigue of glass under stress. *Nature.* 154: 341–343.

33. Stokes, R. J. and C. H. Li. 1963. *Fracture of Solids.* New York: Wiley.

34. Rehbinder, P. A., Shchukin, E. D., and L. Ya. Margolis. 1964. On the mechanical strength of porous disperse bodies. *Doklady AN SSSR.* 154: 695.

35. Rehbinder, P. 1947. New physico-chemical phenomena in the deformation and mechanical treatment of solids. *Nature.* 159: 866–867.

36. Yushenko, V. S., Polukarova, Z. M., and E. D. Shchukin. 1971. The adsorption strength lowering in non-eutectic systems. *Fiz.-Him. Meh. Mater.* 7: 85.

37. Lichtman, V. I., Rehbinder, P. A., and G. V. Karpenko. 1954. *The Effect of Surface-Active Medium on the Processes of Deformation of Metals.* Moscow, Russia: Izd. AN SSSR.

38. Shchukin, E. D., Kochanova, L. A., and A. V. Pertsov. 1963. On the temperature transition from brittleness to plasticity under the condition of adsorption strength lowering. *Kristallografiya.* 8:69.

39. Rozhanskiy, V. N., Pertsov, N. V., Shchukin, E. D., and P. A. Rehbinder. 1957. The effect of thin coatings of mercury on the strength of single crystals of metals. *Doklady AN SSSR.* 116: 769–771.

40. Eremenko, V. and Naydich, Yu. (Eds.), 1968. *Surface Phenomena in Melts.* Kiev, Ukraine: Naukova Dumka.

41. Skvortsova, Z. N., Traskin, V. Yu., Bryuhanova, L. S., and N. V. Pertsov. 1992. Mechanics of fracture of cohesive boundaries with different concentrations of foreign inclusions. In *The Successes of Colloid Chemistry and Physical-Chemical Mechanics,* E. D. Shchukin (Ed.), pp. 222–228. Moscow, Russia: Nauka.

42. Traskin, V. Y., Pertsov, N. V., Skvortsova, Z. N., Shchukin, E. D., and P.A. Rehbinder. 1970. Adsorption strength lowering of alkali halide crystals. *Doklady AN SSSR.* 191: 876–880.

43. Traskin, V. Y. and Z. N. Skvortsova. 1988. Thermodynamic activity of water in electrolyte solutions and their impact on the strength of solids. In *Surface Water Films in Dispersed Structures,* E. D. Shchukin (Ed.), pp. 197–202. Moscow, Russia: Izd. MGU.

44. Kuchumova, V. M., Savenko, V. I., and E. D. Shchukin. 2007. Effect of aqueous solutions of cetyltrimethylammonium bromide on a surface strength of the quartz glass under various pH. *Kolloidnyi Zh.* 69: 357–363.

45. Kochanova, L. A., Kuchumova, V. M., Savenko, V. I., and E. D. Shchukin. 1992. Environment effect on sclerometric brittleness of ionic crystals. *J. Mater Sci.* 27: 5516–5522.

46. Savenko, V. I. and E. D. Shchukin. 2007. On the influence of the surface energy on the dynamics of dislocations in ionic crystals. *Kolloidnyi Zh.* 69: 834–838.

47. Ovcharenko, F.D. 1981. *Physical-Chemical Mechanics and Lyophilic Properties of Disperse Systems.* Kiev, Ukraine: Naukova Dumka.

48. Shchukin, E. D., Pertsov, N. V., Osipov, V. I., and R. I. Zlochevskaya. (Eds.). 1985. *Physical-Chemical Mechanics of Natural Disperse Systems.* Moscow, Russia: Izd. MGU.

49. Pertsov, N. V. 1985. Physical-chemical effects of liquid phase at the destruction of rocks. In *Physical-Chemical Mechanics of Natural Disperse Systems,* E. D. Shchukin, N. V. Pertsov, V. I. Osipov, and R. I. Zlochevskaya (Eds.), pp. E107–E117. Moscow, Russia: Izd. MGU.

50. Shchukin, E. D. (Ed.). 1992. *The Successes of Colloid Chemistry and Physical-Chemical Mechanics.* Moscow, Russia: Nauka.

51. Yushchenko, V. S., Dukarevich, M. V., Chuvaev, V. F., and E. D. Shchukin. 1969. Adsorption lowering of the strength of magnesium oxide hydrate in water vapor. *Zh. Fizicheskoy Himii.* 53: 1556–1559.
52. Pertsov, N. V. and P. A. Rehbinder. 1958. On the surface activity of the liquid metal coatings and their effect on the strength of metals. *Doklady AN SSSR.* 123: 1068–1070.
53. Skvortsov, A. G., Sinevich, E. A., Pertsov, N. V., Shchukin, E. D., and P. A. Rehbinder. 1970. Determination of the surface energy of crystals of naphthalene by splitting. *Doklady AN SSSR.* 193: 76–79.
54. Pertsov, N. V., Sinevich, E. A., and E. D. Shchukin. 1968. Adsorption-induced strength lowering of molecular crystals. *Doklady AN SSSR.* 179: 633.
55. Bartenev, G. M. and Yu. S. Zuyev. 1964. Strength and destruction of highly elastic materials. Khimiya, Moscow.
56. Tynniy, A. N. and A. I. Soshko. 1965. On the mechanism of destruction of brittle bodies when exposed to surface-active media. *Fiz.-Him. Meh. Mater.* 1: 312–316.
57. Shchukin, E. D., Soshko, A. I., Mikityuk, O. A., and A. N. Tynniy. 1971. The selectivity of the effects of lowering of the strength of polymeric materials under the action of surface-active media. *Fiz.-Him. Meh. Mater.* 7: 33–39.
58. Volynskiy A. L. and N. F. Bakeev. 1995. *Solvent Crazing of Polymers.* Amsterdam, the Netherlands: Elsevier.
59. Videnskiy, I. V., Shchukin, E. D., Savenko, V. I., Petrova, I. V., and Z. M. Polukarova. 1999. Liquid metal embrittlement in the absence of liquid metal phase: In studies of surface damageability and in hard materials machining. *Colloids Surf.* 156: 349–355.
60. Rozhanskiy, V. N. 1958. On the mechanism of formation of initiating cracks in the crystals during plastic deformation. *Doklady AN SSSR.* 123: 648–651.
61. Lichtman, V. I. and E. D. Shchukin. 1958. Physical-chemical phenomena during the deformation of metals. *Uspehi Fiz. Nauk.* 66: 213–245.
62. Zhurkov, S. N. and A. V. Savitskiy. 1959. On the mechanism of fracture of solids. *Doklady AN SSSR.* 129: 91.
63. Shchukin, E. D. and A. V. Pertsov. 2007. Thermodynamic criterion of spontaneous dispersion. In *Colloid and Interface Science Series*, Tadros, Th. (Ed.), Vol. 1, pp. 23–47. New York: Wiley-VCH.
64. Shchukin, E. D. 2004. Conditions of spontaneous dispersion and formation of thermodynamically stable colloid systems. *J. Dispers. Sci. Technol.* 25: 875–893.
65. Pertsov, A. V. 1967. The studies of dispersion processes under conditions of the strong decrease in the free interfacial energy. Canadian Science thesis. Moscow, Russia: Izd. MGU.
66. Pertsov, A. V., Mirkin, L. I., Pertsov, N. V., and E. D. Shchukin. 1964. On the spontaneous dispersion under conditions of the strong decrease in the free interfacial energy. *Doklady AN SSSR.* 158: 1166–1168.
67. Weiler, S. Y. and V. I. Lichtman. 1960. *Action of Lubricants in Metal Pressure.* Moscow, Russia: Izd. AN SSSR.
68. Shchukin, E. D., Kochanova, L. A., and V. I. Savenko. 1993. Electric surface effects in solids plasticity and strength. In *Modern Aspects of Electrochemistry*, R. E. White, B. E. Conway, and J. O'M. Bockris (Eds.), New York: Plenum Press, pp. 245–298.
69. Shchukin, E. D., Videnskiy, I. V., and V. I. Savenko. 1998. Microscratching in electrochemical cell: The effect of gallium on surface deformation in aluminum. *Colloids Surf.* 142: 175–181.
70. Shchukin, E. D., Videnskiy, I. V., Mihalske, T. A., and V. I. Savenko. 1998. Microsclerometry in electrochemical cell. Gallium Effect on surface deformation in the aluminum. *Fiz.-Him. Meh. Mater.* 3: 49–54.
71. Yushchenko, V. S., Grivtsov, A. G., and E. D. Shchukin. 1974. Numerical simulation of a molecular crystal deformation. *Doklady AN SSSR.* 215: 148–151.
72. Yushchenko, V. S., Grivtsov, A. G., and E. D. Shchukin. 1974. Numerical simulation of Rehbinder effect. *Doklady AN SSSR.* 219: 162–165.
73. Yushchenko, V. S. and E. D. Shchukin. 1978. Molecular dynamics study of the elementary processes of destruction under constant load. *Doklady AN SSSR.* 242: 653–656.
74. Shchukin, E. D. and V. S. Yushchenko. 1981. Molecular dynamics simulation of mechanical behaviour. *J. Mater. Sci.* 16: 313–330.
75. Haile, J. M. 1997. *Molecular Dynamics Simulation: Elementary Methods.* New York: Wiley.
76. Frenkel, D. and B. Smit. 2001. *Understanding Molecular Simulation: From Algorithms to Applications*, 2nd edn. London, U.K.: Academic Press.
77. Griebel, M., Knapek, S., and G. Zumbusch. 2007. *Numerical Simulation in Molecular Dynamics: Numerics, Algorithms, Parallelization, Applications.* Berlin, Germany: Springer.

78. Marx, D. and J. Hutter. 2009. *Ab Initio Molecular Dynamics: Basic Theory and Advanced Methods.* Cambridge, U.K.: Cambridge University Press.

79. Car, R. and M. Parrinello. 1985. Unified approach for molecular dynamics and density-functional theory. *Phys. Rev. Lett.* 55: 2471.

80. Grivtsov, A. G. and E. E. Shnol. 1971. *Numerical Experiments on Simulation of Molecular Motion.* Moscow, Russia: Izd. IPM AN SSSR.

81. Grivtsov, A. G. 1987. *The Numerical Simulation and Dynamics of Microheterogeneous Systems.* Moscow, Russia: Nauka.

82. Tovbin, Y. K. (Ed.). 1996. . *Method of Molecular Dynamics in Physical Chemistry.* Moscow, Russia: Nauka.

83. Shchukin, E. D., Summ, B. D., and Yu. V. Goryunov. 1966. On the role of interatomic interactions in adsorption-induced strength lowering of metals. *Doklady AN SSSR.* 167: 631.

84. Yushchenko, V. S., Ponomareva, T. P., and E. D. Shchukin. 1992. Environment influence on the mechanical strength of chemical bonds in solids—*ab initio* quantum-mechanical calculations. *J. Mater. Sci.* 27: 1659–1662.

85. Shchukin, E. D., Kochanova, L. A., and Z. M. Zanozina. 1965. Loss of strength of the glass under the action of microscopic defects applied to its surface. *Fiz.-Him. Meh. Mater.* 2: 128–133.

86. Frenkel, Y. I. 1952. The theory of reversible and irreversible cracks in brittle solids. *Zh. Tekh. Fiz.* 22: 1857–1866.

87. Elliott, H. A. 1957. An analysis of the conditions for rupture due to Griffith cracks. *Proc. Phys. Soc.* 59: 208.

88. Barenblatt, G. I. 1959. On equilibrium cracks formed in brittle fracture. *Doklady AN SSSR.* 127: 47.

89. Read, W. T. 1953. *Dislocations in Crystals.* New York: McGraw-Hill.

90. Weertman, J. and J. R. Weertman. 1992. *Elementary Dislocation Theory.* Oxford, U.K.: Oxford University Press.

91. Nabarro, F. R. N. and J. P. Hirth. (Eds.). 2007. *Dislocations in Solids.* Amsterdam, the Netherlands: Elsevier.

92. Hull, D. and D. J. Bacon. 2001. *Introduction to Dislocations,* 4th edn. Liverpool, U.K.: University of Liverpool Press.

93. Aleksandrov, A. P. and S. N. Zhurkov. 1933. *The Phenomenon of Brittle Fracture.* Moscow, Russia: GTTI.

94. Aslanova, M. S. and P. A. Rehbinder. 1954. The influence of adsorption on retarded elastic aftereffect in glass. *Doklady AN SSSR.* 96: 299.

95. Bowden, F. P. and D. Tabor. 1954. *Friction and Lubrication of Solids.* Oxford, U.K.: Oxford University Press.

96. Derjaguin, B. V. 1963. *What is Friction,* 2nd edn. Moscow, Russia: Izd. AN SSSR.

97. Kragelskiy, I. V. 1968. *Friction and Wear.* Moscow, Russia: Mashinostroenie.

98. ASM International. 1992. Friction, lubrication, and wear technology. In *ASM Handbook,* Vol. 18, American Society of Metals.

99. Savenko, V. I. and E. D. Shchukin. 1992. A method for evaluating anti-wear lubricant effects. *Tribol. Int.* 25: 367–370.

100. Rabinowicz, E. 1995. *Friction and Wear of Materials.* New York: Wiley.

101. Persson, B. N. J. 2000. *Sliding Friction: Physical Principles and Applications* (Nanoscience and Technology). Berlin, Germany: Springer.

102. Kragelskiy, I. V. and V. V. Alisin (Eds.). 2005. *Tribology: Lubrication, Friction and Wear* (Tribology in Practice Series). New York: Wiley.

103. Mate, C. M. 2008. Tribology on the Small Scale*: A Bottom up Approach to Friction, Lubrication, and Wear* (Mesoscopic Physics and Nanotechnology). Oxford, U.K.: Oxford University Press.

104. Blau, P. J. 2009. *Friction Science and Technology: From Concepts to Applications,* 2nd edn. Boca Raton, FL: CRC Press.

105. Popov, V. L. 2010. *Contact Mechanics and Friction: Physical Principles and Applications.* Berlin, Germany: Springer.

106. Savenko, V. I. and E. D. Shchukin. 1996. New applications of the Rehbinder effect in tribology. A review. *Wear.* 194: 86–94.

107. Shchukin, E. D. 1999. Physical-chemical mechanics in studies of Peter A. Rehbinder and his school. *Colloids Surf.* 49: 529–537.

108. Videnskiy, I. V., Petrova, I. V., and E. D. Shchukin. 1993. Electro-chemo-mechanical treatment: Facilitating steel grinding under electrochemical reduction of active cations. *J. Mater. Res.* 8: 2224–2227.

109. Shchukin, E. D., Mihalske, T. A., Grin, R. E. et al. 1998. Microsclerometry in studying the effect of environment on the mechanical properties of metals. *Fizika i Himiya obrabotki Mater.* 3: 99–104.

110. Shchukin, E. D., Kontorovich, S. I., and B. V. Romanovskiy. 1993. Porous materials sintering under conditions of catalytic reactions. *J. Mater. Sci.* 28: 1937–1930.

111. Shchukin, E. D. 1997. Mutual influence of the solid phase and the environment in the process of heterogeneous catalysis. *Himicheskaya Promyshlennost.* 6: 28–35.

112. Shchukin, E. D., Bessonov, A. I., Kontorovich, S. I. et al. 2003. Physical and chemical mechanics of catalysts: Strength and durability of fine-porous materials in active media. *Fiz.-Him. Meh. Mater.* 39: 28–43.

113. Shchukin, E. D. 1965. On some problems in the physical-chemical theory of strength of the fine-disperse porous bodies—Catalysts and adsorbents. *Kinetika i Kataliz.* 6: 641–650.

114. Shchukin, E. D., Dukarevich, M. V., Kontorovich, S. I., and P. A. Rehbinder. 1967. On the adsorption decrease in strength of a magnesium oxide catalyst in the process of catalysis. *Doklady AN SSSR.* 175: 882–884.

115. Shchukin, E. D., Dukarevich, M. V., Kontorovich, S. I., and P. A. Rehbinder. 1968. On the strength and durability of a magnesium oxide catalyst in the process of catalysis. *Doklady AN SSSR.* 182: 394–397.

116. Schweiger, H., Raybaud, P., Kresse, G., and H. Toulhoat. 2002. Shape and edge sites modification of MoS_2 catalytic nanoparticles induces by working conditions: A theoretical study. *J. Catal.* 207: 76–87.

117. Shchukin, E. D. and E. A. Amelina. 1979. Contact interactions in disperse systems. *Adv. Colloid Interface Sci.* 11: 235–287.

118. Shchukin, E. D. 1996. Some colloid-chemical aspects of the small particles contact interactions. In *Fine Particles Science and Technology*, E. Pelizzetti (Ed.), pp. 239–253. Amsterdam, the Netherlands: Kluwer Academic Publishers.

119. Shchukin, E. D. 2002. Surfactant effects on the cohesive strength of particle contacts: Measurements by the cohesive force apparatus. *J. Colloid Interface Sci.* 25: 159–167.

120. Shchukin, E. D., Amelina, E. A., and S. I. Kontorovich. 1992. Formation of contacts between particles and development of internal stresses during hydration processes. In *Materials Sciences of Concrete*, J. Skalny (Ed.), pp. 1–35. Westerville, OH: ACS.

121. Shchukin, E. D. and E. A. Amelina. 2003. Surface modification and contact interaction of particles. *J. Dispers. Sci. Technol.* 24: 377–395.

122. Shchukin, E. D., Bessonov, A. I., and S. A. Paranskiy. 1971. *Mechanical Testing of Catalysts and Sorbents.* Moscow, Russia: Nauka.

123. Shchukin, E. D., Dukarevich, M. V., Kontorovich, S. I., and P. A. Rehbinder. 1966. On the adsorption decrease in strength of the fine-disperse porous structures. *Doklady AN SSSR.* 167: 1109–1112.

124. Dollimore, D. and G. R. Heal. 1961. The effect of various vapors on the strength of compacted silica. *J. Appl. Chem.* 11: 459.

125. Wittman, F. H. 1968. Influence of the surface energy on the strength of a porous solid. *Z. Angew. Phys.* 25: 160–163.

126. Slepneva, A. T., Kontorovich, S. I., Lipkind, B. A., Amelina, E. A., and E. D. Shchukin. 1969. Effect of the addition of highly-dispersed SiO_2 and Al_2O_3 on the adsorption strength decrease in clays and granulated zeolites. *Kolloidnyi Zh.* 31: 281–284.

127. Slepneva, A. T., Lipkind, B. A., Dukarevich, M. V., Kontorovich, S. I., and E. D. Shchukin. 1970. Regularities in the strength decrease of zeolite pellets under action of water and benzene vapors. *Kolloidnyi Zh.* 32: 251–254.

128. Suzdaltseva, S. F., Skvortsova, E. I., Margolis, L. Y., Shchukin, E. D., and P. A. Rehbinder. 1971. Effect of chemisorption on the strength of porous dispersed catalysts. *Doklady AN SSSR.* 201: 415–418.

129. Zhurkov, S. N. and T. P. Sanfirova. 1955. Temperature and time dependence of the tensile strength of pure metals. *Doklady AN SSSR.* 101: 237–240.

130. Grigoryan, V. A., Zhuhovitskiy, A. A., Shvindlerman, L. S., and M. I. Chikonosova. 1966. Dynamic surface effect in the catalytic reaction. *Zh. Fizicheskoy Himii.* 40: 1144–1147.

131. Muhlenov, I. P., Dobkina, E. I., Deryuzhkina, V. I., and V. E. Soroko. 1989. *Catalysts Technology.* Leningrad, Russia: Himiya.

132. Massoth, F. E. 1973. Molybdenum-alumina catalysts. *J. Catal.* 30: 204–217.

133. Roginskiy, S. Z., Tretyakov, I. I., and A. B. Shehter. 1953. Catalytic corrosion. *Doklady AN SSSR.* 91: 881–884.

134. Iwasawa, Y. 1987. Chemical design surfaces for active solid catalysts. *Adv. Catal.* 35: 187–264.

135. Lyubovskiy, M. P. and V. V. Barelko. 1994. Dynamics of catalytic corrosion on the surface of platinum catalyst in the oxidation of ammonia. *Kinetika i Kataliz.* 35: 376–381.

136. Boreskov, G. K. 1987. *Catalysis: Problems of Theory and Practice.* Novosibirsk, Russia: Nauka.

137. Rao, C. N. R. and J. Gopalakrishnan. 1976. *New Directions in Solid State Chemistry*. Cambridge, U.K.: Cambridge University Press.

138. Whitesides, G. M. 1983. Relations between fracture and coordination chemistry. In *Atomistics in Fracture*, R. M. Latanision and J. R. Pickens (Eds.), pp. 1–23. New York: Plenum Press.

139. Shchukin, E. D. 1983. Interactions between adsorbed species and a strained crystal. In *Atomistics in Fracture*, R. M. Latanision and J. R. Pickens (Eds.), pp. 421–423. New York: Plenum Press.

140. Korzunov, V. A., Chuvylkin, I. Y., Zhidomirov, G. M., and V. B. Kazanskiy. 1981. Semi-empirical quantum-mechanical calculations of intermediate complexes in catalytic reaction. The stages of chemisorption of ethanol at the aluminum γ-oxide. *Kinetika i Kataliz*. 22: 930–936.

141. Berestetskaya, V. and P. Y. Butyagin. 1981. Mechanical-chemical activation of the surface of magnesium oxide. *Doklady AN SSSR*. 260: 361–364.

142. Rozovskiy, A. Y. 1988. *The Catalyst and the Reaction Medium*. Moscow, Russia: Nauka.

143. Johannesen, O. 1989. *Selected Topics in High Temperature Chemistry: Defect Chemistry of Solids* (Studies in Inorganic Chemistry). Amsterdam, the Netherlands: Elsevier Science.

144. Van Hove, M. A. and G. A. Somorjai. 1994. Adsorption and adsorbate-induced restructuring: A LEED perspective. *Surf. Sci*. 299–300: 487–501.

145. Kotelnikov, G. R., Patanov, V. A., Shchukin, E. D., and L. N. Kozina. 1975. Assessment of physical-mechanical properties of catalysts used in suspended layer. *Kolloidnyi Zh*. 37: 875–877.

146. Lankin, Y. I., Kontorovich, S. I., Amelina, E. A., and E. D. Shchukin. 1982. Experimental study of the formation of accretion contacts between silica particles in supersaturated solutions of silicic acid. *Kolloidnyi Zh*. 42: 649–652.

147. Kotelnikov, G. R., Shchukin, E. D., Strunnikova, L. V. et al. 1972. On durability of chromium-nickel-calcium phosphate olefinic hydrocarbon dehydrogenation catalyst. *Kinetika i Kataliz*. 13: 1307–1310.

148. Kontorovich, S. I., Lavrova, K. A., Plavnik, G. M. et al. 1971. Investigation of internal stresses in the filled aluminosilicate xerogels. *Doklady AN SSSR*. 199: 1360–1363.

149. Grechenko, A. N., Golosman, E. Z., Kontorovich, S. I. et al. 1982. Features of structure formation of mixed cement containing catalysts during the hydrothermal treatment. *Neorg. Mater*. 18: 1032–1037.

150. Shchukin, E. D. and L. Y. Margolis. 1982. The mechanism of destruction of the catalyst as a result of the mutual influence of the solid phase and the reaction medium. *Poverhnost*. 8: 1–8.

151. Sokolova, L. N., Shchukin, E. D., Burenkova, L. N., and B. V. Romanovskiy. 1998. Sintering of the porous zirconium oxide in the course of catalytic reaction. *Doklady AN, Himiya*. 360: 782–783.

152. Sokolova, L. N., Shchukin, E. D., Burenkova, L. N., and B. V. Romanovskiy. 2000. Effect of strengthening of porous materials in catalytic reactions. *Doklady AN, Himiya*. 373: 491–492.

153. Romanovskiy, B. V., Shchukin, E. D., Burenkova, L. N., and L. N. Sokolova. 2002. Influence of catalysis on the strength of porous materials with a globular structure. *Zhurnal Fizicheskoy Himii*. 76: 1044–1047.

154. Abukais, A., Burenkova, L. N., Zhilinskaya, E. A. et al. 2003. The influence of the catalytic conversion of the alcohol on the strength of porous materials ZrO_2 and $ZrO_2 + Y_2O_3$. *Neorg. Mater*. 39: 602–608.

155. Shchukin, E. D., Kontorovich, S. I., Girenkova, N. I. et al. 1991. On sintering of the metal porous samples under condition of catalytic reaction. *Doklady AN SSSR*. 318: 1417–1421.

156. Kotelnikov, G. R., Shchukin, E. D., Patanov, V. A., Bushin, A. N., and R. K. Mihaylov. 1973. On the strength of the industrial dehydrogenation catalysts intended for use in suspended layer. *Kolloidnyi Zh*. 35: 362–364.

157. Kontorovich, S. I. 1990. Physical-chemical regularities of strengthening of the fine-disperse porous structures—Catalysts and adsorbents. Science thesis, Institute of Physics and Chemistry, Russian Academic Science, Moscow, Russia.

158. Polukarova, Z. M., Shatalova, I. G., Yusupov, R. K. et al. 1968. Increase of the mechanical strength of a molybdenum-bismuth catalyst for the ammoxidation of propylene. *Neftehimiya*. 8: 899–903.

159. Polukarova, Z. M., Shatalova, I. G., Yusupov, R. K., and E. D. Shchukin. 1968. Use of vibration compacting for increasing the strength of compacts. *Powder Metal*. 6: 54–56.

160. Kontorovich, S. I., Malikova, Z. G., and E. D. Shchukin. 1970. Internal stresses in structures of hydration hardening of mineral binders. *Kolloidnyi Zh*. 32: 224–228.

Index

Milton Keynes UK
Ingram Content Group UK Ltd.
UKHW050259161024
449569UK00043B/1445